The Aim and Structure of Physical Theory

THE AIM AND STRUCTURE OF PHYSICAL THEORY

BY PIERRE DUHEM

TRANSLATED FROM THE FRENCH
BY PHILIP P. WIENER

PRINCETON, NEW JERSEY
PRINCETON UNIVERSITY PRESS

Published by Princeton University Press, 41 William Street,
Princeton, New Jersey 08540
In the United Kingdom, Princeton University Press, Oxford

Library of Congress Card No. 53-6383

ISBN 0-691-02524-X (paperback)
Translated from the second edition, published in 1914 by
Marcel Rivière & Cie., Paris, under the title
La Théorie Physique: Son Objet, Sa Structure

First Princeton paperback printing for the Princeton
Science Library, 1991

Princeton University Press books are printed on acid-free paper,
and meet the guidelines for permanence and durability of the
Committee on Production Guidelines for Book Longevity of the
Council on Library Resources
10 9 8 7 6 5 4 3 2 1 (pbk.)
Printed in the United States of America by Princeton University
Press, Princeton, New Jersey

FOREWORD

Pierre Duhem's Life and Work

Born in Paris on June 10, 1861 and passing away in his country home at Cabrespine (Aude) on September 14, 1916 at the age of fifty-five, Pierre Duhem was one of the most original figures of French theoretic physics a half-century ago. Apart from his strictly scientific works which were brilliant indeed, notably in the domain of thermodynamics, he acquired an extremely extensive knowledge of the history of the physico-mathematical sciences and, after having given much thought to the meaning and scope of physical theories, he shaped a very arresting opinion concerning them, expounding it in various forms in numerous writings. Thus, an excellent theoretician of physics and historian of the sciences, possessing enormous erudition, he also made for himself a great name in scientific philosophy.

Very gifted in mathematics and physics, Pierre Duhem at the age of twenty entered the École Normale Supérieure on the Rue d'Ulm in Paris; in this outstanding institution of higher education which has given France so many great teachers of literature and science, he was a brilliant student, and his attention was turned very quickly toward the study of thermodynamics and its applications, a domain, furthermore, which he was never to cease cultivating.

Reflecting on the works of Thomson (Lord Kelvin), Clausius, Massieu, Gibbs and the other great originators of thermodynamic conceptions, he was especially struck by the analogy between the methods of Lagrange's analytical mechanics and those of thermodynamics. These reflections led him at the age of twenty-three to introduce in a quite general way the notion of thermodynamic potential and to publish soon afterward a book, *Le Potentiel thermodynamique et ses applications à la mécanique chimique et à la théorie des phénomènes électriques* [Paris, 1886—Translator].

Having received first place in 1885 in the competitive examinations for teaching physics, Duhem, already known in scientific circles, became two years later lecturer in the Faculty of Sciences of Lille University, where he taught with brilliance hydrodynamics, elasticity, and acoustics. Very soon after his marriage in Lille his wife died, leaving him an only daughter with whom he was to spend the rest of his life. At thirty-two he became full professor in the Faculty of Sciences of Bordeaux University, and kept this post until his death.

All his life Pierre Duhem retained in his scientific works his initial orientation. His preoccupation with regard to theory was the construction of a kind of general energetics (including classical analytical mechanics as a special case) and abstract thermodynamics. Essentially a systematic mind, he was attracted by axiomatic methods which lay down exact postulates in order to derive by rigorous reasoning unassailable conclusions; he prized their solidity and rigor, and was far from repulsed by their dryness and abstractness. He rejected, it might be said, with horror, the idea of substituting for the formal arguments of energetics the uncertain images or models furnished by atomic theories; he had no inclination to follow Maxwell, Clausius, and Boltzmann in the construction of a kinetic theory of matter permitting a concrete interpretation of the abstract conceptions of thermodynamics. If he admired Willard Gibbs for the rigor of his purely thermodynamic arguments and for the algebraic elegance of his demonstration of the phase rule, he certainly did not follow the great American thinker when the latter tried to base the atomic interpretation of thermodynamics on general statistical mechanics. From his *Commentaires sur la Thermodynamique*, his youthful work, to his great *Traité d'Énergétique générale*, which in his maturity crowned his works on matter, Duhem pursued his efforts at axiomatization and rigorous deduction. He sifted out all the fundamental notions admitted by thermodynamics; for example, he gave a purely mathematical definition of the quantity of heat and thus deprived it of any physical intuitive meaning in order to avoid any begging of the question. This constant effort at abstraction gives the theoretical work of Duhem a rather austere appearance which, despite the very remarkable results it has brought, may not please all minds.

It is fair to insist on the fact that Duhem, though he was constantly preoccupied with the establishment of an impeccable axiomatic system in the theories he developed, never lost sight of the problems of application. Notably in the domain of physical chemistry, familiar to him from his youth, he came to grips with the applications of theory to experiment by examining in detail all the consequences of the often difficult ideas of Willard Gibbs, whose presentation he knew how to make precise, and he was one of the first to spread them in France.

Duhem also occupied himself a great deal with hydrodynamics and with the theory of elasticity, branches of science which his conceptions led him to consider, besides, as particular chapters of general energetics. His works on the propagation of waves in fluids, notably on waves of impact, have retained all their validity. It seems his researches on electromagnetism were less happy, for he always had a great hostility toward Maxwell's theory and preferred Helmholtz' ideas, which are quite forgotten today. His deep antipathy with regard to all pictorial models prevented him, moreover, from understanding the importance of the Lorentz theory of electrons, then in full development, and rendered him as unjust as he was shortsighted about the rise of atomic physics, then in its beginnings.

Pierre Duhem was also a great historian of the sciences belonging to the domains, familiar to him, of mechanics, astronomy, and physics. Very conscious of the continuous evolution which manifests itself in the development of science, justly persuaded that all the great innovators have had forerunners, he demonstrated strongly that the great revival of mechanics, astronomy, and physics at the time of the Renaissance and in modern times has its roots deep in the intellectual work of the Middle Ages, a work whose importance from the scientific point of view had been too often unrecognized prior to Duhem's researches. In several of his writings, and particularly in his important three-volume work, *Léonard de Vinci, ceux qu'il a lus et ceux qui l'ont lu*, he insisted on the part played by the scholars of the medieval universities, and particularly by those of the University of Paris, from the thirteenth to the sixteenth century. He showed that a reaction took place after the death of Saint Thomas Aquinas against the ideas of Aristotle and the Aristotelians, and that this was at the origin of the movement of ideas which, rejecting the Greek philosopher's conceptions of motion, was going to end with the principle of inertia, with the work of Galileo, and with modern mechanics. He established that John Buridan, Rector of the Sorbonne about 1327, had the first idea of the principle of inertia and introduced under the Latin name of *impetus* a magnitude which, though not too well defined, is closely related to what we today call kinetic energy and quantity of motion. He analyzed the important progress due, a little later, to the works of Albert of Saxony and Nicholas Oresme. The latter especially accomplished considerable work, for with his ideas on the solar system he was the precursor of Copernicus, and with his first attempts at analytical geometry he was the forerunner of Descartes. He was even acquainted with the form of the laws of uniformly accelerated motion, so important in the study of weight. Then Duhem shows us Leonardo da Vinci, that admirable and many-sided man of genius, assimilating and pursuing the work of his predecessors and preparing the road on which, after various scientific scholars of the sixteenth century, Galileo and his continuators were definitively to begin modern mechanics.

Through writings of this sort and notably through a valuable sketch of the history of mechanics, Pierre Duhem, who had also studied closely the science of the seventeenth century and brought to light the often unrecognized contributions of Father Mersenne and Malebranche, was classed in the first rank of contemporary historians of the sciences. In his maturity he undertook, it is said with numerous anonymous collaborators, a colossal work: the history of cosmogonic doctrines, i.e. of conceptions about the system of the world from antiquity to the modern period. At his death he had already written eight volumes of this work, but only five have been published: the publication of the last three, whose manuscripts had been entrusted to the Academy of Sciences of Paris, having been postponed as a consequence of the financial difficulties of publication. It is a work of profound erudition, a mine of precious documents concerning the history of ideas and of philosophy in ancient times and in the Middle Ages at least as much as what is properly called the history of science. It would be immensely desirable for subscriptions abroad to help complete the publication of this vast synthesis which the author nearly had time, despite his premature death, to bring to its completion.

A theoretic physicist of indisputable value, possessing an enormous erudition in the history of the sciences, accustomed through this twofold intellectual formation to reflect on the growth, development, and scope of physical theories, Pierre Duhem naturally turned toward the philosophy of science. An essentially systematic mind, he worked out for the meaning of the theories of physics a very precise opinion which he expounded in numerous publications. The most important of these is his book entitled *La Théorie Physique: Son Objet, Sa Structure*, which enjoyed a great success in France and which the present volume offers in an English translation for American (and other English-speaking) readers. It is a capital work whose clarity and often impassioned tone are an exact reflection of the mind that created it. Without wishing to analyze completely a work so rich in substance, we should like to underscore rapidly a few essential points.

Pierre Duhem held firmly to separating physics from metaphysics: he saw in the history of physical theories, whether they were based on continuous or discontinuous images, or whether they were of the field or atomic type of physics, a proof of our radical inability to reach the depths of reality. It was not that Pierre Duhem, a convinced Catholic, rejected the value of metaphysics; he wished to separate it completely from physics and to give it a very different basis, the religious basis of revelation. This preoccupation with a complete separation of physics from metaphysics led him, as a logical but curious consequence, to be ranked, at least with respect to the interpretation of physical theories, among positivists with an energetistic or phenomenological tendency. In fact, he summarized his opinion concerning physical theories in the following conclusion: "A physical theory is not an explanation; it is a system of mathematical propositions whose aim is to represent as simply, as completely, and as exactly as possible a whole group of experimental laws."

Physical theory would then be merely a method of classification of physical phenomena which keeps us from drowning in the extreme complexity of these phenomena. And Duhem, arrived at this positivist and pragmatist conception of nature bordering closely on the conventionalism (*commodisme*) of Henri Poincaré, was in complete agreement with the positivist Mach in proclaiming that physical theory is above all an "economy of thought." For him all hypotheses based on images are transitory and infirm; only relations of an algebraic nature which sound theories have established among phenomena can stand imperturbably. Such, in the main, is the essential idea which Duhem produced about physical theory. It certainly pleased the physicists of the school of energetics, his contemporaries; it certainly is also favored by a great number of quantist physicists of the present day. Others were already finding it or will still find it a little narrow, and will reproach it for diminishing too much the knowledge of the depth of reality which the progress of physics can procure for us.

We must be fair and emphasize the fact that Duhem did not fall into the extremes to which his views might perhaps have led him. He believed instinctively, as all physicists do, in the existence of a reality external to man, and did not wish to allow himself to be dragged into the difficulties raised by a thoroughgoing "idealism." Hence, taking a position which is a very personal one at that, and separating himself on this point from pure phenomenalism, he declared that the mathematical laws of theoretical physics, without informing us what the deep reality of things is, reveal to us nonetheless certain appearances of a harmony which can only be of an ontological order. In perfecting itself physical theory progressively takes on the character of a "natural classification" of phenomena, and he made precise the meaning of the adjective "natural" by saying: "The more theory is perfected, the more we apprehend that the logical order in which it arranges experimental laws is the reflection of an ontological order." In this manner, it seems, he had been led to mitigate the rigor of his scientific positivism because he felt, and we think justifiably so, the force of the following objection: "If physical theories are only a convenient and logical classification of observable phenomena, how does it come about that they can anticipate experiment and foresee the existence of phenomena as yet unknown?" In order to answer this objection he really felt that we must attribute to physical theories a deeper bearing than that of a mere methodical classification of facts already known. In particular, he was clearly aware, and some passages of his book show this to be so, that the analogy of the formulas employed by physical theories bearing on different phenomena most often do not reduce to a mere formal analogy but may correspond to deep connections among diverse appearances of reality.

Such in the main is the conception which Duhem propounded concerning the scope of physical theories—an idea more subtly nuanced in the end than one might first believe. It is possible, however, to think that despite the subtlety of his doctrine brought about by the idea of a natural classification, Duhem, led on by the uncompromising tendency of his mind, often maintained judgments that were too absolute. Thus, inspired by a genuine horror of all mechanical or pictorial models, he kept on combatting atomism and, faithful to the school of energetics, he never became interested in the interpretation of the abstract concepts of classical thermodynamics, though it was so instructive and fruitful, which statistical mechanics furnished in his own lifetime. Thus preparing himself for perhaps too easy a success, he attacked the simplistic representation of atoms by small, hard, and elastic corpuscles; he attacked the ideas, at times somewhat naïve, of Lord Kelvin on the representation of natural phenomena by gears or vortices. He does not seem to have been aware of the tremendous revival which the atomic theory in its present form was to bring to physics, nor to have had any presentiment of the prodigious developments it was to have in a half-century. The passages in which he exposes almost to derision the notion of the electron and its introduction into science have since received cruel refutation inflicted by the extraordinary advances of microphysics.

Other parts of his book bear some of the marks of its age. Thus, when he compares, using great psychological penetration, narrow and deep minds with ample and weak minds, he is perhaps right in mentioning Napoleon as an example of the latter, but is he also right in putting into the same category all physicists of the English school? His opinion is no doubt explained by the times in which the book was written, in the aftermath of the brilliant works of William Thomson, whose strong personality appeared to symbolize all contemporary English physics. But it takes one by surprise today when nobody, I think, would have any notion of saying that Mr. Dirac is preoccupied merely with concrete representations! Moreover, by his parallel contrast between deep minds and ample minds, Duhem also appears to me to have been unjust toward the "pictorial" theoreticians of the second category whose contribution to the progress of physics, after all, has undoubtedly been greater than has been that of theoreticians solely preoccupied with axiomatization and perfectly rigorous logical deduction.

Despite these reservations, the work of Duhem on physical theory deserves great admiration because, based on the great personal experience of the author and on the acuteness of judgment of a remarkably strong mind, it contains views which are very often correct and profound, and which, even in the cases where we cannot adopt them without restrictions, are nonetheless still interesting and supply ample matter for thought. I shall give as an example the penetrating reflections devoted by Duhem to the so-called crucial experiment (Bacon's *experimentum crucis*). According to Duhem, there are no genuine crucial experiments because it is the ensemble of a theory forming an indivisible whole which has to be compared to experiment. The experimental confirmation of one of its consequences, even when selected among the most characteristic ones, cannot bring a crucial proof to the theory; for indeed nothing permits us to assert that other consequences of the theory will not be contradicted by experiment, or that another theory yet to be discovered will not be able to interpret as well as the preceding one the observed facts. And with much perspicacity Duhem cites as an instance the famous experiment in which Foucault, with the help of his method of a rotating mirror, demonstrated, a century ago now, that the speed of propagation of light in water is less than the speed of propagation of light in a vacuum. It was thought, at the time Duhem was writing, that this experiment contributed a crucial proof in favor of the wave theory of light and compelled us to reject any corpuscular conception of this physical entity. Very correctly Duhem declared that the experiment of Foucault is by no means crucial, for if its result is easily interpreted by Fresnel's theory and is in contradiction with Newton's corpuscular theory, nothing permits one to assert that another corpuscular theory resting on other postulates than the old form of this doctrine may not enable us to interpret Foucault's result. And the choice made by Duhem in giving this example turns out to be a particularly happy one as a result of the evolution of our ideas about light which he surely had not foreseen. We know, in fact, that the same year in which Duhem was writing his book (1905), Einstein introduced into science the idea of a "quantum of light," the photon, and that today the existence of photons is not in doubt. No matter in what way we finally interpret the double aspect of light, its corpuscular and wave appearances whose reality can no longer be doubted, it will of course be necessary to reconcile the existence of photons with Foucault's result. This shows us the profundity of Duhem's remarks on crucial experiments and the skill with which he knew instinctively how to choose his example. We cannot therefore deny that Duhem's analyses are very often marked by a great penetration and great scope.

Pierre Duhem, although he was kind and affable, had an uncompromising character and did not always spare adversaries of his ideas. A convinced Catholic, conservative in politics, he asserted his opinions with a sincerity which was at times not exempt from an aggressive vivacity. Everybody paid tribute to the rectitude of his character, but some did not appreciate its harshness. He had enemies and that no doubt explains why this eminent scientist and scholar, philosopher and historian, did not obtain what in a centralized country like France is the natural crown of every fine scientific career: a chair in a large institution of higher education in Paris. It must be said that he did nothing to procure it, and one day when he was approached to find out whether he would accept an appointment to teach the history of science at the Collège de France, he answered that he was a physicist and did not wish to be classified as a historian. Three years before his death, he had a satisfaction which consoled him for many injustices: the Academy of Sciences of Paris called him to become a non-resident member.

An indefatigable worker, Pierre Duhem, dying prematurely at fifty-five, left an enormous contribution in theoretical physics, in philosophy, and in the history of science. The value of his strictly scientific researches, the profundity of his thought, and the incredible extent of his erudition make him one of the most remarkable figures of French science of the end of the nineteenth and beginning of the twentieth centuries.

LOUIS DE BROGLIE

Paris, 1953

INTRODUCTION

Preface

The Aim and Structure of Physical Theory was favorably received when it first appeared in 1906, and at the time of the second edition in 1914.* W.v.O. Quine appropriated Pierre Duhem's thesis in a justly famous article,[1] and, in so doing, gave it a second life, but not without interpreting it and changing its meaning and implication.[2] Here the English version of Duhem's work will be analyzed in a manner faithful to its original meaning, without the mixing in of foreign elements that the modern reader almost inevitably associates with it.

In the Preface to the second edition, the author mentions adding two articles, dated 1905 and 1907, to the collection that appeared in 1904–1905 in the *Revue de philosophie.* He forgets to remind us that Chapter IV—a chapter common to both editions—develops ideas first articulated in an 1893 essay ("L'Ecole anglaise et les theories physiques") published by the *Revue des questions scientifiques.* Several questions of classificatory coherence posed by this chapter need to be examined first. Since Duhem constantly draws upon the history of science in support of his theses, and since it is the peripatetic doctrine which, by his own admission, offers the best analogy to his own conception, a few reflections on the "Duhemian Aristotle" will be added to this examination, as represented by *The Aim and Structure of Physical Theory* [hereafter PT].

Having clarified these questions of internal coherence and history, we will try to isolate in PT a thesis that is apt to be favorably received today—logically weaker than the one Duhem arrived at, and derived from the dual character of knowledge proper to physics in its symbolic and approximate aspects.

Duhem's full thesis goes further. We will examine by way of conclusion the nature of its governing principle, as well as certain questions it poses for philosophical reflection.

* All references to PT are to the French edition.

[1] "Two Dogmas of Empiricism," *Philosophical Review* 1951: 60, 20–43.

[2] Jules Vuillemin, "On Duhem's and Quine's Theses," in *The Philosophy of W. V. Quine*, ed. L. E. Hahn and P. A. Schilpp, The Library of Living Philosophers, vol. XLVIII (LaSalle: Open Court, 1986), 594–618.

I. Internal Coherence and History in Physical Theory

Duhem presents two different classifications of physical theories.[3] The first distinguishes four types of theories which, in increasing order of dogmatism and metaphysical commitment, constitute the four great cosmological schools: the hylomorphist Peripatetic School; the Newtonian school, which appeals to forces and mutual actions; the Atomist school, which searches for indivisible corpuscles; and the Cartesian school, which asserts the identity of matter and space. The second opposes abstract theories to mechanical models. Whereas each of the four types of theory outlined above is to some degree explanatory (even if the explanation is reduced to its minimum in Peripateticism), the two theories here opposed are both purely classificatory.[4] Abstract and purely algebraic classifications are favored on the Continent, especially in France and Germany,[5] while models are preferred by the English, especially Thomson (Lord Kelvin), despite the fact that Newtonian theory is abstract.[6] Corresponding to these geographical differences, Duhem believes to have found two different kinds of minds at work: the deep but narrow minds that require the discipline of logic to imagine a multitude of empirical laws, and the broad but weak minds that are not put off by such a diversity of laws, as long as they dispose of a sufficient number of images that speak directly to the senses. The opposition of these two types of minds appears to coincide with the opposition between the faculties of understanding and imagination,[7] which in turn fix the two kinds of relations possible between theory and logic: understanding demands the compatibility and *a fortiori* the unity of the diverse hypotheses that make up physical theory, and as a constitutive principle depends upon it, and as a regulative principle does not declare itself satisfied until the possibility of such a thing is asserted, in keeping with the dictates of reason;[8] the imagination, on the other hand, relies upon the juxtaposition of diversity, so long as this is plainly presented, free from any pretense of being subordinated to logic.

The coherence of Duhem's doctrine becomes clear once a relation-

[3] PT, 9–17; *Le mixte et la combinaison chimique, Essai sur l'évolution d'une idée* (Paris: Naud, 1902), 1.

[4] PT, 117, 107–9.

[5] PT, 99–100. Carried away by his taste for national character traits, Duhem would soon separate French and German physicists, here joined. The events of the First World War explain without excusing "Quelques reflexions sur la science allemande," *Revue des deux mondes* 25 (1915), 657–86.

[6] PT, 104.

[7] PT, 142, 152, 451.

[8] PT, 455, 507.

ship is established between the theory of models and, on the one hand, the operations that form the basis of the theoretical structure, and, on the other hand, Henri Poincaré's own conception of physics. First, let it be noted that models put into action various theoretically necessary operations:[9] the definition and measurement of physical magnitudes, the choice of hypotheses, mathematical (especially algebraic) development, and comparison with experience.[10] For the choice of hypotheses, even if it supposes their non-contradiction in the case of a given theory,[11] tolerates contradiction on the condition that contradictory theories are not combined. But these necessary conditions of physical theory, which Poincaré and others in France also find sufficient, are indeed enough for the classification of empirical laws. When paired with imaginative models, they can even yield a knowledge in no way mixed up with metaphysics[12] and the coherence typical of it.[13] But they arrive, and can only hope to arrive, at an artificial classification. If Duhem suggests occasionally that mechanical models take the place of explanation, while the abstract aim of logical unity excludes it,[14] he condemns them in general, as he must, not for their theoretical excess, but for their deficiency. A complete theory must be rationalistic—and this is why it cannot be English!—by demanding coherence as one of its requisites. Coherence, and only coherence (as an idea, not a given) ensures that the proposed classification of empirical laws is not artificial, but natural, and it authorizes us to see the reflection of ontological order in the logical order so described.[15]

Thus a complete and autonomous physical theory, as described by Duhem, occupies the second of six levels within a six-tiered system of classification. The first two give physical theory the sole object of the classification of empirical laws, but the first contents itself with an artificial classification, which the second converts into a natural classification by postulating the unity and logical coherence of hypotheses. The last four levels are explanatory and metaphysical. They go beyond what a physicist considers legitimate to assert, and arise not out of physical theory but out of cosmology. The most humble of these four, however, deserves to be examined more closely.

Whether it is a question of the attraction of iron to a magnet,[16] or the chemical conception of an element[17] or gravity,[18] Aristotle's doc-

[9] PT, 26.
[10] PT, 90, 109, 111.
[11] PT, 25.
[12] PT, 107.
[13] PT, 133.
[14] PT, 152.
[15] PT, 35.
[16] PT, 9–10.
[17] PT, 189.
[18] PT, 339–40.

trine—metaphysical in that it claims to disclose ultimate substantial forms or absolute elements—must be rejected. Like all explanatory theories, it subordinates physics to metaphysics and in effect prevents the possibility of agreement among physicists belonging to the different cosmological schools.[19] It is nevertheless sufficient to amend this dogmatic system by retaining no more than the opposition of form to matter, both relative to the state of our experience and our symbolic arrangements.[20] Then we see that Peripateticism stands apart from the cosmological theories: Atomists, Cartesians, and Newtonians all lay down principles to which they pretend to submit nature and which can therefore always enter into conflict with the dogmas of metaphysics and of faith.[21] Peripateticism, on the contrary, escapes this possible conflict by virtue of its capacity to absorb as many primary qualities as natural classification requires.[22]

The natural classification that physical theory strives for (and whose possibility is preserved by the postulate of coherence and logical unity[23] without ever actually furnishing reason with a dogmatic basis on which it could rest)[24] limits and corrects the positivism immanent in physical theory[25] by giving a foundation to the analogy between cosmology and metaphysics, on the one hand, and cosmology and physical theory on the other. The sort of rational faith expressed by the unity of the theory is especially facilitated when one examines the analogy that suggests itself between general thermodynamics—the normal result of abstract theory—and Peripateticism interpreted in terms of equilibrium and entropy.[26] The Aristotelian analogy, correctly interpreted, represents exactly the surplus that is typical of natural classification and that was lacking in mechanical models and pure positivism.

Duhem's doctrine is thus coherent as much in its internal economy as in its historic, albeit at first glance strange, relation with Aristotelianism.

I will not examine the questions posed by this relation, for the author of Book XII of the *Metaphysics* and of Books VII and VIII of *Physics* did after all give movement and physical theory a definition which orders the motion of the heavens and earth in imitation of the Prime Mover. I will limit myself to questioning the virtue of the logical co-

[19] PT, 47–48.
[20] PT, 444.
[21] PT, 430–31.
[22] PT, 164, 170ff.
[23] PT, 450.
[24] PT, 433; example: the interpretation of the conservation of energy (432).
[25] PT, 455, 454, 509.
[26] PT, 464–71.

herence required of physical theory, seen from the point of view of late nineteenth century physics. It would be impossible, in fact, to estimate the value of PT without taking into account the reality it sought to understand.

However, when Duhem claims as his own the opinion of an observer who declared at the time of publication of PT that "theoretical physics in no way presents us with a group of divergent or contradictory theories,"[27] does he faithfully describe the contemporary experience of both theoreticians and experimentalists? Does not Lorentz's admission of impotence ("at a loss how to reach the end of contradiction") find an echo in Michelson's fruitless attempts to detect a vibration in the aether?[28] Do not analogous difficulties face statistical mechanics with contradictions that result in the "ultraviolet catastrophe"? These instances of conflict, it is true, lead back to metaphysical suppositions of Mechanicism (i.e., the philosophy of the science of motion) and Atomism. But could it not be said that what one physicist—confined by the logical unity of thermodynamical phenomenology—might interpret as the symptoms of explanatory delirium, another scientist might use to remedy a defective explanation? If one denies experience the power to decide between two physical theories, how can one allow it the right to judge between two philosophical theories of physical theory?

On the one hand, Duhem reproaches Poincaré and the English school for turning up their noses at the logical unity requisite to physical theory.[29] Pressed for an explanation of the nature of this logical unity, he shows it at work in energetics and gives his scientific works the form of final synthesis in his *Treatise on Energetics or General Thermodynamics*.[30] As against Mechanicism and kinetic theory, he proceeds from a very abstract formulation of phenomenological thermodynamics to which he subordinates mechanics without, however, quite incorporating electro-magnetism.[31] The logical unity desired and

[27] Abel Rey, quoted by Duhem, PT, 448.

[28] G. Holton, *Thematic Origins of Scientific Thought, Kepler to Einstein*, (Cambridge, Mass.: Harvard University Press 1973), 266. Michelson's experiment took place in 1881: Lorentz's text is taken from a letter to Rayleigh dated August 18, 1882.

[29] PT, 137, 148–49.

[30] Two volumes (Paris: Gauthiers-Villars, 1911). For a good analysis of the philosophical implications of energetics and, in general, Duhem's doctrine, see Anastase Brenner, *Duhem, Science, réalité et apparence* (Paris: Vrin, 1990), especially chapter II.

[31] Brenner, 1990, 80n.2.

achieved is an algebraic correspondence by which the most diverse physical phenomena are expressed through an equation of the same form. It is therefore an analogical unity,[32] which does not entail ontological commitment, but which guarantees that common sense, thus satisfied, has the natural character of classification.[33]

On the other hand, without denying the importance and omnipresence of these analogies (about which it could in fact be asked whether nature imposes them on classification or whether the constraints of simplification and classificatory routine suggest them),[34] other physicists have a stricter notion of the unity of physics. Phenomenological analogy does not satisfy them. They insist that the common equation obeyed by different phenomena unify them not in the way in which the diffusion equation permits calculation of, for example, the propagation of heat, or the diffusion of ions in a gas or of neutrons in graphite, but in the way in which Newtonian gravity applies to both the fall of an apple to the ground and the fall of the moon toward the earth. Let us not be deceived by the language of physicists who, in speaking of mechanisms and atoms, end up preferring the incoherence of theories to formal unity![35] Incoherence is the mark of research carried out to turn upside down the concepts essential to physics: space, time, magnitude, physical state. When Duhem speaks ironically of Thomson's investigations concerning the size of atoms,[36] and when he appoints common sense as the judge fit to decide which hypotheses should be abandoned,[37] history does not hesitate to pronounce its verdict. With regard to a question of the highest importance for physical theory until the formulation of its inequalities by Heisenberg, it is Thomson who shows himself possessed of common sense—at least if, as Duhem would have it, common sense cannot be separated from the evolution of ideas in physics.[38]

Although coherent in its internal composition and in its relation to Peripatetic cosmology, PT pays for its jealous regard for logical unity and historical continuity by its insensitivity to the crisis that leads the physics of the period to the double revolution of relativity and quantum theory.

[32] PT, 141–42.
[33] PT, 153.
[34] R. Feynman, R. B. Leighton, M. Sands, *The Feynman Lectures on Physics*, (Reading, Mass.: Addison-Wesley, 1970), vol. II, chap. 12.
[35] PT, 446–47, 461–62.
[36] PT, 107–8.
[37] PT, 331.
[38] PT, 394, 397.

II. Symbolism and Approximation: Their Consequences for Physical Theory

Duhem, initially a partisan of the inductive method, became aware of its limitations while teaching thermodynamics, and finally renounced the attempt to coordinate theoretical postulates with experiments term by term and to give a synthetic account of the relation between them.[39] The import of PT, specifically with regard to magnitudes, experiments, and physical laws, can be most easily understood by keeping in mind the preeminence of thermodynamics. Two characteristics stand out: symbolism and approximation, which pick out these elements and oppose them to attributes, perceptions, and laws of common sense.

It is plainly shown by the analysis of magnitudes (such as mass, the length of an object, or the duration of an event) and the operations to which they lend themselves (in particular the correspondence between the concatenation of bodies or events and the addition of appropriate magnitudes) that all physical magnitudes are the result of a symbolic abstraction by which we place a number or a set of numbers into correspondence with a physical attribute.[40] But thermodynamics has the advantage of drawing upon the irreducible vocabulary of science for certain magnitudes, such as temperature, which distinguish themselves from quantities in that they cannot be added (even though they figure in polynomial expressions and lend themselves to the relations of equality, greater than and less than, and to operations of derivation).[41] They are intensities, symbols of primary qualities that a phenomenological theory counts among its givens,[42] just as chemical analysis posits elements on the basis of their power to dissociate.

To measure a physical magnitude is to conduct an experiment. To conduct an experiment is to obtain the approximate value of a magnitude. Scientific fact is thus divided into two disparate facts: the observed, and the theoretical act which assigns an exact value to the symbol of that magnitude. As a consequence of this disparity, it can easily be seen that any given theoretical fact is consistent with an infinite

[39] Brenner, 1990, 29–53. [40] PT, 160ff. [41] PT, 166.

[42] PT, 164–78, 185–86. Duhem carefully notes, p. 175, that the notion of addition "comes up again when one studies the quantitative effect that furnishes a scale capable of discerning the diverse intensities of a quality." This is an essential remark that limits the reach of the author's remarks on the intrinsic importance of intensities, on the supposed link between explanation and the reduction of quantity (166), and on the immunity of mathematical physics to the reduction of quality to quantity.

number of distinct practical facts, and that an infinite number of incompatible theoretical facts can be made to correspond to any given practical fact.[43] The definition of physical magnitudes is thus not possible without this complex coordination between theory and experiment, which involves scientific fact in the interpretation of language and which superimposes upon the actual use of observational instruments its own abstract use, inevitably tied to the correction of errors.[44] These remarks apply to geometric magnitudes—lengths and volumes. The consideration of thermodynamic magnitudes—temperatures and pressures—and of highly theoretical instruments such as thermometers and manometers that permit their indirect measurement, adds another sort of evidence.[45]

Is it surprising then that all physical law is symbolic and approximate?

All laws are abstract. The laws of common sense (e.g., all men are mortal) are no less abstract than the laws of physiology. But Duhem appears to hesitate over physical laws. On the one hand, when he defines physical theory as a classification of experimental laws, he treats these as an assumption of the theory, and treats as an induction the economy of thought they permit with regard to the potential infinity of observations comprehended.[46] When he criticizes the Newtonian method, he demonstrates that Newton did not advance from Kepler's empirical laws to the principles of gravity and mechanics through induction, yet he constantly supposes that it is by induction that Kepler arrived at his three laws, incidentally designated *phenomena* by Newton.[47] By contrast, when he attributes a symbolic and not merely abstract character to physical laws, he does not expressly exclude experimental laws from those laws which are inevitably theoretical.[48] Furthermore, he bases his analysis on the crudest and most primitive of laws, the Mariotte-Boyle law.[49] This law, rudimentary though it may be, belongs to thermodynamics. If PT's coherence is to be preserved, what then can be concluded from Duhem's *ex silentio* argument, except that certain laws of celestial kinematics can be regarded as experimental, and that it is not unreasonable, despite their elaborate math-

[43] PT, 229–30, 201–2.

[44] PT, 222, 227, 235, 246, 241.

[45] PT, 251. On the nature of thermodynamical experiments see C. Truesdell, "The Disastrous Effects of Experiment Upon the Early Developments of Thermodynamics," in *Scientific Philosophy Today*, ed. J. Agassi and R. S. Cohen, BSPS (Dordrecht: Reidel, 1982), 415–23.

[46] PT, 27–28.

[47] PT, 290–91, 295.

[48] PT, 251, 254.

[49] PT, 251.

ematical expression, to construe them as simply abstract in view of their inductive basis, although all the laws of thermodynamics are symbolic and theoretical. Since finally this science, by virtue of the hypothesis of energetics, furnishes the whole of physics with its ultimate foundation; its symbolic and theoretical principles therefore will not fail to confer analogically some of their virtue upon empirical laws, nor will its opposition to veritable physical laws lose, in the final analysis, its decisive and definitive character.

The emphasis placed upon the characteristic approximation of all physical laws confirms this conclusion. Newton, in fact, generalizes Kepler's law of elliptical trajectories to conic sections. He modifies the third law by introducing into consideration the masses of the planet and its satellite, respectively.[50]

Physics is distinguished from geometry, or, more generally, from mathematics, by its approximate character.[51] This assertion is not unique to Duhem. We find it in Hume. It is common to rationalists, empiricists, and conventionalists. Among geometers, those who distinguish between a mathematics of precision and a mathematics of approximation[52] hasten to make the point that when a mathematician develops a number or a function in series and terminates his estimation at the nth term, he determines with exactitude the rest that he ignores, and, in consequence, completely eliminates any error committed. Disagreement arises over the question of knowing which among the geometries of constant curvature the physicist can or must choose. Does the choice of Euclidian geometry, presupposed by optics in the case of a light ray, depend on a convention rendered irresistible by its simplicity and convenience, as Poincaré maintains? Or is every convention—and equally, by implication, the choice of a physical geometry—subject in principle to revision, in view of the totality of implications flowing from its connection with a given physical theory, as Duhem would have it?[53] Here lies a decision which, finally, depends on the proposed analysis of physical law, according to whether one favors the inductive aspect or emphasizes its fully symbolic nature. Neither option, in any case, casts doubt upon the contrast between geometry and physics: between the system which from exact hypotheses deduces equally exact consequences, and the theory whose mag-

[50] PT, 294, 296.

[51] PT, 269, 304, 311, 404, 407, 410.

[52] As does Felix Klein; see *Elementarmathematik vom höheren Standpunkte aus*, 3 vol., 4th ed. (Berlin: Springer, 1968), vol. 1 and vol. 3 *passim*.

[53] PT, 317–22, 323–29.

nitudes, laws, and principles can pretend to no more than an approximate status.

Those who, today, reject any abrupt demarcation between the analytic and the synthetic, and who extend the Duhemian thesis above and beyond physics by attaching it to mathematics as well as to all empirical knowledge, do not appear to have considered specifically how approximation distinguishes a physical law from a geometrical theorem no less than from a generalization of common sense. Duhem, a historian of physics, instinctively perceives what in the cumulative continuity of mathematical physics (once it has been separated from the missteps and backward steps due to the intrusion of metaphysics)[54] contrasts with the cumulative continuity of mathematics. Euclid's theorems are and remain true. Physical laws, because they are approximate, do not retain validity except insofar as the physicist contents himself with the order of precision that theory associates with it. The difference in the historical evolution of the two sciences thus reflects the difference between their laws. The mathematician studies functions and curves. Motivated by the disparity between theory and experiment, the physicist calculates his functions to this or that nearest decimal point. He represents them not by curves, but by bands of an infinity of curves.[55]

Does this opposition permit one to conclude with Duhem that a physical law is neither true nor false, but only approximate and therefore tentative and relative?[56] It is more prudent to limit oneself to expressions that not only are used by physics texts but moreover are supported by Duhem's commentary.[57] Let us return to Mariotte's law. It is, we are told, true and exact for a gas such as air "if one studies its compressibility between one and two atmospheres by methods of measurement within which relative errors can be on the order of a thousandth."[58] It is added at once that the order of magnitude of precision is not fixed once and for all: Mariotte's law is inexact when measurement attains a precision of one ten-thousandth, or when, holding this precision at one-thousandth, the atmospheric pressure is varied by several tenths of atmospheres. Physical laws therefore remain true or false, as long as their truth value is made to depend on the stipulation concerning the margin of approximation that fixes this value. To incorporate approximation into a law as a condition of its truth value is ex-

[54] PT, 384.

[55] PT, What Klein called *Bandkurven*; PT. 258–59.

[56] PT, 254, 260–61.

[57] PT, 263, 266–67.

[58] G. Bruhat, *Cours de physique générale, Mécanique*, 6th ed., revised by A. Foch (Paris: Masson, 1964), 261.

plicitly to grant that the law is relative. But it is also to qualify what is excessive about the word *tentative.* For when it has been stipulated that Mariotte's law is true under such and such a condition, it is no longer provisionally true but absolutely true. Determining the margin of approximation is therefore as important for physical knowledge as the knowledge of the law itself. This determination is lacking, however, for all laws that have yet to be disproved.

As for developing and testing the theory, the symbolism and approximation of magnitudes, experiments, and laws entail four principal consequences:

1. One cannot hope to base theoretic hypotheses on induction[59] or what might be called enlightened induction,[60] or, in general, on some particular principle adequate to justify them in isolation. The formulation of hypotheses, which is never without its arbitrary side, takes time, and only with time can the consequences of hypotheses become known and their fecundity established.
2. Its strictly theoretical character prevents a hypothesis from arising directly from inductive experiment. Verification is never the base, but the crown of a theory,[61] and it is not the postulates themselves, but their consequences, that are tested.[62]
3. A physical theory is therefore checked in a collective manner.[63] Only the entire system of physical theory can be compared with the whole of experimental laws:[64] the unity of a theory is not that of a clock, but of a living body.[65]
4. The *experimentum crucis* is but a myth, suggested to physics by the false geometrical analogy of reduction to the absurd. With what assurance, in fact, does one suppose that the opposition between two rival theories constitutes a rigorous dilemma, or that, in Aristotelian terms, the two branches are contradictory and not simply contrary[66] (as Duhem prophetically remarks, regarding the confrontation of wave and particle theories)?[67]

III. Does the Structure of Physical Theory Unequivocally Determine its Object? The Philosophical Postulate of Phenomenalism

All human endeavor is analyzed in terms of goals—that is, of ends represented as intentions—and of means. Physical theory does not es-

[59] PT, 290.
[60] PT, 330, 398ff.
[61] PT, 311.
[62] PT, 314.
[63] PT, 278, 277, 282–85.
[64] PT, 304.
[65] PT, 285.
[66] PT, 288, 314.
[67] PT, Louis de Broglie, Preface to the English translation of PT, xi.

cape this rule, though Duhem, in order to stress that it is a collective enterprise which aims at representing nature, substitutes for the words *goals* and *means* (laden with subjective connotation) the more specific words *aim* and *structure*.[68] The division of PT into two parts corresponds to these. The aim of physical theory is natural classification, where "natural" recalls Aristotelian analogy, and its structure a set of operations characterized by symbolism and approximation. To inquire after the validity of PT is to question the adequacy of structure in relation to aim.

Duhem demonstrates that if the aim of physical theory is the natural classification of experimental laws, then the operations constituting the structure of theory must be symbolic and approximate, which in fact they are. Of the six possible levels of physical theory, symbolism excludes the first, by imposing a logical unity incompatible with the incoherence inherent in artificial classifications, and invokes the third by analogy, since one can view Peripateticism as the dogmatic and cosmological image of natural classification. As for the three remaining metaphysical theories, corresponding to the Newtonian, Atomist, and Cartesian traditions, they claim by explaining nature to reveal its absolute and irrevocable principle, contrary to the approximation built into the structure of theory.

Is the demonstration convincing? The reservations previously noted about the critique of mechanical models pertain more to a particular phase of crisis of physical theory than to its ordinary state. Moreover, Duhem's blindness before the two sorts of difficulty that were then leading to the theory of relativity and to quantum theory calls into question his perspicacity as a physicist rather than his wisdom as an epistemologist. For apart from the fact that these two theories were developed by elaborating concepts and postulates far removed from intuition and inductive generalization, and in such a way that they appear to complete the assault on Newtonian cosmology, neither of them can be said to lend support to the images, much less to the principles, that had sustained the Cartesians and the Atomists. Furthermore, as is shown by the research and polemics of contemporary physicists, agreement has not been reached on the postulate of the reduction of the wave packet within quantum mechanics, nor has the question of the relation between quantum mechanics and relativity been resolved. In short, the two most metaphysical theories in Duhem's classification

[68] For the date of this change in vocabulary (between 1896 and 1905), see Brenner, 1990, 30n.4.

have emerged victorious, but only in a Pyrrhic sense, since they triumph only by sacrificing the core of their metaphysical claims.

There nevertheless remains a double obscurity in the demonstration, affecting symbolization (as distinct from simple abstraction) no less than explanation (as distinct from simple approximation). For it is necessary to clarify the nature of symbolization's advance, in going beyond abstraction, to justify the exclusion of artificial categories, and of the advance of explanation, in going beyond approximation, to justify the exclusion of cosmologies. In both cases, it must be asked in what the logical or formal unity of physical theory consists.

One will first be inclined to search for the archetype of classification in the classifications of species which from Aristotle to Linnaeus and Cuvier dominated biology until they were supplanted by evolutionist doctrines, which subordinated classification to explanation. But in PT Duhem expressly rejects physiological theories as overly simplistic.[69] The classification he has in mind is chemical classification. This classification shows itself to be natural whenever theory yields a formula for a multitude of substances, anticipating their composition and principal properties and permitting their realization through synthesis.[70] The criterion remains the same for physics properly so called. Thus Poisson deduced from Fresnel's theory a "strange" consequence verified by Arago: what authorizes us to believe that wave optics is a natural classification is that the observed results "accord precisely with the unlikely predictions of the calculus."[71] Duhem borrows his example from a theory that "explains" light by the vibrations of a hypothetical aether, which, unlike acoustics, traces the sensible vibrations of a medium.[72] He could have just as easily borrowed from the English school, as he does in the case of Maxwell's electromagnetic theory.[73]

Are these diverse examples nonetheless specific to a natural classification in the Duhemian sense? It may be doubted. They simply testify to the power of prediction implicitly contained in the equations—undecipherable at first glance and therefore full of surprises owing to the unexpected consequences that can be derived from them. But all that this power requires is that physical theory be mathematized.[74] "Artificial" classifications and explanations are, in this respect, on the same level, with the condition that they be written in the language of equations. What appears to differentiate the physical theories in this

[69] PT, 274.
[70] PT, 39.
[71] PT, 40.
[72] PT, 72, 4–5.
[73] PT, 192–93.
[74] As Duhem would like it to be; see PT, 157–58.

respect is not the ambition of their ontological commitment, but their capacity for calculation and prediction, which is not the privilege of energetics. The line separating abstraction and symbolization passes not between artificial classification and natural classification, but between a physics that does not measure and a physics that does. If the capacity for prediction is retained as a criterion, then one would group under the same rubric Aristotle's hylomorphism, Gilbert's dynamics, the atoms of Democritus and Gassendi, and vortexes of Descartes. And one would consider not only general thermodynamics as symbolic, but equally the mechanics of Newton, statistical mechanics, and Maxwell's vortexes.[75] The effect of symbolization is one thing, its scope another. And it is far from certain that even with regard to the extent of the territory covered, general thermodynamics, by contenting itself with a logical unity of analogy between phenomena, acquired for itself an advantage over the competing piecemeal theories.[76]

The excess of explanation versus approximation condemns more surely explanatory theories when one requires, as does Duhem, that the explanation reveal the ultimate cause of empirical laws. There is no doubt that it is this strong and metaphysical endorsement that physicists gave to explanation at the time of the triumph of rational mechanics. The realization that physical theory is approximate is enough for physics to reject questions of a speculative (even if apparently mathematical) nature whose precision exceeds the given margin of approximation. In particular when this margin merges initial indiscernible conditions with functions susceptible of rapid divergence, long-range predictions rapidly lose all meaning. Duhem limits his reflections to extrapolations concerning the stability of systems. The first principles of the doctrine of chaos[77] emerged from the celebrated work done on this problem by Poincaré, who pitilessly criticized all cosmological constructions,[78] and who drew conclusions that bore upon the inevitably limited import of Laplacian determinism.

[75] It happened that certain theories foreign to measurement (such as Plato's and Aristotle's) imposed on certain phenomena conditions of simplicity and perfection which, while mathematically illusory, were so precise that they permitted quantifiable predictions. Thus the principle of the uniform circularity of elementary celestial movements permitted Eudoxus to build his system of homocentric spheres.

[76] See the criticisms Planck addresses to the analogies drawn by Mach from the notion of potential in M. Planck, *Initiations a la physique*, trans. from the German by J. du Plessis de Grenedan (Paris: Flammarion, 1941), chap. III.

[77] P. Bergé, Y. Pomeau, Ch. Vidal, *L'ordre dans le chaos* (Paris: Hermann, 1984), chap. IV.

[78] *Leçons sur les hypothèses cosmogoniques*, ed. Vergne, 2d edition (Paris: Hermann, 1913).

Must physical theory, for all that, renounce all claims to explanation? Another look at the Duhemian conception of its relation to experiments shows that this is not the case. First, allowances must be made for polemical exaggeration when the author declares "that the principles of physical theory are propositions that are relative to certain mathematical signs stripped of any objective existence,"[79] since a classification is only natural in its indirect and analogical relation, but incontestable with this existence. Let us also accept the global schema of the refutation of a theoretical set of hypotheses by experiment,[80] but without going so far as to ruin the possibility of such a refutation on the pretext that one cannot "leave at the laboratory door *the* theory that one wants to test"[81]: the instruments of the experiment itself suppose certain theoretical principles, not the totality of the theory submitted to testing. Consider then the system of Eudoxus's homocentric spheres, faced with the following observations: Venus and Mars change in brightness, the apparent diameter of the Moon varies from 11 to 12 times a certain magnitude, the central eclipses of the Sun are at times total and at times ring-shaped.[82] What does Duhem conclude? "The astronomy of homocentric spheres did not—could not—resolve the problem posed by Plato; it did not preserve, indeed was condemned never to preserve the totality of celestial phenomena."[83]

PT could be contrasted to *Le Système du Monde*. Indeed, the reported observations do not negate the isolated hypothesis of the constancy of Earth-Sun, Earth-Moon, Earth-Planet radius-vectors, among other things, since they suppose the validity of the law of terrestrial geometric optics (according to which apparent brightness and magnitude vary according to distance). Certainly. But to deny this law of optics would be to doubt the alleged observations themselves, observations which, on the contrary, in no way suppose the constancy of radius-vectors. The logical unity required by natural classification therefore excludes neither the relative independence of hypotheses,

[79] PT, 431; see de Broglie, *Foreword*, ix–x, where we find another example of this exaggeration in the famous declaration: "If Newton's theory is exact, then Kepler's laws are necessarily false" (PT, 293), a formulation that completely ignores the notion of approximation. For more on this point, see Erhard Scheibe, "Die Erklärung der Keplerschen Gesetze durch Newtons Gravitationsgesetz," in *Einheit und Vielheit*, Festschrift für C. F. v. Weizsäcker, ed. E. Scheibe and G. Sussman (Göttingen, 1973), 98–118.

[80] PT, 282.

[81] PT, 277.

[82] Sosigenes, as related by Simplicius: Pierre Duhem, *Le système du Monde*, vol. 1 (Paris: Hermann, 1954), 400–402.

[83] *Ibid.*, 399.

nor the possibility of invalidating one in isolation.[84] Moreover the refutation thus obtained remains uniquely negative. Can the homocentric postulate of elementary uniform circular movements be preserved, while admitting epicycles or eccentrics? Or, on the contrary, must the constancy of radius-vectors be abandoned? The experiment itself does not decide the matter, but nonetheless the optic explanation requires the renunciation of the homocentric system.

Just how extensive are the explanations tolerated by physical theory, once ultimate explanations have been eliminated? Physical realism, which Duhem rejects as metaphysical,[85] has, in general, but one modal meaning, and therefore does not involve any difference in theoretical or experimental content.[86] How then distinguish the logical unity demanded by PT and the unity of integration associated with explanations? The latter, difficult to distinguish from simplicity, has as its principal effect the reduction in number of hypotheses and the elimination of "explanation *ad hoc*." A follower of Fourier, Duhem constructs a theory indifferent to hypotheses that reduce heat to a flowing liquid or agitated molecules. In the tradition of Osiander and Bellarmin, he resorts to the argument of equivalence of hypotheses to disqualify their supposedly metaphysical differences. But it is one thing to translate the observed movement of a planet in the systems of Ptolemy, Copernicus, and Tycho Brahe.[87] It is quite another, as Kepler noted,[88] to diminish the number of kinematic parameters of the solar system by removing from each observed movement the common distortion due to the movement of the observatory, and to explain loops of retrograde motion by comparing the speeds of the Earth and the planet in their orbits.

We have seen Duhem accept the reduction of optics to electromagnetism and to refuse that of thermodynamics to kinetic theory. Referring to chemical elements,[89] he specifies that the reduction is linked

[84] One could say the same for the refutation of the hypothesis used by Galileo to explain tides (PT, 365).

[85] PT, 7.

[86] It so happens that physical realism can be formulated using terms that permit experimental prevision (the most widely spread interpretation thus gives J. Bell the credit for having formulated the EPR paradox in just this way: this is then the transformation of a problem posed in philosophical terms into a physical problem).

[87] PT, 437–41.

[88] Johannes Kepler, the beginning of *Mysterium Cosmographicum*; Jules Vuillemin, "La méthodologie de Kepler," *Traditionen und Perspektiven der analytischen Philosophie*, Festschrift für R. Haller, ed. Gombocz, Rude, and Sauer (Vienna: HPT, 1989), 24–34.

[89] PT, 190.

to our technical ability. Its legitimacy is a matter of theoretical choice. Thus a theory can be, for science, more revolutionary in its indirect effects than in its principles: in setting into motion the center of heavy bodies, Copernicanism destroys the Aristotelian doctrine of gravity.[90] Did Duhem, partisan of continuity and evolution, worry when faced with the complete recasting of concepts unified by explanation, and of which the recent history of physics furnishes so many examples? This recasting not only collided with the Mechanicism proclaimed by kinetic theory, but even more so with the renewal of the Mechanicism by probabilities and statistics.

This is very likely why PT tolerates no explanations: either the theoretical simplicity owed to explanations does not distinguish itself essentially from the logical unity aimed at by natural classification, or else the conceptual speculations that are linked to them, and which often mark the passage from one margin of approximation to another, do not actually form a part of the theory, except heuristically, subjectively, and provisionally. A theory will eliminate them at a later stage and replace them with other concepts. What remains true will be limited to equations relating to the old margins. Attention to approximation entails conceptual indifference. We know how to measure and predict nature, but we do not know how to understand it.[91] Insofar as they are objectives, our concepts remain classificatory; they cannot be explanatory.

If symbolization, then, is by itself powerless to set aside a physical theory conceived as artificial classification, approximation that eliminates absolute explanations does not suffice to eliminate all explanations, contrary to the wishes of natural classification. In order to complete Duhem's demonstration, it is necessary also to search beyond the structure of physical theory for an additional philosophical principle, one that is manifest in the peripatetic analogy that crowns natural classification. The two exclusions that the analysis of theoretical structure is not quite able to provide a foundation for can be readily understood when one defines in Duhem's manner the connection between physics and metaphysics.[92] Either physics is subordinate to metaphysics, as the principles of explanatory theories would have it; or it affirms its autonomy, as does a theory limited to classification (but, in the latter case, autonomy can signify self-sufficiency and exclusion, as hap-

[90] PT, 392.

[91] Quine arrives at a similar conclusion, although he bases his on the indeterminacy of translation, not approximation.

[92] PT, 6ff.

pens in artificial classifications); or else, having as its intrinsic object the logical unity of phenomena, it designates, however ideally, their ontological unity by a metaphysical analogy characteristic of the connection established between energetics and Peripateticism. By adding therefore to the two accepted features of the structure of theory (symbolization and approximation) the unique philosophical postulate of phenomenalism, one determines the necessary and sufficient condition of the object assigned to physical theory by Duhem: it is—it must be—a natural classification.

Conclusion

Like his adversaries—physicists tolerant of incoherence—Duhem defends the autonomy of physical theory. What divides them, despite the distorting presentation of PT, is the appearance of an unequal ontological commitment that would make skeptics of his opponents, as against Duhem's well-founded phenomenalism. The true opposition is between physical theories that are closed in on themselves, and that do not concede any jurisdiction foreign to that of science, and phenomenalism, which certainly takes care not to save phenomena by appealing to metaphysical entities, but which does not exclude, indeed even claims, an analogical connection to these entities. The adversary Duhem aims at is less the pluralist than the naturalist. The contrast is therefore between phenomenalism and naturalism.

Duhem does not express his phenomenalism in the most felicitous form. Whatever sympathy we feel for the work of Aristotle, whatever majesty we recognize in his cosmology, it is difficult to accord the latter any other than a historical significance from the point of view of knowledge. For the Peripatetic analogy of which Duhem speaks seems to belong to the jurisdiction of knowledge, which is furthermore not physical, and this knowledge is, in any case, quite mysterious. Duhem does not cite Kant, who believed he had demonstrated its impossibility.

Let us leave Aristotle. Let us leave phenomenalism when it ventures to affirm that physical knowledge does not exhaust the knowledge of nature.

Naturalism does not content itself with simply limiting the knowledge of nature to physical knowledge. It places within the bounds of this knowledge the very relation between reason and nature. In particular, and in order to restrict us to physical theory, it undertakes to extract from man observed as part of nature what is necessary and sufficient to constitute this theory considered not only as a body of symbolic and approximate sentences, but also as a common enterprise of physicists.

The hypotheses used by the naturalist to explain the history of the sciences are speculative, as are those used by the phenomenalist for describing it. Duhem justly reminds us that experimental criticism of hypotheses is subordinate to moral conditions.[93] Let us translate into contemporary terms the categorical imperative of the physicist: "Seek out the disconfirmable." To explain the survival of that behavior, the naturalist invokes success and interest. The phenomenalist, while more hopeful that nature can be controlled, remains, at the same time, less optimistic about the possibility; he sees those qualities or irreducible forms of moral conduct which, by the impartiality and loyalty necessary for judging hypotheses, make men citizens of another world.

We will let the reader draw his own conclusions, according to his tastes, as to the validity of the philosophical postulate necessary to complete the Duhemian conception of physical theory. But let him nevertheless not be mistaken about the advantage that would appear to belong to naturalism, under the pretext that the latter defends the unity of scientific methodology. Phenomenalism is the equal of naturalism in the fields of physics and natural sciences, the only fields over which science has actually gained control. Where the human element comes into play, for example in history of science, and specifically when it is a question of the ethics of science, the supposedly immanent methodological unity of which naturalism avails itself remains an act of faith, as does the expressly transcendent postulate of practical reason.

Jules Vuillemin
Collège de France
October 1990

TRANSLATED BY TRISHKA WATERBURY
AND MALCOLM DEBEVOISE

[93] PT, 332.

AUTHOR'S PREFACE TO THE SECOND EDITION

THE FIRST EDITION of this book bears the date 1906; the chapters bring together articles published serially in 1904 and 1905 by the *Revue de Philosophie*. Since that time a number of controversies concerning physical theory have been raging among philosophers, and a number of new theories have been proposed by physicists. Neither these discussions nor these discoveries have revealed to us any reasons for casting doubt on the principles we had stated. Indeed we are rather more confident than ever that these principles should be firmly held. It is true that certain schools have affected scorn for them; free from the constraint these schools might have felt on account of these principles, they think they can run all the more easily and quickly from one discovery to another; but this frantic and hectic race in pursuit of a novel idea has upset the whole domain of physical theories, and has turned it into a real chaos where logic loses its way and common sense runs away frightened.

Hence it has not seemed to us idle to recall the rules of logic and to vindicate the rights of common sense; it has not been apparent to us that there was no use in repeating what we had said nearly ten years ago; and so this second edition reproduces the text of all the pages of the first edition.

If the years gone by have not brought any reasons to cause us to doubt our principles, time has given us opportunities to make them precise and to develop them. These opportunities have led us to write two articles: one, "Physique de Croyant," has been published by the *Annales de Philosophie Chrétienne*; the other, "La Valeur de la Théorie Physique," has received the hospitality of the *Revue générale des Sciences pures et appliquées*. Since the reader may perhaps find it somewhat worth while to peruse the clarifications and additions given to our book by these two articles, we have reproduced them in the appendix at the end of this new edition.

CONTENTS

	Page
Foreword by Louis de Broglie	v
Introduction to the 1991 Edition, by Jules Vuillemin	xv
Author's Preface to the Second Edition	xxxiv
Introduction	3

PART I

THE AIM OF PHYSICAL THEORY

Chapter I. Physical Theory and Metaphysical Explanation 7
1. Physical theory considered as explanation. 2. According to the foregoing opinion, theoretical physics is subordinate to metaphysics. 3. According to the foregoing opinion, the value of a physical theory depends on the metaphysical system one adopts. 4. The quarrel over occult causes. 5. No metaphysical system suffices in constructing a physical theory.

Chapter II. Physical Theory and Natural Classification 19
1. What is the true nature of a physical theory and the operations constituting it? 2. What is the utility of a physical theory? Theory considered as an economy of thought. 3. Theory considered as classification. 4. A theory tends to be transformed into a natural classification. 5. Theory anticipating experiment.

Chapter III. Representative Theories and the History of Physics 31
1. The role of natural classifications and of explanations in the evolution of physical theories. 2. The opinions of physicists on the nature of physical theories.

Chapter IV. Abstract Theories and Mechanical Models 55
1. Two kinds of minds: ample and deep. 2. An example of the ample mind: the mind of Napoleon. 3. The ample mind, the supple mind, and the geometrical mind. 4. Ampleness of mind and the English mind. 5. English

physics and the mechanical model. 6. The English school and mathematical physics. 7. The English school and the logical coordination of a theory. 8. The diffusion of English methods. 9. Is the use of mechanical models fruitful for discoveries? 10. Should the use of mechanical models suppress the search for an abstract and logically ordered theory?

PART II

THE STRUCTURE OF PHYSICAL THEORY

Chapter I. Quantity and Quality 107

1. Theoretical physics is mathematical physics. 2. Quantity and measurement. 3. Quantity and quality. 4. Purely quantitative physics. 5. The various intensities of the same quality are expressible in numbers.

Chapter II. Primary Qualities 121

1. On the excessive multiplication of primary qualities. 2. A primary quality is a quality irreducible in fact, not by law. 3. A quality is never primary, except provisionally.

Chapter III. Mathematical Deduction and Physical Theory 132

1. Physical approximation and mathematical precision. 2. Mathematical deductions physically useful and those not. 3. An example of mathematical deduction that can never be utilized. 4. The mathematics of approximation.

Chapter IV. Experiment in Physics 144

1. An experiment in physics is not simply the observation of a phenomenon; it is, besides, the theoretical interpretation of this phenomenon. 2. The result of an experiment in physics is an abstract and symbolic judgment. 3. The theoretical interpretation of phenomena alone makes possible the use of instruments. 4. On criticism of an experiment in physics; in what respects it differs from the examination of ordinary testimony. 5. Experiment in physics is less certain but more precise and detailed than the non-scientific establishment of a fact.

Chapter V. Physical Law 165

1. The laws of physics are symbolic relations. 2. A law of physics is, properly speaking, neither true nor false but approximate. 3. Every law of physics is provisional and

relative because it is approximate. 4. Every physical law is provisional because it is symbolic. 5. The laws of physics are more detailed than the laws of common sense.

Chapter VI. Physical Theory and Experiment 180

1. The experimental testing of a theory does not have the same logical simplicity in physics as in physiology. 2. An experiment in physics can never condemn an isolated hypothesis but only a whole theoretical group. 3. A "crucial experiment" is impossible in physics. 4. Criticism of the Newtonian method. First example: celestial mechanics. 5. Criticism of the Newtonian method (continued). Second example: electrodynamics. 6. Consequences relative to the teaching of physics. 7. Consequences relative to the mathematical development of physical theory. 8. Are certain postulates of physical theory incapable of being refuted by experiment? 9. On hypotheses whose statement has no experimental meaning. 10. Good sense is the judge of hypotheses which ought to be abandoned.

Chapter VII. The Choice of Hypotheses 219

1. What the conditions imposed by logic on the choice of hypotheses reduce to. 2. Hypotheses are not the product of sudden creation, but the result of progressive evolution. An example drawn from universal attraction. 3. The physicist does not choose the hypotheses on which he will base a theory; they germinate in him without him. 4. On the presentation of hypotheses in the teaching of physics. 5. Hypotheses cannot be deduced from axioms provided by common-sense knowledge. 6. The importance in physics of the historical method.

APPENDIX

Physics of a Believer 273

1. Introduction. 2. Our physical system is positivist in its origins. 3. Our physical system is positivist in its conclusions. 4. Our system eliminates the alleged objections of physical science to spiritualistic metaphysics and the Catholic faith. 5. Our system denies physical theory any metaphysical or apologetic import. 6. The metaphysician

should know physical theory in order not to make an illegitimate use of it in his speculations. 7. Physical theory has as its limiting form a natural classification. 8. There is an analogy between cosmology and physical theory. 9. On the analogy between physical theory and Aristotelian cosmology.

The Value of Physical Theory 312

Translator's Index 337

The Aim and Structure of Physical Theory

INTRODUCTION

We shall in this book offer a simple logical analysis of the method by which physical science makes progress. Perhaps certain readers will wish to extend the reflections put forth here to sciences other than physics; perhaps, also, they will desire to draw consequences transcending the proper aim of logic; but so far as we are concerned, we have scrupulously avoided both sorts of generalization. We have imposed narrow limits on our researches in order to explore more thoroughly the restricted domain we have assigned to our inquiry.

Before an experimenter makes use of an instrument for the study of a phenomenon, in his concern to be certain he will dismount the instrument, examine each portion, study the function and play of each part, and subject it to varied tests. He then knows exactly how reliable the readings of the instrument are, and what their limits of precision are; he may then use it with confidence.

Thus have we gone about the analysis of physical theory. We have sought, first of all, to determine precisely its *object* or *aim*. Then, knowing the end to which it is ordered, we have examined its *structure*. We have studied in successive order the mechanism of each one of the operations which go to make up a theory of physics, and have noted how each of them contributes to realizing the aim of the theory.

We have made a deliberate effort to clarify each of our assertions by means of examples, fearing above all things any locutions which fail to bring us into immediate contact with reality.

Furthermore, the doctrine put forth in this work is not a logical system resulting solely from the contemplation of general ideas; it has not been constructed through the sort of meditation that is hostile to concrete detail. It was born and matured in the daily practice of the science.

There is scarcely a chapter of physical theory that we have not had to teach in every detail, and we have more than once tried to contribute to the progress of almost every such topic. The summary ideas on the aim and structure of physical theory now presented are the fruit of this labor, prolonged over a period of twenty years. This long testing period has made us confident that the ideas were correct and fruitful.

PART I

The Aim of Physical Theory

CHAPTER I

PHYSICAL THEORY AND METAPHYSICAL EXPLANATION

1. *Physical Theory Considered as Explanation*

THE FIRST QUESTION we should face is: What is the aim of a physical theory? To this question diverse answers have been made, but all of them may be reduced to two main principles:

"A physical theory," certain logicians have replied, "has for its object the *explanation* of a group of laws experimentally established."

"A physical theory," other thinkers have said, "is an abstract system whose aim is to *summarize* and *classify logically* a group of experimental laws without claiming to explain these laws."

We are going to examine these two answers one after the other, and weigh the reasons for accepting or rejecting each of them. We begin with the first, which regards a physical theory as an explanation.

But, first, what is an explanation?

To explain (explicate, *explicare*) is to strip reality of the appearances covering it like a veil, in order to see the bare reality itself.

The observation of physical phenomena does not put us into relation with the reality hidden under the sensible appearances, but enables us to apprehend the sensible appearances themselves in a particular and concrete form. Besides, experimental laws do not have material reality for their object, but deal with these sensible appearances, taken, it is true, in an abstract and general form. Removing or tearing away the veil from these sensible appearances, theory proceeds into and underneath them, and seeks what is really in bodies.

For example, string or wind instruments have produced sounds to which we have listened closely and which we have heard become stronger or weaker, higher or lower, in a thousand nuances productive in us of auditory sensations and musical emotions; such are the acoustic facts.

These particular and concrete sensations have been elaborated by our intelligence, following the laws by which it functions, and have provided us with such general and abstract notions as intensity, pitch, octave, perfect major or minor chord, timbre, etc. The experimental laws of acoustics aim at the enunciation of fixed relations among these and other equally abstract and general notions. A law, for example, teaches us what relation exists between the dimensions of two strings of the same metal which yield two sounds of the same pitch or two sounds an octave apart.

But these abstract notions—sound intensity, pitch, timbre, etc.—depict to our reason no more than the general characteristics of our sound perceptions; these notions get us to know sound as it is in relation to us, not as it is by itself in sounding bodies. This reality whose external veil alone appears in our sensations is made known to us by theories of acoustics. The latter are to teach us that where our perceptions grasp only that appearance we call sound, there is in reality a very small and very rapid periodic motion; that intensity and pitch are only external aspects of the amplitude and frequency of this motion; and that timbre is the apparent manifestation of the real structure of this motion, the complex sensation which results from the diverse vibratory motions into which we can analyze it. Acoustic theories are therefore explanations.

The explanation which acoustic theories give of experimental laws governing sound claims to give us certainty; it can in a great many cases make us see with our own eyes the motions to which it attributes these phenomena, and feel them with our fingers.

Most often we find that physical theory cannot attain that degree of perfection; it cannot offer itself as a *certain* explanation of sensible appearances, for it cannot render accessible to the senses the reality it proclaims as residing underneath those appearances. It is then content with proving that all our perceptions are produced *as if* the reality were what it asserts; such a theory is a hypothetical explanation.

Let us, for example, take the set of phenomena observed with the sense of sight. The rational analysis of these phenomena leads us to conceive certain abstract and general notions expressing the properties we come across in every perception of light: a simple or complex color, brightness, etc. Experimental laws of optics make us acquainted with fixed relations among these abstract and general notions as well as among other analogous notions. One law, for instance, connects the intensity of yellow light reflected by a thin plate with the thickness of the plate and the angle of incidence of the rays which illuminate it.

Of these experimental laws the vibratory theory of light gives a hypothetical explanation. It supposes that all the bodies we see, feel, or weigh are immersed in an imponderable, unobservable medium called the ether. To this ether certain mechanical properties are attributed; the theory states that all simple light is a transverse vibration, very small and very rapid, of this ether, and that the frequency and amplitude of this vibration characterize the color of this light and its brightness; and, without enabling us tò perceive the ether, without putting us in a position to observe directly the back-and-forth motion of light vibration, the theory tries to prove that its postulates entail consequences agreeing at every point with the laws furnished by experimental optics.

2. *According to the Foregoing Opinion, Theoretical Physics Is Subordinate to Metaphysics*

When a physical theory is taken as an explanation, its goal is not reached until every sensible appearance has been removed in order to grasp the physical reality. For example, Newton's research on the dispersion of light has taught us to decompose the sensation we experience of light emanating from the sun; his experiments have shown us that this light is complex and resolvable into a certain number of simpler light phenomena, each associated with a determinate and invariable color. But these simple or monochromatic light data are abstract and general representations of certain sensations; they are sensible appearances, and we have only dissociated a complicated appearance into other simpler appearances. But we have not reached the real thing, we have not given an explanation of the color effects, we have not constructed an optical theory.

Thus, it follows that in order to judge whether a set of propositions constitutes a physical theory or not, we must inquire whether the notions connecting these propositions express, in an abstract and general form, the elements which really go to make up material things, or merely represent the universal properties perceived.

For such an inquiry to make sense or to be at all possible, we must first of all regard as certain the following affirmation: Under the sensible appearances, which are revealed in our perceptions, there is a reality distinct from these appearances.

This point granted, and without it the search for a physical explanation could not be conceived, it is impossible to recognize having reached such an explanation until we have answered this next question: What is the nature of the elements which constitute material reality?

Now these two questions—Does there exist a material reality distinct from sensible appearances? and What is the nature of this reality?—do not have their source in experimental method, which is acquainted only with sensible appearances and can discover nothing beyond them. The resolution of these questions transcends the methods used by physics; it is the object of metaphysics.

Therefore, *if the aim of physical theories is to explain experimental laws, theoretical physics is not an autonomous science; it is subordinate to metaphysics.*

3. *According to the Foregoing Opinion, the Value of a Physical Theory Depends on the Metaphysical System One Adopts*

The propositions which make up purely mathematical sciences are, to the highest degree, universally accepted truths. The precision of language and the rigor of the methods of demonstration leave no room for any permanent divergences among the views of different mathematicians; over the centuries doctrines are developed by continuous progress without new conquests causing the loss of any previously acquired domains.

There is no thinker who does not wish for the science he cultivates a growth as calm and as regular as that of mathematics. But if there is a science for which this wish seems particularly legitimate, it is indeed theoretical physics, for of all the well-established branches of knowledge it surely is the one which least departs from algebra and geometry.

Now, to make physical theories depend on metaphysics is surely not the way to let them enjoy the privilege of universal consent. In fact, no philosopher, no matter how confident he may be in the value of the methods used in dealing with metaphysical problems, can dispute the following empirical truth: Consider in review all the domains of man's intellectual activity; none of the systems of thought arising in different eras or the contemporary systems born of different schools will appear more profoundly distinct, more sharply separated, more violently opposed to one another, than those in the field of metaphysics.

If theoretical physics is subordinated to metaphysics, the divisions separating the diverse metaphysical systems will extend into the domain of physics. A physical theory reputed to be satisfactory by the sectarians of one metaphysical school will be rejected by the partisans of another school.

Consider, for example, the theory of the action exerted by a magnet on iron, and suppose for a moment that we are Aristotelians.

What does the metaphysics of Aristotle teach us concerning the real nature of bodies? Every substance—in particular, every material substance—results from the union of two elements: one permanent (matter) and one variable (form). Through its permanence, the piece of matter before me remains always and in all circumstances the same piece of iron. Through the variations which its form undergoes, through the *alterations* that it experiences, the properties of this same piece of iron may change according to circumstances; it may be solid or liquid, hot or cold, and assume such and such a shape.

Placed in the presence of a magnet, this piece of iron undergoes a special alteration in its form, becoming more intense with the proximity of the magnet. This alteration corresponds to the appearance of two poles and gives the piece of iron a principle of movement such that one pole tends to draw near the pole opposite to it on the magnet and the other to be repelled by the one designated as the like pole on the magnet.

Such for the Aristotelian philosopher is the reality hidden under the magnetic phenomena; when we have analyzed all these phenomena by reducing them to the properties of the magnetic quality of the two poles, we have given a complete explanation and formulated a theory altogether satisfactory. It was such a theory that Niccolo Cabeo constructed in 1629 in his remarkable work on magnetic philosophy.[1]

If an Aristotelian declares he is satisfied with the theory of magnetism as Father Cabeo conceives it, the same will not be true of a Newtonian philosopher faithful to the cosmology of Father Boscovich.

According to the natural philosophy which Boscovich has drawn from the principles of Newton and his disciples,[2] to explain the laws of the action which the magnet exerts on the iron by a magnetic alteration of the substantial form of the iron is to explain nothing

[1] Nicolaus Cabeus, S. J., *Philosophia magnetica, in qua magnetis natura penitus explicatur et omnium quae hoc lapide cernuntur causae propriae afferuntur, multa quoque dicuntur de electricis et aliis attractionibus, et eorum causis* (Cologne: Joannem Kinckium, 1629).

[2] P. Rogerio Josepho Boscovich, S. J., *Theoria philosophiae naturalis redacta ad unicam legem virium in natura existentium* (Vienna, 1758).

at all; we are really concealing our ignorance of reality under words that sound deep but are hollow.

Material substance is not composed of matter and form; it can be resolved into an immense number of points, deprived of extension and shape but having mass; between any two of these points is exerted a mutual attraction or repulsion proportional to the product of the masses and to a certain function of the distance separating them. Among these points there are some which form the bodies themselves. A mutual action takes place among the latter points, and as soon as the distances separating them exceed a certain limit, this action becomes the universal gravitation studied by Newton. Other points, deprived of this action of gravity, compose weightless fluids such as electric fluids and calorific fluid. Suitable assumptions about the masses of all these material points, about their distribution, and about the form of the functions of the distance on which their mutual actions depend are to account for all physical phenomena.

For example, in order to explain magnetic effects, we imagine that each molecule of iron carries equal masses of south magnetic fluid and north magnetic fluid; that the distribution of the fluids about this molecule is governed by the laws of mechanics; that two magnetic masses exert on one another an action proportional to the product of those masses and to the inverse square of the distance between them; finally, that this action is a repulsion or an attraction according to whether the masses are of the same or of different kinds. Thus was developed the theory of magnetism which, inaugurated by Franklin, Oepinus, Tobias Mayer, and Coulomb, came to full flower in the classical memoirs of Poisson.

Does this theory give an explanation of magnetic phenomena capable of satisfying an atomist? Surely not. Among some portions of magnetic fluid distant from one another, the theory admits the existence of actions of attraction or repulsion; for an atomist such actions at a distance amount to appearances which cannot be taken for realities.

According to the atomistic teachings, matter is composed of very small, hard, and rigid bodies of diverse shapes, scattered profusely in the void. Separated from each other, two such corpuscles cannot in any way influence each other; it is only when they come in contact with one another that their impenetrable natures clash and that their motions are modified according to fixed laws. The magnitudes, shapes, and masses of the atoms, and the rules governing their impact alone provide the sole satisfactory explanation which physical laws can admit.

In order to explain in an intelligible manner the various motions which a piece of iron undergoes in the presence of a magnet, we have to imagine that floods of magnetic corpuscles escape from the magnet in compressed, though invisible and intangible, streams, or else are precipitated toward the magnet. In their rapid course these corpuscles collide in various ways with the molecules of the iron, and from these collisions arise the forces which a superficial philosophy attributed to magnetic attraction and repulsion. Such is the principle of a theory of the magnet's properties already outlined by Lucretius, developed by Gassendi in the seventeenth century, and often taken up again since that time.

Shall we not find more minds, difficult to satisfy, who condemn this theory for not explaining anything at all and for taking appearances for reality? Here is where the Cartesians appear.

According to Descartes, matter is essentially identical with the extended in length, breadth, and depth, as the language of geometry goes; we have to consider only its various shapes and motions. Matter for the Cartesians is, if you please, a kind of vast fluid, incompressible and absolutely homogeneous. Hard, unbreakable atoms and the empty spaces separating them are merely so many appearances, so many illusions. Certain parts of the universal fluid may be animated by constant whirling or vortical motions; to the coarse eyes of the atomist these whirlpools or vortices will look like individual corpuscles. The intermediary fluid transmits from one vortex to the other forces which Newtonians, through insufficient analysis, will take for actions at a distance. Such are the principles of the physics first sketched by Descartes, which Malebranche investigated further, and to which W. Thomson, aided by the hydrodynamic researches of Cauchy and Helmholtz, has given the elaboration and precision characteristic of present-day mathematical doctrines.

This Cartesian physics cannot dispense with a theory of magnetism; Descartes had already tried to construct such a theory. The corkscrews of "subtle matter" with which Descartes, not without some naïveté, in his theory replaced the magnetic corpuscles of Gassendi were succeeded, among the Cartesians of the nineteenth century, by the vortices conceived more scientifically by Maxwell.

Thus we see each philosophical school glorifying a theory which reduces magnetic phenomena to the elements with which it composes the essence of matter, but the other schools rejecting this theory, in which their principles do not let them recognize a satisfactory explanation of magnetism.

4. *The Quarrel over Occult Causes*

There is one form of criticism which very often occurs when one cosmological school attacks another school: the first accuses the second of appealing to "occult causes."

The great cosmological schools, the Aristotelian, the Newtonian, the atomistic, and the Cartesian, may be arranged in an order such that each admits the existence in matter of a smaller number of essential properties than the preceding schools are willing to admit.

The Aristotelian school composes the substance of bodies out of only two elements, matter and form; but this form may be affected by qualities whose number is not limited. Each physical property can thus be attributed to a special quality: a *sensible* quality, directly accessible to our perception, like weight, solidity, fluidity, heat, or brightness; or else an *occult* quality whose effects alone will appear in an indirect manner, as with magnetism or electricity.

The Newtonians reject this endless multiplication of qualities in order to simplify, to a high degree, the notion of material substance: in the elements of matter they leave only masses, mutual actions, and shapes, when they do not go as far as Boscovich and several of his successors, who reduce the elements to unextended points.

The atomistic school goes further: its material elements preserve mass, shape, and hardness. But the forces through which the elements act on one another, according to the Newtonian school, disappear from the domain of realities; they are regarded merely as appearances and fictions.

Finally, the Cartesians push to the limit this tendency to strip material substances of various properties: they reject the hardness of atoms and even the distinction between plenum and void, in order to identify matter, as Leibniz said, with "completely naked extension and its modification."[3]

Thus each cosmological school admits in its explanations certain properties of matter which the next school refuses to take as real, for the latter regards them as mere words designating more deeply

[3] G. W. Leibniz, *Oeuvres*, ed. Gerhardt, IV, 464. (Translator's note: See Leibniz, *Selections* [Charles Scribner's Sons, 1951], pp. 100ff.).

hidden realities without revealing them; it groups them, in short, with the occult qualities created in so much profusion by scholasticism.

It is hardly necessary to recall that all the cosmological schools other than the Aristotelian have agreed in attacking the latter for the arsenal of qualities which it stored in substantial form, an arsenal which added a new quality each time a new phenomenon had to be explained. But Aristotelian physics has not been the only one obliged to meet such criticisms.

The Newtonians who endow material elements with attractions and repulsions acting at a distance seem to the atomists and Cartesians to be adopting one of those purely verbal explanations usual with the old Scholasticism. Newton's *Principia* had hardly been published when his work excited the sarcasm of the atomistic clan grouped around Huygens. "So far as concerns the cause of the tides given by Mr. Newton," Huygens wrote Leibniz, "I am far from satisfied, nor do I feel happy about any of his other theories built on his principle of attraction, which to me appears absurd."[4]

If Descartes had been alive at that time, he would have used a language similar to that of Huygens. In fact, Father Mersenne had submitted to Descartes a work by Roberval[5] in which the author adopted a form of universal gravitation long before Newton. On April 20, 1646 Descartes expressed his opinion as follows:

"Nothing is more absurd than the assumption added to the foregoing; the author assumes that a certain property is inherent in each of the parts of the world's matter and that, by the force of this property, the parts are carried toward one another and attract each other. He also assumes that a like property inheres in each part of the earth considered in relation with the other parts of the earth, and that this property does not in any way disturb the preceding one. In order to understand this, we must not only assume that each material particle is animated, and even animated by a large number of diverse souls that do not disturb each other, but also that these souls of material particles are endowed with knowledge of a truly divine sort, so that they may know without any

[4] Christian Huygens to G. W. Leibniz, Nov. 18, 1690, *Oeuvres complètes de Huygens, Correspondance*, 10 vols. (The Hague, 1638-1695), IX, 52. (Translator's note: The complete edition of Huygens' Collected Works was published in twenty-two volumes by the Holland Society of Sciences [Haarlem, 1950].)

[5] *Aristarchi Samii "De mundi systemate, partibus et motibus ejusdem, liber singularis"* (Paris, 1643). This work was reproduced in 1647 in Volume III of the *Cogitata physico-mathematica* by Marin Mersenne. [See pp. 242-243, below.]

medium what takes place at very great distances and act accordingly."[6]

The Cartesians agree, then, with the atomists when it comes to condemning as an occult quality the action at a distance which Newtonians invoke in their theories; but turning next against the atomists, the Cartesians deal just as harshly with the hardness and indivisibility attributed to corpuscles by the atomists. The Cartesian Denis Papin wrote to the atomist Huygens: "Another thing that bothers me is . . . that you believe that perfect hardness is of the essence of bodies; it seems to me that you are there assuming an inherent quality which takes us beyond mathematical or mechanical principles."[7] The atomist Huygens, it is true, did not deal less harshly with Cartesian opinion: "Your other difficulty," he replied to Papin, "is that I assume hardness to be of the essence of bodies whereas you and Descartes admit only their extension. By which I see that you have not yet rid yourself of that opinion which I have for a long time judged very absurd."[8]

5. *No Metaphysical System Suffices in Constructing a Physical Theory*

Each of the metaphysical schools scolds its rivals for appealing in its explanations to notions which are themselves unexplained and are really occult qualities. Could not this criticism be nearly always applied to the scolding school itself?

In order for the philosophers belonging to a certain school to declare themselves completely satisfied with a theory constructed by the physicists of the same school, all the principles used in this theory would have to be deduced from the metaphysics professed by that school. If an appeal is made, in the course of the explanation of a physical phenomenon, to some law which that metaphysics is powerless to justify, then no explanation will be forthcoming and physical theory will have failed in its aim.

Now, no metaphysics gives instruction exact enough or detailed enough to make it possible to derive all the elements of a physical theory from it.

In fact, the instruction furnished by a metaphysical doctrine concerning the real nature of bodies consists most often of negations.

[6] R. Descartes, *Correspondance*, ed. P. Tannery and C. Adam, Vol. IV (Paris, 1893), Letter CLXXX, p. 396.

[7] Denis Papin to Christian Huygens, June 18, 1690, *Oeuvres complètes de Huygens* . . . , IX, 429.

[8] Christian Huygens to Denis Papin, Sept. 2, 1690, *ibid.*, IX, 484.

The Aristotelians, like the Cartesians, deny the possibility of empty space; the Newtonians reject any quality which is not reducible to a force acting among material points; the atomists and Cartesians deny any action at a distance; the Cartesians do not recognize among the diverse parts of matter any distinctions other than shape and motion.

All these negations are appropriately argued when it is a matter of condemning a theory proposed by an adverse school; but they appear singularly sterile when we wish to derive the principles of a physical theory.

Descartes, for example, denied that there is anything else in matter than extension in length, breadth, depth, and its diverse modes—that is to say, shapes and motions; but with these data alone, he could not even begin to sketch the explanation of a physical law.

At the very least, before attempting the construction of any theory, he would have had to know the general laws governing diverse motions. Hence, he proceeded from his metaphysical principles to attempt first of all to deduce a dynamics.

The perfection of God requires him to be immutable in his plans; from this immutability the following consequence is drawn: God preserves as constant the quantity of motion that he gave the world in the beginning.

But this constancy of the quantity of motion in the world is still not a sufficiently precise or definite principle to make it possible for us to write any equation of dynamics. We must state it in a quantitative form, and that means translating the hitherto very vague notion of "quantity of motion" into a completely determined algebraic expression.

What, then, will be the mathematical meaning to be attached by the physicist to the words "quantity of motion"?

According to Descartes, the quantity of motion of each material particle will be the product of its mass—or of its volume, which in Cartesian physics is identical with its mass—times the velocity with which it is animated, and the quantity of motion of all matter in its entirety will be the sum of the quantities of motion of its diverse parts. This sum should in any physical change retain a constant value.

Certainly the combination of algebraic magnitudes through which Descartes proposed to translate the notion of "quantity of motion" satisfies the requirements imposed in advance by our instinctive knowledge of such a translation. It is zero for a whole at rest, and always positive for a group of bodies agitated by a certain motion;

its value increases when a determined mass increases the velocity of its movement; it increases again when a given velocity affects a larger mass. But an infinity of other expressions might just as well have satisfied these requirements: for the velocity we might notably have substituted the square of the velocity. The algebraic expression obtained would then have coincided with what Leibniz was to call "living force"; instead of drawing from divine immutability the constancy of the Cartesian quantity of motion in the world, we should have deduced the constancy of the Leibnizian living force.

Thus, the law which Descartes proposed to place at the base of dynamics undoubtedly agrees with the Cartesian metaphysics; but this agreement is not necessary. When Descartes reduced certain physical effects to mere consequences of such a law, he proved, it is true, that these effects do not contradict his principles of philosophy, but he did not give an explanation of the law by means of these principles.

What we have just said about Cartesianism can be repeated about any metaphysical doctrine which claims to terminate in a physical theory; in this theory there are always posited certain hypotheses which do *not* have as their grounds the principles of the metaphysical doctrine. Those who follow the thought of Boscovich admit that all the attractions or repulsions which are observable at a perceptible distance vary inversely with the square of the distance. It is this hypothesis which permits them to construct three systems of mechanics: celestial, electrical, and magnetic; but this form of law is dictated to them by the desire to have their explanations agree with the facts and not by the requirements of their philosophy. The atomists admit that a certain law governs the collisions of corpuscles; but this law is a singularly bold extension to the atomic world of another law which is permissible only when masses big enough to be observed are considered; it is not deduced from the Epicurean philosophy.

We cannot therefore derive from a metaphysical system all the elements necessary for the construction of a physical theory. The latter always appeals to propositions which the metaphysical system has not furnished and which consequently remain mysteries for the partisans of that system. At the root of the explanations it claims to give there always lies the unexplained.

CHAPTER II

PHYSICAL THEORY AND NATURAL CLASSIFICATION

1. *What Is the True Nature of a Physical Theory and the Operations Constituting It?*

WHILE we regard a physical theory as a hypothetical explanation of material reality, we make it dependent on metaphysics. In that way, far from giving it a form to which the greatest number of minds can give their assent, we limit its acceptance to those who acknowledge the philosophy it insists on. But even they cannot be entirely satisfied with this theory since it does not draw all its principles from the metaphysical doctrine from which it is claimed to be derived.

These thoughts, discussed in the preceding chapter, lead us quite naturally to ask the following two questions:

Could we not assign an aim to physical theory that would render it *autonomous*? Based on principles which do not arise from any metaphysical doctrine, physical theory might be judged in its own terms without including the opinions of physicists who depend on the philosophical schools to which they may belong.

Could we not conceive a method which might be *sufficient* for the construction of a physical theory? Consistent with its own definition the theory would employ no principle and have no recourse to any procedure which it could not legitimately use.

We intend to concentrate on this aim and this method, and to study both.

Let us posit right now a definition of physical theory; the sequel of this book will clarify it and will develop its complete content: A physical theory is not an explanation. It is a system of mathematical propositions, deduced from a small number of principles, which aim to represent as simply, as completely, and as exactly as possible a set of experimental laws.

In order to start making this definition somewhat more precise, let us characterize the four successive operations through which a physical theory is formed:

1. Among the physical properties which we set ourselves to repre-

sent we select those we regard as simple properties, so that the others will supposedly be groupings or combinations of them. We make them correspond to a certain group of mathematical symbols, numbers, and magnitudes, through appropriate methods of measurement. These mathematical symbols have no connection of an intrinsic nature with the properties they represent; they bear to the latter only the relation of sign to thing signified. Through methods of measurement we can make each state of a physical property correspond to a value of the representative symbol, and vice versa.

2. We connect the different sorts of magnitudes, thus introduced, by means of a small number of propositions which will serve as principles in our deductions. These principles may be called "hypotheses" in the etymological sense of the word for they are truly the grounds on which the theory will be built; but they do not claim in any manner to state real relations among the real properties of bodies. These hypotheses may then be formulated in an arbitrary way. The only absolutely impassable barrier which limits this arbitrariness is logical contradiction either among the terms of the same hypothesis or among the various hypotheses of the same theory.

3. The diverse principles or hypotheses of a theory are combined together according to the rules of mathematical analysis. The requirements of algebraic logic are the only ones which the theorist has to satisfy in the course of this development. The magnitudes on which his calculations bear are not claimed to be physical realities, and the principles he employs in his deductions are not given as stating real relations among those realities; therefore it matters little whether the operations he performs do or do not correspond to real or conceivable physical transformations. All that one has the right to demand of him is that his syllogisms be valid and his calculations accurate.

4. The various consequences thus drawn from the hypotheses may be translated into as many judgments bearing on the physical properties of the bodies. The methods appropriate for defining and measuring these physical properties are like the vocabulary and key permitting one to make this translation. These judgments are compared with the experimental laws which the theory is intended to represent. If they agree with these laws to the degree of approximation corresponding to the measuring procedures employed, the theory has attained its goal, and is said to be a good theory; if not, it is a bad theory, and it must be modified or rejected.

Thus a true theory is not a theory which gives an explanation of physical appearances in conformity with reality; it is a theory which represents in a satisfactory manner a group of experimental laws. A false theory is not an attempt at an explanation based on assumptions contrary to reality; it is a group of propositions which do not agree with the experimental laws. *Agreement with experiment is the sole criterion of truth for a physical theory.*

The definition we have just outlined distinguishes four fundamental operations in a physical theory: (1) the definition and measurement of physical magnitudes; (2) the selection of hypotheses; (3) the mathematical development of the theory; (4) the comparison of the theory with experiment.

Each one of these operations will occupy us in detail as we proceed with this book, for each of them presents difficulties calling for minute analysis. But right now it is possible for us to answer a few questions and to refute a few objections raised by the present definition of physical theory.

2. *What Is the Utility of a Physical Theory? Theory Considered as an Economy of Thought*

And first, of what use is such a theory?

Concerning the very nature of things, or the realities hidden under the phenomena we are studying, a theory conceived on the plan we have just drawn teaches us absolutely nothing, and does not claim to teach us anything. Of what use is it, then? What do physicists gain by replacing the laws which experimental method furnishes directly with a system of mathematical propositions representing those laws?

First of all, instead of a great number of laws offering themselves as independent of one another, each having to be learnt and remembered on its own account, physical theory substitutes a very small number of propositions, viz., fundamental hypotheses. The hypotheses once known, mathematical deduction permits us with complete confidence to call to mind all the physical laws without omission or repetition. Such condensing of a multitude of laws into a small number of principles affords enormous relief to the human mind, which might not be able without such an artifice to store up the new wealth it acquires daily.

The reduction of physical laws to theories thus contributes to that "intellectual economy" in which Ernst Mach sees the goal and directing principle of science.[1]

[1] E. Mach, "Die ökonomische Natur der physikalischen Forschung," *Populär-wissenschaftliche Vorlesungen* (3rd ed.; Leipzig, 1903), Ch. XIII, p. 215.

The experimental law itself already represented a first intellectual economy. The human mind had been facing an enormous number of concrete facts, each complicated by a multitude of details of all sorts; no man could have embraced and retained a knowledge of all these facts; none could have communicated this knowledge to his fellows. Abstraction entered the scene. It brought about the removal of everything private or individual from these facts, extracting from their total only what was general in them or common to them, and in place of this cumbersome mass of facts it has substituted a single proposition, occupying little of one's memory and easy to convey through instruction: it has formulated a physical law.

"Thus, instead of noting individual cases of light-refraction, we can mentally reconstruct all present and future cases, if we know that the incident ray, the refracted ray, and the perpendicular lie in the same plane and that $\sin i/\sin r = n$. Here, instead of the numberless cases of refraction in different combinations of matter and under all different angles of incidence, we have simply to note the rule above stated and the values of n—which is much easier. The economical purpose here is unmistakable."[2]

The economy achieved by the substitution of the law for the concrete facts is redoubled by the mind when it condenses experimental laws into theories. What the law of refraction is to the innumerable facts of refraction, optical theory is to the infinitely varied laws of light phenomena.

Among the effects of light only a very small number had been reduced to laws by the ancients; the only laws of optics they knew were the law of the rectilinear propagation of light and the laws of reflection. This meager contingent was reinforced in Descartes' time by the law of refraction. An optics so slim could do without theory; it was easy to study and teach each law by itself.

Today, on the contrary, how can a physicist who wishes to study

---

(Translator's note: Translated by T. J. McCormack, "The Economical Nature of Physical Research," Mach's *Popular Scientific Lectures* [3rd ed.; La Salle, Ill.: Open Court, 1907], Ch. XIII.)

See also E. Mach, *La Mécanique; exposé historique et critique de son développement* (Paris, 1904), Ch. IV, Sec. 4: "La Science comme économie de la pensée," p. 449. (Translator's note: Translated from the German 2nd ed. by T. J. McCormack, *The Science of Mechanics: a Critical and Historical Account of Its Development* [Open Court, 1902], Ch. IV, Sec. iv: "The Economy of Science," pp. 481-494.)

[2] E. Mach, *La Mécanique* . . . , p. 453. (Translator's note: Translated in *The Science of Mechanics* . . . , p. 485.)

optics, as we know it, acquire even a superficial knowledge of this enormous domain without the aid of a theory? Consider the effects of simple refraction, of double refraction by uniaxial or biaxial crystals, of reflection on isotropic or crystalline media, of interference, of diffraction, of polarization by reflection and by simple or double refraction, of chromatic polarization, of rotary polarization, etc. Each one of these large categories of phenomena may occasion the statement of a large number of experimental laws whose number and complication would frighten the most capable and retentive memory.

Optical theory supervenes, takes possession of these laws, and condenses them into a small number of principles. From these principles we can always, through regular and sure calculation, extract the law we wish to use. It is no longer necessary, therefore, to keep watch over the knowledge of all these laws; the knowledge of the principles on which they rest is sufficient.

This example enables us to take firm hold of the way the physical sciences progress. The experimenter constantly brings to light facts hitherto unsuspected and formulates new laws, and the theorist constantly makes it possible to store up these acquisitions by imagining more condensed representations, more economical systems. The development of physics incites a continual struggle between "nature that does not tire of providing" and reason that does not wish "to tire of conceiving."

3. *Theory Considered as Classification*

Theory is not solely an economical representation of experimental laws; it is also a *classification* of these laws.

Experimental physics supplies us with laws all lumped together and, so to speak, on the same plane, without partitioning them into groups of laws united by a kind of family tie. Very often quite accidental causes or rather superficial analogies have led observers in their research to bring together different laws. Newton put into the same work the laws of the dispersion of light crossing a prism and the laws of the colors adorning a soap bubble, simply because of the colors that strike the eye in these two sorts of phenomena.

On the other hand, theory, by developing the numerous ramifications of the deductive reasoning which connects principles to experimental laws, establishes an order and a classification among these laws. It brings some laws together, closely arranged in the same group; it separates some of the others by placing them in two groups very far apart. Theory gives, so to speak, the table of contents and the chapter headings under which the science to be studied will be methodically divided, and it indicates the laws which are to be arranged under each of these chapters.

Thus, alongside the laws which govern the spectrum formed by a prism it arranges the laws governing the colors of the rainbow; but the laws according to which the colors of Newton's rings are ordered go elsewhere to join the laws of fringes discovered by Young and Fresnel; still in another category, the elegant coloration analyzed by Grimaldi is considered related to the diffraction spectra produced by Fraunhofer. The laws of all these phenomena, whose striking colors lead to their confusion in the eyes of the simple observer, are, thanks to the efforts of the theorist, classified and ordered.

These classifications make knowledge convenient to use and safe to apply. Consider those utility cabinets where tools for the same purpose lie side by side, and where partitions logically separate instruments not designed for the same task: the worker's hand quickly grasps, without fumbling or mistake, the tool needed. Thanks to theory, the physicist finds with certitude, and without omitting anything useful or using anything superfluous, the laws which may help him solve a given problem.

Order, wherever it reigns, brings beauty with it. Theory not only renders the group of physical laws it represents easier to handle, more convenient, and more useful, but also more beautiful.

It is impossible to follow the march of one of the great theories of physics, to see it unroll majestically its regular deductions starting from initial hypotheses, to see its consequences represent a multitude of experimental laws down to the smallest detail, without being charmed by the beauty of such a construction, without feeling keenly that such a creation of the human mind is truly a work of art.

4. *A Theory Tends to Be Transformed into a Natural Classification*[3]

This esthetic emotion is not the only reaction that is produced by a theory arriving at a high degree of perfection. It persuades us also to see a natural classification in a theory.

Now first, what is a natural classification? For example, what

[3] We have already noted natural classification as the ideal form toward which physical theory tends in "L'Ecole anglaise et les théories physiques," Art. 6, *Revue des questions scientifiques*, October 1893.

does a naturalist mean in proposing a natural classification of vertebrates?

The classification he has imagined is a group of intellectual operations not referring to concrete individuals but to abstractions, species; these species are arranged in groups, the more particular under the more general. In order to form these groups the naturalist considers the diverse organs—vertebral column, cranium, heart, digestive tube, lungs, swim-bladder—not in the particular and concrete forms they assume in each individual, but in the abstract, general, schematic forms which fit all the species of the same group. Among these organs thus transfigured by abstraction he establishes comparisons, and notes analogies and differences; for example, he declares the swim-bladder of fish analogous to the lung of vertebrates. These homologies are purely ideal connections, not referring to real organs but to generalized and simplified conceptions formed in the mind of the naturalist; the classification is only a synoptic table which summarizes all these comparisons.

When the zoologist asserts that such a classification is natural, he means that those ideal connections established by his reason among abstract conceptions correspond to real relations among the associated creatures brought together and embodied in his abstractions. For example, he means that the more or less striking resemblances which he has noted among various species are the index of a more or less close blood-relationship, properly speaking, among the individuals composing these species; that the cascades through which he translates the subordination of classes, of orders, of families, and of genera reproduce the genealogical tree in which the various vertebrates are branched out from the same trunk and root. These relations of real family affiliation can be established only by comparative anatomy; to grasp them in themselves and put them in evidence is the business of physiology and of paleontology. However, when he contemplates the order which his methods of comparison introduce into the confused multitude of animals, the anatomist cannot assert these relations, the proof of which transcends his methods. And if physiology and paleontology should someday demonstrate to him that the relationship imagined by him cannot be, that the evolutionist hypothesis is controverted, he would continue to believe that the plan drawn by his classification depicts real relations among animals; he would admit being deceived about the nature of these relations but not about their existence.

The neat way in which each experimental law finds its place in the classification created by the physicist and the brilliant clarity imparted to this group of laws so perfectly ordered persuade us in an overwhelming manner that such a classification is not purely artificial, that such an order does not result from a purely arbitrary grouping imposed on laws by an ingenious organizer. Without being able to explain our conviction, but also without being able to get rid of it, we see in the exact ordering of this system the mark by which a natural classification is recognized. Without claiming to explain the reality hiding under the phenomena whose laws we group, we feel that the groupings established by our theory correspond to real affinities among the things themselves.

The physicist who sees in every theory an explanation is convinced that he has grasped in light vibration the proper and intimate basis of the quality which our senses reveal in the form of light and color; he believes in an ether, a body whose parts are excited by this vibration into a rapid to-and-fro motion.

Of course, we do not share these illusions. When, in the course of an optical theory, we talk about luminous vibration, we no longer think of a real to-and-fro motion of a real body; we imagine only an abstract magnitude, i.e., a pure, geometrical expression. It is a periodically variable length which helps us state the hypotheses of optics, and to regain by regular calculations the experimental laws governing light. This vibration is to our mind a *representation*, and not an *explanation*.

But when, after much groping, we succeed in formulating with the aid of this vibration a body of fundamental hypotheses, when we see in the plan drawn by these hypotheses a vast domain of optics, hitherto encumbered by so many details in so confused a way, become ordered and organized, it is impossible for us to believe that this order and this organization are not the reflected image of a real order and organization; that the phenomena which are brought together by the theory, e.g., interference bands and colorations of thin layers, are not in truth slightly different manifestations of the same property of light; and that phenomena separated by the theory, e.g., the spectra of diffraction and of dispersion, do not have good reasons for being in fact essentially different.

Thus, physical theory never gives us the explanation of experimental laws; it never reveals realities hiding under the sensible appearances; but the more complete it becomes, the more we apprehend that the logical order in which theory orders experimental laws is the reflection of an ontological order, the more we suspect that the relations it establishes among the data of observation correspond to real relations among things,[4] and the more we feel that theory tends to be a natural classification.

The physicist cannot take account of this conviction. The method at his disposal is limited to the data of observation. It therefore cannot prove that the order established among experimental laws reflects an order transcending experience; which is all the more reason why his method cannot suspect the nature of the real relations corresponding to the relations established by theory.

But while the physicist is powerless to justify this conviction, he is nonetheless powerless to rid his reason of it. In vain is he filled with the idea that his theories have no power to grasp reality, and that they serve only to give experimental laws a summary and classificatory representation. He cannot compel himself to believe that a system capable of ordering so simply and so easily a vast number of laws, so disparate at first encounter, should be a purely artificial system. Yielding to an intuition which Pascal would have recognized as one of those reasons of the heart "that reason does not know," he asserts his faith in a real order reflected in his theories more clearly and more faithfully as time goes on.

Thus the analysis of the methods by which physical theories are constructed proves to us with complete evidence that these theories cannot be offered as explanations of experimental laws; and, on the other hand, an act of faith, as incapable of being justified by this analysis as of being frustrated by it, assures us that these theories are not a purely artificial system, but a natural classification. And so, we may here apply that profound thought of Pascal: "We have an impotence to prove, which cannot be conquered by any dogmatism; we have an idea of truth which cannot be conquered by any Pyrrhonian skepticism."

5. *Theory Anticipating Experiment*

There is one circumstance which shows with particular clarity our belief in the natural character of a theoretical classification; this circumstance is present when we ask of a theory that it tell us the results of an experiment before it has occurred, when we give it the bold injunction: "Be a prophet for us."

A considerable group of experimental laws had been established by investigators; the theorist has proposed to condense the laws into a very small number of hypotheses, and has succeeded in doing so;

[4] Cf. H. Poincaré, *La Science et l'Hypothèse* (Paris, 1903), p. 190. (Translator's note: Translated by Bruce Halsted, "Science and Hypothesis" in *Foundations of Science* [Lancaster, Pa.: Science Press, 1905].)

each one of the experimental laws is correctly represented by a consequence of these hypotheses.

But the consequences that can be drawn from these hypotheses are unlimited in number; we can, then, draw some consequences which do not correspond to any of the experimental laws previously known, and which simply represent possible experimental laws.

Among these consequences, some refer to circumstances realizable in practice, and these are particularly interesting, for they can be submitted to test by facts. If they represent exactly the experimental laws governing these facts, the value of the theory will be augmented, and the domain governed by the theory will annex new laws. If, on the contrary, there is among these consequences one which is sharply in disagreement with the facts whose law was to be represented by the theory, the latter will have to be more or less modified, or perhaps completely rejected.

Now, on the occasion when we confront the predictions of the theory with reality, suppose we have to bet for or against the theory; on which side shall we lay our wager?

If the theory is a purely artificial system, if we see in the hypotheses on which it rests statements skillfully worked out so that they represent the experimental laws already known, but if the theory fails to hint at any reflection of the real relations among the invisible realities, we shall think that such a theory will fail to confirm a new law. That, in the space left free among the drawers adjusted for other laws, the hitherto unknown law should find a drawer already made into which it may be fitted exactly would be a marvelous feat of chance. It would be folly for us to risk a bet on this sort of expectation.

If, on the contrary, we recognize in the theory a natural classification, if we feel that its principles express profound and real relations among things, we shall not be surprised to see its consequences anticipating experience and stimulating the discovery of new laws; we shall bet fearlessly in its favor.

The highest test, therefore, of our holding a classification as a natural one is to ask it to indicate in advance things which the future alone will reveal. And when the experiment is made and confirms the predictions obtained from our theory, we feel strengthened in our conviction that the relations established by our reason among abstract notions truly correspond to relations among things.

Thus, modern chemical symbolism, by making use of developed formulas, establishes a classification in which diverse compounds are ordered. The wonderful order this classification brings about in the tremendous arsenal of chemistry already assures us that the classification is not a purely artificial system. The relations of analogy and derivation by substitution it establishes among diverse compounds have meaning only in our mind; yet, we are convinced that they correspond to kindred relations among substances themselves, whose nature remains deeply hidden but whose reality does not seem doubtful. Nevertheless, for this conviction to change into overwhelming certainty, we must see the theory write in advance the formulas of a multitude of bodies and, yielding to these indications, synthesis must bring to light a large number of substances whose composition and several properties we should know even before they exist.

Just as the syntheses announced in advance sanction chemical notation as a natural classification, so physical theory will prove that it is the reflection of a real order by anticipating observation.

Now the history of physics provides us with many examples of this clairvoyant guesswork; many a time has a theory forecast laws not yet observed, even laws which appear improbable, stimulating the experimenter to discover them and guiding him toward that discovery.

The Académie des Sciences had set, as the subject for the physics prize that was to be awarded in the public meeting of March 1819, the general examination of the phenomena of the diffraction of light. Two memoirs were presented, and one by Fresnel was awarded the prize, the commission of judges consisting of Biot, Arago, Laplace, Gay-Lussac, and Poisson.

From the principles put forward by Fresnel, Poisson deduced through an elegant analysis the following strange consequence: If a small, opaque, and circular screen intercepts the rays emitted by a point source of light, there should exist behind the screen, on the very axis of this screen, points which are not only bright, but which shine exactly as though the screen were not interposed between them and the source of light.

Such a corollary, so contrary, it seems, to the most obvious experimental certainties, appeared to be a very good ground for rejecting the theory of diffraction proposed by Fresnel. Arago had confidence in the natural character arising from the clairvoyance of this theory. He tested it, and observation gave results which agreed absolutely with the improbable predictions from calculation.[5]

[5] *Oeuvres complètes d'Augustin Fresnel,* 3 vols. (Paris, 1866-1870), I, 236, 365, 368.

Thus physical theory, as we have defined it, gives to a vast group of experimental laws a condensed representation, favorable to intellectual economy.

It classifies these laws and, by classifying, renders them more easily and safely utilizable. At the same time, putting order into the whole, it adds to their beauty.

It assumes, while being completed, the characteristics of a natural classification. The groups it establishes permit hints as to the real affinities of things.

This characteristic of natural classification is marked, above all, by the fruitfulness of the theory which anticipates experimental laws not yet observed, and promotes their discovery.

That sufficiently justifies the search for physical theories, which cannot be called a vain and idle task even though it does not pursue the explanation of phenomena.

CHAPTER III

REPRESENTATIVE THEORIES AND THE HISTORY OF PHYSICS

1. *The Role of Natural Classifications and of Explanations in the Evolution of Physical Theories*

WE HAVE PROPOSED that the aim of physical theory is to become a natural classification, to establish among diverse experimental laws a logical coordination serving as a sort of image and reflection of the true order according to which the realities escaping us are organized. Also, we have said that on this condition theory will be fruitful and will suggest discoveries.

But an objection immediately arises to the doctrine we are here expounding.

If theory is to be a natural classification, if it is to group appearances in the same way realities are grouped, then is not the surest way to reach this goal to inquire first what these realities are? Instead of constructing a logical system representing experimental laws in as condensed and as exact a form as possible, in the hope that this system will end by being an image of the ontological order of things, would it not make more sense to try to explain these laws and to unveil those hidden things? Moreover, is this not the way in which the masters of science have proceeded? Have they not, by striving for the explanation of physical phenomena, created those fruitful theories whose prophecies have taken hold and aroused our astonishment? Can we do better than imitate their example and return to the methods condemned in our first chapter?

There is no doubt that several of the geniuses to whom we owe modern physics have built their theories in the hope of giving an explanation of natural phenomena, and that some even have believed they had gotten hold of this explanation. But that, nevertheless, is no conclusive argument against the opinion we have expounded concerning physical theories. Chimerical hopes may have incited admirable discoveries without these discoveries embodying the chimeras which gave birth to them. Bold explorations which have contributed greatly to the progress of geography are due to adventurers who were looking for the golden land—that is not a sufficient reason for inscribing "El Dorado" on our maps of the globe.

Hence, if we want to prove that the search for explanations is a truly fruitful method in physics, it is not enough to show that a goodly number of theories has been created by thinkers who strove for such explanations; we have to prove that the search for explanation is indeed the Ariadne's thread which has led them through the confusion of experimental laws and has allowed them to draw the plan of this labyrinth.

Now it is not only impossible to give this proof, but, as we shall see, even a superficial study of the history of physics provides abundant arguments to the contrary.

When we analyze a theory created by a physicist who proposes to explain sensible appearances, we generally do not take long to recognize that this theory is formed of two really distinct parts: one is the simply representative part which proposes to classify laws; the other is the explanatory part which proposes to take hold of the reality underlying the phenomena.

Now, it is very far from being true that the explanatory part is the reason for the existence of the representative part, the seed from which it grew or the root which nourishes its development; actually, the link between the two parts is nearly always most frail and most artificial. The descriptive part has developed on its own by the proper and autonomous methods of theoretical physics; the explanatory part has come to this fully formed organism and attached itself to it like a parasite.

It is not to this explanatory part that theory owes its power and fertility; far from it. Everything good in the theory, by virtue of which it appears as a natural classification and confers on it the power to anticipate experience, is found in the representative part; all of that was discovered by the physicist while he forgot about the search for explanation. On the other hand, whatever is false in the theory and contradicted by the facts is found above all in the explanatory part; the physicist has brought error into it, led by his desire to take hold of realities.

Whence the following consequence: When the progress of experimental physics goes counter to a theory and compels it to be modified or transformed, the purely representative part enters nearly whole in the new theory, bringing to it the inheritance of all the valuable possessions of the old theory, whereas the explanatory part falls out in order to give way to another explanation.

Thus, by virtue of a continuous tradition, each theory passes on to the one that follows it a share of the natural classification it was able to construct, as in certain ancient games each runner handed on the lighted torch to the courier ahead of him, and this continuous tradition assures a perpetuity of life and progress for science.

This continuity of tradition is not visible to the superficial observer due to the constant breaking-out of explanations which arise only to be quelled.

Let us support all we have just said by some examples. They will be provided by the theories about the refraction of light. We shall borrow them from these theories, not, indeed, because they are exceptionally favorable to our thesis, but, on the contrary, because those who study the history of physics superficially might think that these theories owe their principal progress to the search for explanations.

Descartes has given a theory which *represents* the phenomena of simple refraction; it is the principal object of two admirable treatises, *Dioptrique* and *Météores*, to which the *Discours de la méthode* served as a preface. Based on the constant relation between the sine of the angle of incidence and the sine of the angle of refraction, his theory arranges in a very clear order the properties of lenses of diverse shapes and of optical instruments composed of these lenses; it takes account of the phenomena attending vision, and analyzes the laws of the rainbow.

Descartes has also given an explanation of light effects. Light is only an appearance; the reality is a pressure engendered by the rapid motions of incandescent bodies within a "subtle matter" penetrating all bodies. This subtle matter is incompressible, so that the pressure which constitutes light is transmitted in it instantaneously to any distance: no matter how far away a point is from a light source, at the very same instant the latter is lit, the point is lit. This instantaneous transmission of light is an absolutely necessary consequence of the system of physical explanations created by Descartes. Beeckman, who did not wish to admit this proposition and who, in imitation of Galileo, sought to contradict it by means of experiments, rather childish at that, was addressed by Descartes as follows: "To my mind, it [the instantaneous velocity of light] is so certain that if, by some impossibility, it were found guilty of being erroneous, I should be ready to acknowledge to you immediately that I know nothing in philosophy. You have such great confidence in your experiment that you declare yourself ready to hold all of your philosophy false if no lapse of time should separate the instant when one sees the motion of the lantern in the mirror from the instant when one perceives it in his hand; I, on the other hand, declare to you that if this lapse of time could be observed, then my whole philosophy would be completely upset."[1]

Whether Descartes himself created the fundamental law of refraction or borrowed it from Snell, as Huygens insinuated, has been a passionately debated question; the answer is doubtful, but it matters little to us. What is certain is that this law and the representative theory based on it are not offspring of the explanation of light phenomena proposed by Descartes; the Cartesian cosmology had no part in generating them; experiment, induction, and generalization have alone produced them.

Moreover, Descartes never made the attempt to connect the law of refraction with his explanatory theory of light.

It is indeed true that at the beginning of the *Dioptrique*, he developed mechanical analogies concerning this law; he compared the change of direction of the ray which passes from air into water to the change of the path of a ball thrown vigorously and passing from a certain medium into another more resistant one. But these mechanical comparisons, whose logical validity would be exposed to many criticisms, should rather connect the theory of refraction to the doctrine of emission, a doctrine in which a ray of light is compared with a shower of small particles violently projected by the source of light. This explanation, maintained in Descartes' time by Gassendi, and taken up later by Newton, has no analogy with the Cartesian theory of light; it is incompatible with the latter theory.

Thus, the Cartesian explanation of light phenomena and the Cartesian representation of the diverse laws of refraction are simply juxtaposed without any connection or penetration. Hence, the day when the Danish astronomer Römer showed that light is propagated in space with a finite and measurable velocity, the Cartesian explanation of light phenomena collapsed completely; but it did not bring down with it even the slightest part of the doctrine which represents and classifies the laws of refraction; the latter continues, even today, to form the major part of our elementary optics.

A single light ray, travelling from air into the interior of certain crystalline media such as Icelandic spar, provides two distinct refracted rays: one, the ordinary ray, follows Descartes' law, while the other, the extraordinary ray, escapes the confines of this law. This "admirable and unusual refraction of the cleavable crystal from Iceland" had been discovered and studied in 1657 by the Dane

[1] R. Descartes, *Correspondance*, ed. P. Tannery and C. Adam, Vol. I, Letter LVII (August 22, 1634), p. 307.

Erasmus Barthelsen or Bartholinus.[2] Huygens proposed to formulate a theory which represents at the same time the laws of simple refraction, the object of Descartes' works, and the laws of double refraction. He succeeded in doing this in the most felicitous manner. His geometric constructions, after furnishing in amorphous media or in cubic crystals the single ray following Descartes' law, not only trace in non-cubic crystals two refracted rays, but they also determine completely the laws governing these two rays. These laws are so complicated that experiment left to its own resources would not perhaps have unravelled them; but after theory has given the formula for them, experiment verifies them in detail.

Did Huygens draw this beautiful and fruitful theory from the principles of atomistic cosmology, from those "reasons of mechanics" by virtue of which, according to him, "the true philosophy conceives the cause of all nature's effects"? By no means. The consideration of the void, and of atoms and their hardness and motions played no role at all in the construction of this representation. A comparison between the propagation of sound and the propagation of light, the experimental fact that one of the two refracted rays followed Descartes' law while the other did not obey it, a felicitous and bold hypothesis about the form of the surface of the optical wave in media of crystals—such were the steps by which the great Dutch physicist proceeded to reveal the principles of his classification.

Not only did Huygens not draw the theory of double refraction from the principles of atomistic physics, but once this theory was discovered, he did not try to join it to those principles. In fact, in order to take account of the forms of crystals, he imagined the spar, or rock, crystal to be formed by regular piles of spheroidal molecules, thus preparing the way for Haüy and Bravais; but after developing this assumption, he was content to write: "I shall add only that these little spheroids might well contribute to form the spheroids of the light waves assumed above, several being situated in the same way with their axes parallel."[3] In this short sentence we have all that Huygens attempted in order to explain the form

[2] Erasmus Bartholinus, *Experimenta crystalli Islandici disdiaclastici, quibus mira et insolita refractio detegitur* (The Hague, 1657).

[3] Christian Huygens, *Traité de la lumière, où sont expliquées les causes de ce qui luy arrive dans la réflexion et dans la réfraction, et particulièrement dans l'étrange réfraction du cristal d'Islande* (Leyden, 1690), ed. W. Burckhardt (Paris, 1920), p. 71. (Translator's note: See English translation of this work, *Treatise on Light. In Which Are Explained the Causes of That Which Occurs in Reflexion, and in Refraction. And Particularly in the Strange Refraction of Iceland Crystal,* tr. S. P. Thompson [Chicago: Chicago University Press, 1945].)

of the surface of a light wave by attributing to the crystals an appropriate structure.

Thus his theory will remain intact, whereas the diverse explanations of light phenomena will continue to succeed one another, fragile and decrepit as they are, despite the confidence in their enduring value to which their authors will give testimony.

Under the influence of Newton, the emissionist, corpuscular explanation is absolutely contrary to the one Huygens, creator of the wave theory, gave to light phenomena; from this explanation, joined to a cosmology of attraction along the lines of Boscovich's principles, which the great Dutch physicist had considered absurd, Laplace draws a justification of Huygens' constructions.

Not only does Laplace explain by means of the physics of attraction the theory of simple or double refraction, discovered by a physicist who praised quite opposite ideas, not only does he deduce it "from those principles for which we are indebted to Newton, by means of which all the phenomena of the motion of light, traversing any number whatever of transparent media and the atmosphere, have been subjected to rigorous calculation,"[4] but he also thinks that this deduction augments the certainty and precision of the explanation. Undoubtedly, the solutions of the problems of double refraction given by Huygens' construction, "considered as an experimental result, may be placed on the high plane of the most beautiful discoveries of that rare genius. . . . We ought not to waver in placing them among the most certain as well as the most beautiful results in physics." But "hitherto this law has been only a result of observation, approximating truth, within the limits of error where the most precise experiments still fall. But the simplicity of the law of action on which it depends should make it be considered as a rigorous law." Laplace even goes so far, in his confidence in the value of the explanation he proposes, as to declare that this explanation alone could remove the improbabilities of Huygens' theory and render it acceptable to good minds; for "this law has experienced the same fate as Kepler's beautiful laws, which were not appreciated for a long time because of their having been associated with ideas of a system unfortunately permeating all of Kepler's works."

At the very time that Laplace was so disdainful of the optics of

[4] P. S. Laplace, *Exposition du système du monde* I (Paris, 1796), IV, Ch. XVIII: "De l'attraction moléculaire." (Translator's note: See English translations by J. Pond [Dublin, 1809] and by H. H. Harte [Dublin and London, 1830]. Laplace's nebular hypothesis appeared in a note [vii] in his work.)

waves, the latter, promoted by Young and Fresnel, superseded the optics of corpuscular emission; but, thanks to Fresnel, wave optics has undergone a profound change: the vibration of the source of light no longer acts in the direction of the ray but is perpendicular to it. The analogy between sound and light which guided Huygens has disappeared; nevertheless, the new explanation still leads physicists to adopt the construction of rays refracted by a crystal in the way Huygens imagined it.

However, in changing its explanatory part Huygens' doctrine has enriched its representative part; it no longer expresses only the laws governing the path of rays, but also the laws of their state of polarization.

The holders of this theory would now be in a good position to turn back against Laplace the scornful pity he evinced toward their stand; it becomes difficult to read without smiling the following sentences which the great mathematician wrote at the very moment when the optics of Fresnel was triumphing: "The phenomena of double refraction and of the aberration of the stars appeared to me to give to the system of the corpuscular emission of light, if not complete certainty, at least an extremely high probability. These phenomena are inexplicable on the hypothesis of waves of an ethereal fluid. The singular property of a ray polarized by a crystal, and not dividing again in passing through a second crystal parallel to the first, evidently indicates different actions of the same crystal on the diverse faces of a molecule of light."[5]

The theory of refraction given by Huygens did not cover all possible cases; a large category of crystalline bodies, viz., biaxial crystals, offered phenomena which could not enter into its framework. Fresnel proposed to enlarge this framework so that one could classify not only the laws of uniaxial double refraction but also the laws of biaxial double refraction. How did he succeed in doing this? By seeking an explanation of the mode of light propagation in crystals? By no means; he did it by geometrical intuition where there was no room for any hypothesis about the nature of light or about the constitution of transparent bodies. He noticed that all the wave surfaces that Huygens had had to consider could be deduced through a simple geometrical construction from a certain surface of the second degree. This surface was a sphere for singly refracting media, an ellipsoid of revolution for uniaxial doubly refracting media; he imagined that by applying the same construction to three

[5] *loc.cit.*

unequal axes, one could obtain the wave surface suited for biaxial crystals.

This bold intuition was rewarded with the most brilliant success; not only was the theory proposed by Fresnel in scrupulous agreement with all the experimental determinations, but it also made it possible to guess and discover unforeseen and paradoxical results which the experimenter, left to himself, would never have thought of looking for. Such are the two kinds of conical refraction. The great mathematician Hamilton deduced from the form of the wave surface of biaxial crystals the laws of those strange phenomena which the physicist Lloyd subsequently looked for and discovered.

The theory of biaxial double refraction therefore possesses that fruitfulness and predictive power in which we recognize the marks of a natural classification; yet they were not born of any attempt at explanation.

Not that Fresnel did not try to explain the form of wave surface that he had obtained; this attempt aroused him to such an emotional pitch that he did not publish the method which had led him to the discovery; this method became known only after his death, when his first memoir on double refraction was finally released for publication.[6] In the writings that he published, while alive, on double refraction, Fresnel tried constantly to reestablish, by means of hypotheses about the properties of ether, the laws he had discovered; "but these hypotheses from which he had made his principles do not stand a thorough examination."[7] Fresnel's theory is admirable when it is limited to playing the role of natural classification, but becomes untenable as soon as it is given as an explanation.

The same is true of most physical doctrines; what is lasting and fruitful in these is the logical work through which they have succeeded in classifying naturally a great number of laws by deducing them from a few principles; what is perishable and sterile is the labor undertaken to explain these principles in order to attach them to assumptions concerning the realities hiding underneath sensible appearances.

Scientific progress has often been compared to a mounting tide; applied to the evolution of physical theories, this comparison seems to us very appropriate, and it may be pursued in further detail.

Whoever casts a brief glance at the waves striking a beach does

[6] See "Introduction aux oeuvres d'Augustin Fresnel" by É. Verdet, Arts. 11 and 12, *Oeuvres complètes d'Augustin Fresnel*, Vol. I, pp. lxx and lxxvi.

[7] *ibid.*, p. 84.

not see the tide mount; he sees a wave rise, run, uncurl itself, and cover a narrow strip of sand, then withdraw by leaving dry the terrain which it had seemed to conquer; a new wave follows, sometimes going a little farther than the preceding one, but also sometimes not even reaching the sea shell made wet by the former wave. But under this superficial to-and-fro motion, another movement is produced, deeper, slower, imperceptible to the casual observer; it is a progressive movement continuing steadily in the same direction and by virtue of it the sea constantly rises. The going and coming of the waves is the faithful image of those attempts at explanation which arise only to be crumbled, which advance only to retreat; underneath there continues the slow and constant progress whose flow steadily conquers new lands, and guarantees to physical doctrines the continuity of a tradition.

2. *The Opinions of Physicists on the Nature of Physical Theories*

One of the thinkers who have insisted most energetically on the point that physical theories should be regarded as condensed representations and not as explanations, Ernst Mach, has written as follows:

"My idea of the economy of thought was developed out of my experience as a professor, and grew out of my practice in teaching. I possessed the idea as early as 1861 when I began my lectures as Privat Docent, and at the time I believed that I was in exclusive possession of the principle—a conviction which will be found pardonable. But today, on the contrary, I am convinced that at least some presentiment of this idea must have always been a common possession of *all* inquirers who have reflected on the nature of scientific research."[8]

Indeed, since antiquity there have been certain philosophers who have recognized that physical theories are by no means explanations, and that their hypotheses were not judgments about the nature of things, only premises intended to provide consequences conforming to experimental laws.[9]

[8] E. Mach, *La Mécanique; exposé historique et critique de son développement* (Paris, 1904), p. 360. (Translator's note: Translated from the German 2nd ed. by T. J. McCormack, *The Science of Mechanics: a Critical and Historical Account of Its Development* [Open Court, 1902], p. 579.)

[9] Since the first edition of this work we have on two occasions developed the thoughts that follow in the text. First, in a series of articles entitled "*Σώζειν τὰ φαινόμενα* Essai sur la notion de théorie physique de Platon à Galilée," *Annales de Philosophie chrétienne*, 1908. Second, in our work entitled *Le Système du Monde, Histoire des doctrines cosmologiques de Platon à Copernic,*

The Greeks were acquainted, properly speaking, with only one physical theory, the theory of celestial motions; that is why, in dealing with systems of cosmography, they expressed and developed their conception of physical theory. Moreover, other theories that they carried to a certain degree of perfection, and that today emerge again in physics—namely, the theories of equilibrium of the lever and hydrostatics—rested on principles whose nature could not be subject to any doubt. The axioms or demands of Archimedes were plainly propositions of experimental origin which generalization had transformed; the agreement of their consequences with the facts summarized and ordered the latter without explaining them.

The Greeks clearly distinguished, in the discussion of a theory about the motion of the stars, what belongs to the physicist—we should say today the metaphysician—and to the astronomer. It belonged to the physicist to decide, by reasons drawn from cosmology, what the real motions of the stars are. The astronomer, on the other hand, must not be concerned whether the motions he represented were real or fictitious; their sole object was to represent exactly the *relative* displacements of the heavenly bodies.[10]

In his beautiful research on the cosmographic systems of the Greeks, Schiaparelli has brought to light a very remarkable passage concerning this distinction between astronomy and physics. The passage is from Posidonius, was summarized or quoted by Geminus, and has been preserved for us by Simplicius. Here it is: "In an absolute way it does not belong to the astronomer to know what is fixed by nature and what is in motion; but among the hypotheses relative to what is stationary and to what is moving, he inquires as to which ones correspond to the heavenly phenomena. For the principles he has to refer to the physicist."

These ideas, expressing pure Aristotelian doctrine, inspired many a passage by the astronomers of old; Scholasticism has formally adopted them. It is up to physics—that is, to cosmology—to give the reasons for the astronomical appearances by going back to the causes themselves; astronomy deals only with the observation of phenomena and with conclusions that geometry can deduce

5 vols. (Paris, 1913-1917), Vol. II, Part I, Chs. X and XI, pp. 50-179. (Translator's note: The remaining manuscript notes left by Duhem for this work are being published by Hermann, Paris.)

[10] We have borrowed several of the informative items which follow in the text from a very important article by P. Mansion, "Note sur le caractère géométrique de l'ancienne Astronomie," *Abhandlungen zur Geschichte der Mathematik*, IX (Leipzig). See also P. Mansion, *Sur les principes fondamentaux de la Géométrie, de la Mécanique et de l'Astronomie* (Paris, 1903).

from them. Saint Thomas, in commenting on Aristotle's *Physics*, said: "Astronomy has some conclusions in common with physics. But as it is not purely physics, it demonstrates them by other means. Thus the physicist demonstrates that the earth is spherical by the procedure of a physicist, for example, by saying its parts tend equally in every direction towards the center; the astronomer, on the contrary, does this by relying on the shape of the moon in eclipses or the fact that the stars are not seen to be the same from different parts of the world."

It is by furtherance of this conception of the role of astronomy that Saint Thomas, in his commentary on Aristotle's *De Caelo*, expressed himself in the following manner on the subject of the motion of the planets: "Astronomers have tried in diverse ways to explain this motion. But it is not necessary that the hypotheses they have imagined be true, for it may be that the appearances the stars present might be due to some other mode of motion yet unknown by men. Aristotle, however, used such hypotheses relative to the nature of motion as if they were true."

In a passage from the *Summa Theologiae* (I, 32), Saint Thomas showed even more clearly the incapacity of physical method to grasp an explanation that is certain: "We may give reasons for a thing in two ways. The first consists in proving a certain principle in a sufficient way; thus, in cosmology (*scientia naturalis*) we give a sufficient reason to prove that the motion of the heavens is uniform. In the second way, we do not bring in a reason which proves the principle sufficiently, but the principle being posited in advance, we show that its consequences agree with the facts; thus, in astronomy, we posit the hypothesis of epicycles and eccentrics because, by making this hypothesis, the sensible appearances of the heavenly motions can be preserved; but that is not a sufficiently probative reason, for they might perhaps be preserved by another hypothesis."

This opinion concerning the role and nature of astronomical hypotheses agrees very easily with a good number of passages in Copernicus and his commentator Rheticus. Copernicus, notably in his *Commentariolus de hypothesibus motuum caelestium a se constitutis*, simply presents the fixity of the sun and the mobility of the earth as *postulates* which he asks that he be granted: *Si nobis aliquae petitiones . . . concedentur*. It is proper to add that in certain passages of his *De revolutionibus caelestibus libri sex*, he professes an opinion concerning the reality of his hypotheses which is less reserved than the doctrine inherited from Scholasticism and expounded in the *Commentariolus.*

This last doctrine is formally enunciated in the famous preface which Osiander wrote for Copernicus' book *De revolutionibus caelestibus libri sex.* Osiander expresses himself thus: "*Neque enim necesse est eas hypotheses esse veras, imo, ne verisimiles quidem; sed sufficit hoc unum, si calculum observationibus congruentam exhibeant.*"* And he ends his preface with these words: "*Neque quisquam, quod ad hypotheses attinet, quicquam certi ab Astronomia expectet, cum nihil tale praestare queat.*"†

Such a doctrine concerning astronomical hypotheses aroused Kepler's indignation.[11] In his oldest writing, he said:

"Never have I been able to assent to the opinion of those people who cite to you the example of some accidental demonstration in which from false premises a strict syllogism deduces some true conclusion, and who try to prove that the hypotheses admitted by Copernicus may be false and that, nevertheless, true phenomena may be deduced from them as from their proper principles. . . . I do not hesitate to declare that everything that Copernicus gathered *a posteriori* and proved by observation could without any embarrassment have been demonstrated *a priori* by means of geometrical axioms, to an extent that would be a delightful spectacle to Aristotle, were he living."[12]

This enthusiastic and somewhat naïve confidence in the boundless power of the physical method is prominent among the great discoverers who inaugurated the seventeenth century. Galileo did in-

* Translator's note: "Nor is it, to be sure, necessary that these hypotheses be true, or even probable; but this one thing suffices, namely, whether the calculations show agreement with the observations."

† Translator's note: "Nor should anyone, because he holds fast to hypotheses, expect certainty from astronomy, as it cannot be responsible for anything like that." For the English translation of the *Commentariolus* and a brief discussion of Duhem's attitude to Copernicus' view of astronomical hypotheses, see E. Rosen, *Three Copernican Treatises* (New York, 1939), pp. 57-90 and p. 33, respectively.

[11] In 1597, Nicolas Raimarus Ursus published a book in Prague entitled *De hypothesibus astronomicis,* in which he upheld the opinions of Osiander to an exaggerated extent. Three years later, hence in 1600 or 1601, Kepler answered with the following: *Joannis Kepleri "Apologia Tychonis contra Nicolaum Raymarum Ursum"*; this work remained in manuscript in a very incomplete state and was published only in 1858 by Frisch (*Joannis Kepleri astronomi "Opera omnia"* [Frankfort-on-the-Main and Erlangen], I, 215). This work contains lively refutations of Osiander's ideas.

[12] *Prodromus dissertationum cosmographicarum, continens mysterium cosmographicum . . . a M. Joanne Keplero Wirtembergio* (Georgius Gruppenbachius, 1591); see *Joannis Kepleri astronomi "Opera omnia,"* I, 112-153.

deed distinguish between the point of view of astronomy, whose hypotheses have no other sanction than agreement with experience, and the point of view of natural philosophy, which grasps realities. When he defended the earth's motion he claimed to be talking only as an astronomer and not to be giving hypotheses as truths, but these distinctions are in his case only loopholes created in order to avoid the censures of the church; his judges did not consider them sincere, and if they had regarded them as such, these judges would have shown very little insight. If they had thought that Galileo sincerely spoke as an astronomer and not as a natural philosopher or, in their idiom, "physicist," if they had regarded his theories as a system suited to *represent* celestial motions and not as an affirmative doctrine about the *real nature* of astronomical phenomena, they would not have censured his ideas. We are assured of this by a letter which Galileo's principal adversary, Cardinal Bellarmin, wrote to Foscarini on April 12, 1615: "Your Fatherhood and the honorable Galileo will act prudently by contenting yourselves to speak hypothetically, *ex suppositione*, and not absolutely, as Copernicus has always done, I believe; in fact, to say that by supposing the earth mobile and the sun stationary we give a better account of the appearances than we could with eccentrics and epicycles, is to speak very well; there is no danger in that, and it is sufficient for the mathematician."[13] In this passage Bellarmin maintained the distinction, familiar to the Scholastics, between the physical method and the metaphysical method, a distinction which to Galileo was no more than a subterfuge.

The one who contributed most to break down the barrier between physical method and metaphysical method, and to confound their domains, so clearly distinguished in the Aristotelian philosophy, was surely Descartes.

Descartes' method calls into doubt the principles of all our knowledge and leaves them suspended on this methodological doubt until it can reach the point of demonstrating the legitimacy of principles by a long chain of deductions stemming from the famous *Cogito, ergo sum.* Nothing is more contrary than such a method to the Aristotelian conception, according to which a science, such as physics, rests on self-evident principles whose nature is investigated by a metaphysics which cannot increase their certainty.

The first proposition in physics that Descartes established, in pur-

[13] H. Grisar, *Galileistudien: Historische-theologische Untersuchungen über die Urtheile der römischen Congregationen in Galileiprocess* (Regensburg, 1882), Appendix, IX.

suing his method, grasps and expresses the very essence of matter: "The nature of body consists only in the fact that it is a substance having extension in length, width, and depth."[14] The essence of matter thus being known, we shall be able, through the procedures of geometry, to deduce from it the explanation of all natural phenomena. Summarizing the method by which he claimed he dealt with the science of physics, Descartes said: "I accept no principles of physics which are not also accepted in mathematics, for the sake of being able to prove by demonstration everything that I shall deduce from them, and these principles are sufficient, so long as all the phenomena of nature may be explained by means of them."

Such is the audacious formula of Cartesian cosmology: man knows the very essence of matter, namely, extension; he may then logically deduce all the properties of matter from it. The distinction between physics, which studies phenomena and their laws, and metaphysics, which seeks to know the essence of matter insofar as it is the cause of phenomena and the basis of laws, is deprived of any foundation. The mind does not start from the knowledge of phenomena to rise to the knowledge of matter; what it can know from the start is the very nature of matter, and thence the explanation of phenomena.

Descartes pushed this proud principle to its extreme consequences. He was not content with asserting that the explanation of all natural phenomena may be derived completely from this single proposition: "The essence of matter is extension;" he tried to give this explanation in detail. He investigated the question of constructing the world with shape and motion by starting with this definition. And when he reached the end of his work, he stopped to contemplate it, and declared that nothing was missing in it: "That there is no phenomenon in nature not included in what has been explained in this treatise"—so runs the title of one of the last paragraphs of the *Principia Philosophiae*.[15]

Sometimes Descartes seemed for a moment to have been frightened by the boldness of his cosmological doctrine and to have wished to assimilate it to the Aristotelian doctrine. That is what happens in one of the sections of the *Principia*; let us quote this section in its entirety, for it touches closely on the object of our study:

"It may still be retorted to this that, although I may have imagined causes capable of producing effects similar to those we see, we should not for that reason conclude that those we see are produced

[14] R. Descartes, *Principia Philosophiae* (Amsterdam, 1644), Part III, 4.
[15] *ibid.*, Part IV, 199.

by these causes; because, as an industrious watchmaker may make two watches indicating the hour in the same way and without any difference between them in their external appearance, yet without anything similar in the composition of their wheels, so it is certain that God works in an infinity of diverse ways, each of which enables him to make everything in the world appear as it does without making it possible for the human mind to know which of all these ways he has willed to employ. I have no difficulty in agreeing with this. And I believe I shall have done enough if the causes that I have explained are such that all the effects they may produce are similar to those we see in the world without being informed whether there are other ways in which they are produced. I even believe that it is as useful in life to know the causes thus imagined as if we had knowledge of the true causes, for medicine, mechanics, and generally all the arts served by a knowledge of physics, aim only to apply certain observable bodies to one another in such a manner that certain observable effects are produced by a series of natural causes. This could be accomplished just as well by considering the series of a few causes thus imagined, however false they may be, as if they were the true ones, since this series is supposed to be the same so far as the observable effects go. And in order that it may not be imagined that Aristotle ever claimed to do more than that, he himself said, at the beginning of the seventh book of his *Meteors*, that 'concerning things not manifest to the senses, they are sufficiently demonstrated, as much as may be reasonably desired, if it can only be shown that they may be such as are explained.' "[16]

But this sort of concession to the ideas of the schoolmen is manifestly in disagreement with the very method of Descartes. It is simply one of those precautions against any censure by the holy office that the great philosopher took, very much disturbed, as we know, by the condemnation of Galileo. Moreover, it seems that Descartes himself feared that his circumspection might be taken too seriously, for he followed the section we have just quoted by two others entitled "That nevertheless we have a moral certainty that all the things in this world are the same as what is demonstrated here they may be" and "And even that we have more than a moral certainty about them."

The words "moral certainty" do not suffice, indeed, to express the boundless faith Descartes professed in his method. Not only did he believe he had given a satisfactory explanation of all natural

[16] *ibid.*, Part IV, 204.

phenomena, but he thought he had furnished the only possible explanation for them, and could demonstrate it mathematically. On March 16, 1640 he wrote to Mersenne: "As to physics, I should think I knew nothing about it if I could only say how things may be without demonstrating that they cannot be otherwise; for having reduced physics to the laws of mathematics, I know it is possible, and I believe I can do it for all the little knowledge I believe I have; although I did not do it in my *Essais* because I did not want to give my principles there, and I still do not see anything which invites me to give them in the future."[17]

This proud confidence in the boundless power of the metaphysical method was just the thing to cause Pascal to smile disdainfully; when you but admit that matter is nothing but extension in three dimensions, how foolish it is to wish to draw the detailed explanation of the world: "We must say crudely: that is done by shape and motion, for that is true. But to tell more, and to compose the machine —that is ridiculous, for that is useless, and uncertain, and painful."[18]

Pascal's famous rival, Christian Huygens, was not so harsh about the method which claims to derive the explanation of natural phenomena. Of course, Descartes' explanations are untenable on more than one point; but that is because his cosmology which reduces matter to extension is not the sound philosophy of nature, namely, the physics of the atomists. From the latter we may hope to deduce, though with great difficulties, the explanation of natural phenomena:

"Descartes has recognized, better than those before him, that we should never understand anything important in physics except what might be related to principles not going beyond our mind's reach, such as the principles which depend on bodies, considered devoid of qualities, and on their motions. But as the greatest difficulty consists in showing how so many diverse things are brought about by these principles alone, in that respect he has not succeeded in several particular subjects he proposed to examine; one of them, among others, is the subject of weight. This may be judged by the remarks I make in several places about what he has written, to which I could have added others. And yet, I confess that his essays and his insights, though false, have helped me to discover the road to the discoveries I have myself made on the same subject.

"I do not offer it as being exempt from all doubt, nor one to which no objections can be made. It is too hard to go that far in investiga-

[17] R. Descartes, *Correspondance*, ed. P. Tannery and C. Adam, III, 39.

[18] B. Pascal, *Pensées*, ed. Havet, Art. 24. This thought is preceded by these words: "To write against those who go too deeply in the sciences: Descartes."

tions of this nature. Still, I believe that if the principal hypothesis, which I take as basic, is not the true one, there is little expectation that it can be found while staying within the limits of true and sound philosophy."[19]

Between the time Huygens communicated to the Académie des Sciences of Paris his *Essay on the Cause of Weight* and the time he had it published, there appeared the immortal work of Newton: *Philosophiae naturalis principia mathematica.* This work transformed celestial mechanics, and inaugurated opinions on the subject of the nature of physical theories altogether opposed to those of Descartes and Huygens.

Newton expressed clearly what he thought about the construction of physical theories in several passages in his works.

The attentive study of phenomena and their laws permits the physicist to discover by the inductive method appropriate to his science some of the very general principles from which experimental laws may be deduced; thus the laws of all celestial phenomena are found condensed in the principle of universal gravitation.

Such a condensed representation is not an explanation; the mutual attraction that celestial mechanics imagines between any two parts whatsoever of matter permits us to submit all celestial movements to calculation, but the cause itself of this attraction is not laid bare because of that. Must we see in it a primary and irreducible quality of matter? Must we regard it as the result of impulses produced by a certain ether, as Newton was to judge probable at certain times in his life? These are difficult questions whose solution can only be obtained later. In any case, this problem is the task of the philosopher and not of the physicist; whatever the answer may be, the representative theory constructed by the physicist will keep its full value.

Here is the doctrine stated in a few words in the "General Scholium" with which the *Philosophiae naturalis principia mathematica* ends:

"And now we might add something concerning a certain most subtle spirit which pervades and lies hidden in all gross bodies. By the force and action of this spirit the particles of bodies attract one another at near distances, and cohere, if contiguous; electric bodies operate to greater distances, as well repelling as attracting the neighboring corpuscles; and light is emitted, reflected, refracted,

[19] Christian Huygens, *Discours de la cause de la Pesanteur* (Leyden, 1690).

inflected, and heats bodies. All sensation is excited and the members of animal bodies move at the command of the will, namely, by the vibrations of this spirit, mutually propagated along the solid filaments of the nerves, from the outward organs of sense to the brain, and from the brain into the muscles. But these are things that cannot be explained in few words, nor are we furnished with that sufficiency of experiments which is required for an accurate determination and demonstration of the laws by which this electric and elastic spirit operates."

Later, in the famous Query XXXI at the end (the fourth paragraph from the last) of the second edition of his *Optics*, Newton enunciated with great precision his opinion concerning physical theories; he assigned to them as their object the economic condensation of phenomena:

"To tell us that every species of things is endowed with an occult specific quality by which it acts and produces manifest effects, is to tell us nothing; but to derive two or three general principles of motion from phenomena, and afterwards to tell us how the properties and actions of all corporeal things follow from those manifest principles, would be a very great step in philosophy, though the causes of those principles were not yet discovered; and therefore I scruple not to propose the principles of motion above mentioned, they being of very large extent, and leave their causes to be found out."

Those who shared the proud confidence of the Cartesians or atomists could not allow such modest limits to be imposed on the claims of theoretical physics. To limit one's self to giving a geometric representation of phenomena was to their mind not to advance in the knowledge of nature. Those who were content with such vain progress deserved scarcely anything but sarcasm. One Cartesian said:

"Before making use of the principles we have just established, I believe it will not be inappropriate to examine those Mr. Newton used as the foundation of his system. This new philosopher, already distinguished by the rare knowledge he had drawn from geometry, suffered impatiently because a nation foreign to his own could take such advantage of the position it had as to teach other nations and serve as a model for them. Moved by a noble pride and guided by his superior genius, he thought only of freeing his country from the necessity it felt of borrowing from us the art of throwing light on the processes of nature and of following her in her operations. That was still not enough for him. Opposed to all restraint, and feeling that physics would constantly embarrass him, he banished it from his philosophy; and for fear of being compelled to solicit its aid sometimes, he took the trouble to construct the intimate causes of each particular phenomenon in primordial laws; whence every difficulty was reduced to one level. His work did not bear on any subjects except those that could be treated by means of the calculations he knew how to make; a geometrically analyzed subject became an explained phenomenon for him. Thus, this distinguished rival of Descartes soon experienced the singular satisfaction of being a great philosopher by sole virtue of his being a great mathematician."[20]

". . . I therefore return to what I first advanced, and I conclude that by following the method of this great geometer, we can with the greatest of ease develop the mechanism of nature. Do you wish to give an account of a complicated phenomenon? Expound it geometrically, and you will have done everything; whatever remains embarrassing to the physicist will depend, most certainly, either on a fundamental law or on some particular determination."[21]

Newton's disciples, however, did not all adhere to the prudent reserve of their master; several could not remain in the narrow confines assigned to them by his method in physics. Crossing these limits, they asserted, as metaphysicians, that mutual attractions were the real and primary qualities of matter and that a phenomenon reduced to these attractions was truly explained. This was the opinion expressed by Roger Cotes in the famous preface he wrote at the head of the second edition of Newton's *Principia.* This was also the doctrine developed by Boscovich that the Leibnizian metaphysics often inspired.

However, several of Newton's followers, and not the least distinguished ones, adhered to the method that their illustrious predecessor had so well defined.

Laplace professed utmost confidence in the power of the principle of attraction. This confidence, however, is not a blind one; in some places in the *Exposition du système du Monde*, Laplace indicated that this universal attraction, which in the form of gravity or of molecular attraction coordinates all natural phenomena, is not perhaps the ultimate explanation, and that it may itself depend on a higher cause. This cause, it is true, seems to have been relegated

[20] E. S. de Gamaches, *Principes généraux de la Nature appliqués au mécanisme astronomique et comparés aux principes de la Philosophie de M. Newton* (Paris, 1740), p. 67.

[21] *ibid.*, p. 81.

by Laplace to an unknowable domain. In any case, he recognized with Newton that the quest for this cause, if at all possible, constitutes a problem distinct from the one which physical and astronomical theories solve. He asked: "Is this principle a fundamental law of nature? Is it only a general effect of an unknown cause? Here, we are stopped by our ignorance of the intimate properties of matter, depriving us of any hope of answering these questions satisfactorily."[22] Again, he said: "Is the principle of universal gravity a fundamental law of nature or but the general effect of an unknown cause? May we not reduce the attractions to this principle? Newton, more circumspect than several of his disciples, did not pronounce judgment on these matters where our ignorance of the properties of matter does not permit us to give any satisfactory answer."[23]

Ampère, a more profound philosopher than Laplace, saw with perfect clarity the importance of regarding a physical theory as independent of any metaphysical explanation; in fact, that is the way to keep out of physics the divisive quarrels of the diverse cosmological schools. At the same time, physics remains acceptable to minds that profess incompatible philosophical opinions; and yet, very far from blocking the inquiries of those who would lay claim to giving an explanation of phenomena, we expedite their task. We condense in a small number of very general propositions the countless laws they are to explain, so that it suffices for them to explain these few propositions in order to get at anything mysteriously contained in that enormous collection of laws:

"The chief importance of the formulas which are thus immediately concluded from some general facts, given by a number of observations sufficient to make their certainty incontestable, is that they remain independent both of the hypotheses used by their authors in the search for these formulas and of those hypotheses which may be substituted subsequently. The expression of universal attraction deduced from Kepler's laws does not depend on the hypotheses that a few authors have ventured concerning a mechanical cause they wished to assign to it. The theory of heat really rests on general facts immediately given to observation; and the equation deduced from these facts being confirmed by the agreement of the results drawn from the equation with those given by experience, should be regarded as expressing the true laws of the propagation of heat, both by those who attribute heat to a

[22] P. S. Laplace, *Exposition du système* . . . I, IV, Ch. XVII.
[23] *ibid.*, I, V, Ch. V.

radiation of calorific molecules as well as by those who explain the same phenomenon by having recourse to the vibrations of a fluid pervading space. But it is necessary that the former show how the equation in question results from their way of looking at things, and that the latter deduce it from the general formula of vibratory motions, not for the sake of adding anything to the certainty of this equation but to maintain their own respective hypotheses. The physicist who has not taken sides in this regard accepts this equation as the exact representation of the facts without worrying about the way it may result from either one of the above explanations."[24]

Fourier, moreover, shared Ampère's judgment concerning the theory of heat; in fact, here is how he expressed himself in the *Discours Préliminaire* which prefaces his immortal work:

"The fundamental causes are not known to us, but they are subject to simple and constant laws that may be discovered by observation, and the study of these is the object of natural philosophy.

"Heat, like gravity, penetrates every substance in the universe; its rays fill every part of space. The aim of our work is to expound the mathematical laws that this element follows. This theory will henceforth form one of the most important branches of physics.

". . . the principles of this theory are deduced, like those of mechanics, from a small number of fundamental facts whose cause is not considered by mathematicians but which are accepted by them as resulting from common observations and confirmed by all the experiments."[25]

Fresnel did not assign, any more than Ampère or Fourier, any metaphysical explanation as the aim of theory. He saw in theory a powerful means of discovery because it is a summary and classified representation of experimental knowledge: "It is not useless to unite facts under the same viewpoint by tying them to a small number of general principles. That is the means for grasping laws more easily, and I think that efforts of this kind may contribute as much as the observations themselves to the advancement of science."[26]

The rapid development of thermodynamics in the middle of the nineteenth century reinstated to favor the hypotheses Descartes had first formulated concerning the nature of heat; Cartesian and atomistic opinions received renewed vitality; and the hope of con-

[24] A. M. Ampère, *Théorie mathématique des phénomènes electrodynamiques, uniquement déduite de l'expérience*, ed. Hebemann (Paris, 1824), p. 3.

[25] J. B. Fourier, *Théorie analytique de la chaleur*, ed. Darboux (Paris, 1822), pp. xv, xxi. (Translator's note: See English translation, *The Analytical Theory of Heat*, tr. A. Freeman [Cambridge: Cambridge University Press, 1878].)

[26] *Oeuvres complètes d'Augustin Fresnel*, 3 vols. (Paris, 1866-1870), I, 480.

structing explanatory theories was revived in the thought of more than one physicist.

However, some of the more important physicists, creators of the new doctrine, did not let themselves become intoxicated by this hope; among them and of the first rank was Robert Mayer, whom it is appropriate to quote: "Concerning the intimate nature of heat," he wrote to Griesinger, "or of electricity, etc., I know nothing, any more than I know the *intimate nature* of any matter whatsoever, or of anything else."[27]

The first contributions of Macquorn Rankine to the progress of the mechanical theory of heat had been attempts at explanation; but his ideas soon evolved and, in a short paper of his,[28] too little known, he traced very clearly the characteristics which distinguish a representative theory—called by him "abstractive theory"—from an explanatory theory—designated by the name "hypothetical theory."

Let us quote some passages from this work:

"An essential distinction exists between two stages in the process of advancing our knowledge of the laws of physical phenomena. The first stage consists in observing the relations of phenomena, whether of such as occur in the ordinary course of nature, or of such as are artificially produced in experimental investigations, and in expressing the relations so observed by propositions called formal laws. The second stage consists in reducing the formal laws of an entire class of phenomena to the form of a science; that is to say, in discovering the most simple system of principles, from which all the formal laws of the class of phenomena can be deduced as consequences.

"Such a system of principles, with its consequences methodically deduced, constitutes the *physical theory* of a class of phenomena. . . .

"Two methods of framing a physical theory may be distinguished, characterized chiefly by the manner in which classes of phenomena are defined. They may be termed, respectively, the *abstractive* and the *hypothetical* methods.

"According to the *abstractive* method, a class of objects or phenomena is defined by describing, or otherwise making to be understood, and assigning a name or symbol to, that assemblage of properties which is common to all the objects or phenomena composing

[27] Robert Mayer, *Kleinere Schriften und Briefe* (Stuttgart, 1893), p. 181.

[28] J. Macquorn Rankine, *Outlines of the Science of Energetics*, read to the Philosophical Society of Glasgow, May 2, 1855, and published in the *Proceedings* of this society, Vol. III, No. 4. See Rankine, *Miscellaneous Scientific Papers*, p. 209.

the class, as perceived by the senses, without introducing anything hypothetical.

"According to the *hypothetical* method, a class of objects or phenomena is defined, according to a conjectural conception of their nature, as being constituted, in a manner not apparent to the senses, by a modification of some other class of objects or phenomena whose laws are already known. Should the consequences of such a hypothetical definition be found to be in accordance with the results of observation and experiment, it serves as the means of deducing the laws of one class of objects or phenomena from those of another." It is in this way that we shall derive the laws of light or heat from the laws of mechanics.

Rankine thought that hypothetical theories will be gradually replaced by abstractive theories; however, he believed "that a hypothetical theory is necessary, as a first step, in order to put simplicity and order into the expression of phenomena before it is possible to make any progress in the construction of an abstractive theory." We have seen in the preceding paragraph that this assertion was scarcely confirmed by the history of physical theories; we shall have occasion to discuss it again in Chapter IV, Section 9.

Toward the end of the nineteenth century, hypothetical theories which were offered as more or less probable explanations of phenomena were extraordinarily multiplied. The noise of their battles and the fracas of their collapse have wearied physicists and led them gradually back to the sound doctrines Newton had expressed so forcefully. Renewing the interrupted tradition, Ernst Mach has defined theoretical physics as an abstract and condensed representation of natural phenomena.[29] G. Kirchhoff offered as the object of mechanics: "to describe as completely and as simply as possible the motions produced in nature."[30]

Therefore, if some very great physicists could take pride in the powerful method that they employed to the point of exaggerating its scope, if they could believe that their theories would reveal the metaphysical nature of things, many discoverers who excite our

[29] E. Mach, *Die Gestalten der Flüssigkeit* (Prague, 1872); *Die ökonomische Natur der physikalischen Forschung* (Vienna, 1882); *Die Mechanik in ihrer Entwickelung, historisch-kritisch dargestellt* (Leipzig, 1883). This last work has been translated into French by M. Bertrand under the title *La Mécanique; exposé historique et critique de son développement* (Paris, 1904). (Translator's note: English translation by T. J. McCormack, *The Science of Mechanics, a Critical and Historical Account of Its Development* [Open Court, 1902].)

[30] G. Kirchhoff, *Vorlesungen über mathematische Physik; Mechanik* (Leipzig, 1874), p. 1.

admiration have been more modest and more farsighted. They have recognized that physical theory is not an explanation, but a simplified and orderly representation grouping laws according to a classification which grows more and more complete, more and more natural.

CHAPTER IV

ABSTRACT THEORIES AND MECHANICAL MODELS[1]

1. *Two Kinds of Minds: Ample and Deep*

THE CONSTITUTION of any physical theory results from the two-fold work of abstraction and generalization.

In the first place, the mind analyzes an enormous number of concrete, diverse, complicated, particular facts, and summarizes what is common and essential to them in a law, that is, a general proposition tying together abstract notions.

In the second place, the mind contemplates a whole group of laws; for this group it substitutes a very small number of extremely general judgments, referring to some very abstract ideas; it chooses these primary properties and formulates these fundamental hypotheses in such a way that all the laws belonging to the group studied can be derived by deduction that is very lengthy perhaps, but very sure. This system of hypotheses and deducible consequences, a work of abstraction, generalization, and deduction, constitutes a physical theory in our definition; it surely merits the epithet Rankine used to designate it: abstractive theory.

The two-fold work of abstraction and generalization that goes to make up a theory brings about, we have said,[2] a double economy of thought; it is economical when it substitutes a law for a multitude of facts; it is economical again when it substitutes a small group of hypotheses for a vast set of laws.

Will all those who reflect on the methods of physics agree with us in attributing this character of double economy to abstract theory?

To bring directly before the visual imagination a very large number of objects so that they may be grasped simultaneously in their complex functioning and not taken one by one, arbitrarily separated from the whole to which they are in reality attached—this is for most men an impossible or, at least, a very painful operation. A host of laws, all put on the same plane, without any classifica-

[1] The ideas expounded in this chapter are the development of an article entitled "L'École anglaise et les Théories physiques," published in October 1893 by the *Revue des Questions Scientifiques.*

[2] See Ch. II, Sec. 2, above.

tion grouping them, without any system coordinating or subordinating them, appears to such minds as chaotic and frightening to the imagination, as a labyrinth in which their intelligences go astray. On the other hand, they have no difficulty in conceiving of an idea which abstraction has stripped of everything that would stimulate the sensuous memory; they grasp clearly and completely the meaning of a judgment connecting such ideas; they are skillful in following, untiringly and unwaveringly, down to its final consequences, the reasoning which adopts such judgments for its principles. Among these men, the faculty of conceiving abstract ideas and reasoning from them is more developed than the faculty of imagining concrete objects.

For these *abstract minds* the reduction of facts to laws and the reduction of laws to theories will truly constitute intellectual economies; each of these two operations will diminish to a very large degree the trouble their minds will have to take in order to acquire a knowledge of physics.

But not all vigorously developed minds are abstract minds.

There are some minds that have a wonderful aptitude for holding in their imaginations a complicated collection of disparate objects; they envisage it in a single view without needing to attend myopically first to one object, then to another; and yet this view is not vague and confused, but exact and minute, with each detail clearly perceived in its place and relative importance.

But this intellectual power is subject to one condition; namely, the objects to which it is directed must be those falling within the purview of the senses, they must be tangible or visible. The minds possessing this power need the help of sensuous memory in order to have conceptions; the abstract idea stripped of everything to which this memory can give shape seems to vanish like an impalpable mist. A general judgment sounds to them like a hollow formula void of meaning; a long and rigorous deduction seems to them to be the monotonous and heavy breathing of a windmill whose parts turn constantly but crush only the wind. Endowed with a powerful faculty of imagination, these minds are ill prepared to abstract and deduce.

Will such *visualizing minds* regard an abstract physical theory as an intellectual economy? Surely not. They will regard it rather as an undertaking whose painful nature will appear to them more certain than its utility, and they doubtlessly will compose their physical theories on an entirely different type of model.

Physical theory, of the sort we have conceived, will not then be accepted offhand as the true form in which nature is to be represented, except by abstract minds. Pascal did not fail to take notice of this in that fragment in which he characterized so forcefully the two sorts of minds we have just distinguished:

"Two sorts of right sense: the first in a certain order of things and not in other orders where it sees no sense. The first derives consequences straight from a few principles, and that is one sort of right sense. The other derives consequences from things where there are many principles. For example, the former understands well the phenomena of water whose nature has few principles, but whose consequences are so subtle that only an extreme straitness of mind can grasp them; on that account this mind would not be a great geometer because geometry includes a great number of principles, and because the kind of mind that can penetrate a few principles thoroughly may not in the least be able to penetrate things where there are many principles.

"There are, then, two kinds of minds: one kind, able to penetrate quickly and profoundly the consequences of principles, we call the exact mind; the other, able to comprehend a great number of principles without confusing them, we call the geometrical mind. The first has a strong and rigorous incisiveness of mind, the other, a broad scope of mind. Now, one may exist without the other, because the mind is capable of being strong and narrow, but also capable of being broad and weak."[3]

Abstract physical theory, as we have defined it, will surely attract strong but narrow minds; on the other hand, it should expect to repel broad and weak minds. Since, then, we shall have to combat the latter type of mind, let us first become well acquainted with it.

2. *An Example of the Ample Mind: the Mind of Napoleon*

When a zoologist plans to study a certain organ, he discovers, if he is lucky, an animal in which this organ has had an exceptional development, for he can dissect its different parts more easily, see its structure more clearly, and grasp its function better. In the same way, the psychologist who desires to analyze a faculty of the mind has his wish answered if he meets a creature who possesses this faculty to an eminent degree.

Now, history presents us with a man in whom this form of intellect, which Pascal characterized as broad in scope but weak, was developed to an almost monstrous extent: that man was Napoleon.

[3] B. Pascal, *Pensées*, ed. Havet, Art. VII, 2.

If we read again the portrait of Napoleon so profoundly delineated and so curiously documented by Taine,[4] we shall recognize immediately the following two characteristics, which are so salient that they cannot escape the notice of the least perspicacious person: first, an extraordinary power to hold in mind an extremely complex collection of objects, provided these are sensory objects having shape and color that the imagination can visualize; second, an incapacity for abstraction and generalization, even going so far as a deep aversion with regard to these intellectual operations.

Pure ideas, stripped of the drapery of the concrete and particular details which would have made them visible and tangible, had no access to the mind of Napoleon: "From Brienne it was known that he had no disposition for languages and belles-lettres." He not only did not take easily to abstract ideas, but he rejected them with horror: "Madame de Staël said he examined things only with relation to their immediate utility; a general principle displeased him as a bad joke or as an enemy." Those who make use of abstraction, generalization, and deduction as their habitual means of thought appeared to him as incomprehensible, defective, and immature creatures; he treated those he called "ideologists" with deep scorn. "There you have twelve or fifteen ideologists good for drowning in hot water," he said; "they are vermin I have on my clothes."

On the other hand, if his reason refused to take to general principles, if, on the testimony of Stendhal, "he is ignorant of the great truths discovered a hundred years ago," with what power he could see things at a glance, and comprehend clearly the whole, not letting go of any detail of the mass of complex objects and concrete facts!

Bourienne says: "He had a poor memory for proper names, words and dates, but a prodigious one for *facts* and *localities*. I recall that on a trip from Paris to Toulon he mentioned to me ten good places from which to wage battle. . . . That came from memories of the first travels in his youth, and he described to me the disposition of the terrain and designated the positions he would have occupied before we came to the places." Moreover, Napoleon himself took notice of this peculiarity of his memory, so powerful in regard to facts, so weak towards everything not concrete: "I always keep in mind the conditions of my position. I cannot remember enough to retain a line of Alexandrine verse, but I do not forget a syllable of my reports on strategic position. I am going to find them in my

[4] H. Taine, *Les Origines de la France contemporaine. Le Régime Moderne*, Vol. I (Paris, 1891), Book I, Ch. 1, Secs. 2, 3, 4.

room tonight, and shall not go to bed until I have read them."

Just as he was horrified by abstraction and generalization, because these operations were accomplished in him with great difficulty and labor, so he found himself happy in putting his prodigious imaginative faculty to work, like an athlete who takes pleasure in testing the power of his muscles. His curiosity about exact and concrete facts was "insatiable," according to Mollien. "The good position of my armies," he told us himself, "comes from the fact that I concentrate on it every day for an hour or two, and when they send me monthly reports on the conditions of my troops and ships, which form a score of thick booklets, I stop everything else I am doing in order to read them in detail, in order to see the difference there is between one month and the next. I take greater pleasure in reading this material than a girl does in reading a novel."

This imaginative faculty exercised so easily and willingly by Napoleon is remarkable for its flexibility, broad scope, and accuracy. Many examples could be given to help us appreciate the marvelous qualities of this faculty in Napoleon, but the following two are characteristic enough for us to dispense with a long enumeration.

"Monsieur de Ségur, responsible for visiting all the places on the Northern coast, had conveyed his report. ' "I have seen all your reports on location," the First Consul said to me, "they are accurate. However, you forget two cannons at Ostend." And he designated the spot, "a dam opposite the city."—It was true.—I came out, bewildered by astonishment at the fact that among thousands of pieces of cannon spread out among fixed or mobile batteries behind the beach, two pieces did not elude his memory.' "

"Returning from the camp at Boulogne, Napoleon meets a squad of lost soldiers, asks them for the number of their regiment, figures out the day of their departure, the route they have taken, the road they should be on, and says to them: 'You will find your battalion at such and such a stopping point.'—Now, the army then consisted of 200,000 men."[5]

It is by deeds, attitudes, and visible gestures that man is known by his fellows, that he reveals his thoughts, instincts, passions; in such revelation the slightest and most fleeting detail—an imperceptible blush, a faintly outlined curving of the lips—is often the essential sign, throwing light quickly and suddenly on a joy or deception concealed at the bottom of the soul. This minuteness of detail did not escape Napoleon's scrutiny, and his visual memory fixed it once and for all as would an instantaneous photograph. Whence

[5] The quotations are all taken from *ibid.*

his profound knowledge of the men with whom he had to deal: "Such an invisible, psychical power may be judged and approximately measured by its outward manifestation, by a decisive scrutiny of this or that word, accent, or gesture. He takes in these words, gestures, and accents; he perceives intimate thoughts through their outer expression, he pictures to himself the inner man through such and such a physiognomic trait and manner of speaking, through a summary and typical little scene, through samplings and foreshortened views chosen so well and under such circumstances that they summarize the whole indefinite line of analogous cases. In that way the vague and fleeting object is suddenly grasped, harnessed, and weighed."[6] The surprising psychology of Napoleon was entirely the result of his power of imagining with accuracy, in the large and in detail, visible and palpable objects, men of flesh and bones.

And this faculty is also what rendered his intimate talk so lively and colorful: he used no abstract terms or general judgments but images which immediately strike the eye or ear. "I am not satisfied with the management of the customs house on the Alps; it shows no signs of life; we don't hear the clink of its gold coins pouring into the public treasury."

Everything in the mind of Napoleon—his disgust with ideology, his administrative and tactical vision, his deep knowledge of social circles and of men, the often trivial vigor of his talk—proceeded from this same essential character: breadth and weakness of mind.

3. *The Ample Mind, the Supple Mind, and the Geometrical Mind**

In studying the mind of Napoleon we were able to observe all the characteristics of the ample mind, and we have seen them wonderfully magnified as in a microscope. It will henceforth be easy to recognize them everywhere we meet them, diversified by the various objects to which the mind they characterize applies itself.

We shall recognize them, first of all, wherever we find *the supple mind* which, as Pascal described it, consists essentially in the apti-

[6] *ibid.*, p. 35.

* Translator's note: There are no exact English equivalents for Pascal's "l'esprit de finesse et l'esprit géométrique," adopted by Duhem. "Ample, supple, subtle, broad, diplomatic, imaginative" are some of the connotations of "l'esprit de finesse" in contrast to the narrower, straiter, logically rigorous, abstract, and stronger "esprit géométrique" (not found in those mathematicians who merely compute or measure).

tude to see clearly a very large number of concrete notions, and to grasp simultaneously the whole and the details. "In the supple mind, principles are what common usage and the whole world accept. One has but to look around and not do violence to one's self. It is just a question of having a good view of things; but it must be a good one, for principles are so pervasive and so numerous that it is nearly impossible that they should escape detection. Now, the omission of a principle leads to error; hence, the view must be very clear in order to see all the principles. . . . They are hardly seen, they are felt rather than seen; it is infinitely difficult to make others feel them if they do not feel them themselves. It requires a very delicate and very clear sense to feel such delicate and numerous things, and to judge correctly according to this feeling, most often without being able to demonstrate the things in an order of a geometrical sort, because we do not acquire principles in that way, and because it would be an infinite task to undertake it. The thing must be seen all of a sudden in a single glance, and not by progressive reasoning to any degree.

". . . Minds of this sort, being thus accustomed to judge in a single glance, are so astonished when they are presented with propositions which they do not understand and which require formulation in definitions and sterile principles such as they have not usually seen in detail, that they are repulsed and disgusted by them. . . . Those that are exclusively minds of finesse cannot have the patience to descend to the first principles of speculative things of the imagination which they have never seen in the world and which are out of the usual run of things."[7]

Thus, it is ampleness of mind which gives birth to the finesse of the diplomat, skillful in noticing the smallest facts, and the slightest gestures and attitudes of the man with whom he negotiates, while wishing to penetrate through any dissimulation. Such is the finesse of a Talleyrand gathering thousands of very small bits of information which will help him guess the ambitions, the vanities, the grudges, the jealousies, the hatreds of all the ambassadors at the Congress of Vienna, and which will permit him to play with these men like marionettes whose strings he would hold.

This ampleness of mind we find in the historian preserving in his writings the detailed facts and attitudes of men; in a Saint-Simon, leaving us in his *Memoires* "the portraits of four hundred rascals no two of whom resemble each other." It is the essential instru-

[7] B. Pascal, *op.cit.*, Sec. 7.

ment of the great novelist: it enabled Balzac to create the multitude of characters who people the *Comédie humaine*; to plant each of them down before us in flesh and bones; to carve out of this flesh the wrinkles, the warts, the grimaces which threw into relief each of the passions, vices, and ridiculous aspects of the soul; to dress these bodies, give them living attitudes and gestures, surround them with the things which will be their environment; in a word, to make of them men who live in a stirring world.

It is this ampleness of mind which gives color and warmth to the style of a Rabelais, filling him with visible, palpable, tangible, concrete images to the point of caricature, images which are as full of life as a noisy, moving crowd. Hence the ample mind is the opposite of that classical mind that Taine depicted, of that mind which is in love with abstract notions. It is the opposite of the order and simplicity which speaks so naturally in the style of Buffon, who always chose the most general term in order to express an idea.

We have ample minds in all those who can unroll in their visual imagination a clear, exact, detailed picture where a multitude of objects are in operation. Ample is the mind of the financial speculator who from a mass of telegrams infers the condition of the wheat or wool market all over the world, and, with one glance, judges whether he is to gamble when the market is high or low. Ample is the mind of the military head of a state capable of thinking out a plan of mobilization by which millions of men will arrive at the place of combat on time as necessity demands, without a hitch and without confusion.[8] Ample also is the mind of the chessplayer who, even without looking at the chessboards, can hold matches against five opponents simultaneously.

It is again ampleness of mind which constitutes the peculiar genius of many a geometer and algebraist. More than one reader of Pascal, perhaps, will not fail to be astonished on seeing him sometimes place mathematicians among the number of ample but weak minds. This cross-classification is not one of the lesser proofs of his penetration.

Undoubtedly, every branch of mathematics deals with concepts which are abstract to the highest degree. It is abstraction which furnishes the notions of number, line, surface, angle, mass, force, and pressure; it is abstraction and philosophical analysis which disentangle and make precise the fundamental properties and

[8] Ampleness of mind was almost as characteristic of Caesar as it was of Napoleon. We recall that Caesar dictated at the same time to four secretaries letters composed in four languages.

postulates. It is the most rigorous deduction which makes sure that these postulates are compatible and independent, and which patiently, with impeccable order, unrolls the long chain of theorems which are contained in the postulates. To this mathematical method we owe the most perfect masterpieces, whose logical accuracy and intellectual depth have enriched mankind ever since Euclid's *Elements* and Archimedes' treatises on the lever and on floating bodies.

But precisely because this method calls for an almost exclusive use of the logical faculties of intelligence, because it requires in the highest degree a strong and accurate mind, this method appears extremely laborious and painful to those who have an ample but weak mind. Hence mathematicians have created procedures which substitute for this purely abstract and deductive method another method in which the imaginative faculty plays a greater part than the power of reasoning. Instead of studying directly the abstract notions with which they are concerned, instead of considering them by themselves, they take advantage of the simplest properties of the notions in order to represent them by numbers; that is, in order to *measure* them. Then, instead of linking the properties themselves of these notions, they submit the numbers furnished by measurement to manipulations performed according to the fixed rules of algebra; instead of deducing, they *calculate*. Now this manipulation of algebraic symbols (which we may call calculus, in the largest meaning of the word) presupposes, on the part of the creator as well as of the one who uses it, much less power to abstract and much less skill in arranging one's thoughts in order than aptitude for expressing diverse and complicated combinations. These may be formed with certain visible and traceable signs in order to see offhand the transformations permitting one to pass from one combination to another. The author of certain algebraic discoveries—a Jacobi, for example—has nothing in him of the metaphysician; he is much more like the player who brings about a sure checkmate with a castle or knight. On many occasions the mathematical mind will take a place next to the supple mind (*esprit de finesse*) among the ample but weak minds.

4. *Ampleness of Mind and the English Mind*

In every nation we find some men who have the ample type of mind, but there is one people in whom this ampleness of mind is endemic; namely, the English people.

Let us look, in the first place, among the written works produced by the English genius for the two marks of the ample and weak mind: one, an extraordinary facility for imagining very complicated collections of concrete facts; and two, an extreme difficulty in conceiving abstract notions and formulating general principles.

What is it that strikes the French reader when he opens an English novel, a masterpiece of a great novelist like Dickens or George Eliot, or a first attempt by a young authoress aspiring to literary fame? What strikes one is the lengthy, minute character of the descriptions. At first he feels his curiosity piqued by the picturesqueness of each object, but he soon loses sight of the whole. The numerous images that the author has evoked for him flow confusedly into one another, while new images pour in constantly only to increase this disorder; before you are a quarter of the way through the description, you have forgotten the beginning of it, and you turn the pages without reading them, fleeing from this nightmarish series of concrete things. What this deep but narrow sort of French mind wants are the descriptions of a Loti, abstracting and condensing in three lines the essential idea, the soul of a whole landscape. The Englishman has no such requirements. All those visible, palpable, tangible things that the novelist enumerates and describes minutely are seen by his compatriots, without any trouble, as a whole: each thing in its place and with all its characteristic details. The English reader sees a charming picture where we French perceive nothing but a chaos importuning us.

This opposition between the French mind, strong enough to be unafraid of abstraction and generalization but too narrow to imagine anything complex before it is classified in a perfect order, and the ample but weak mind of the English will come home to us constantly while we compare the written monuments raised by these two peoples.

Do we wish to verify this in the works of the dramatists? Take one of Corneille's heroes, Auguste, hesitating between revenge and mercy, or Rodrigue deliberating between his filial piety and his love. Two feelings wage a dispute for his heart; but what a perfect order there is in their discussion! Each takes the floor in turn, like two lawyers before the bar expounding in perfectly finished briefs their reasons why they will win the case, and when the reasons on both sides have been clearly expounded, the will of man puts an end to the debate through a precise decision, resembling a judicial decree or a conclusion in geometry.

And now, opposite the Auguste or Rodrigue of Corneille place the Lady Macbeth or Hamlet of Shakespeare: what a mess of confused, imperfect thoughts, with vague, incoherent outlines, dominating and being dominated at the same time! The French spectator, shaped by our classical theatre, tries in vain to *understand* such characters; that is, to deduce clearly from a definite setting that multitude of attitudes and of inexact and contradictory words. The English spectator does not assume this undertaking; he does not seek to understand these characters, to classify and to arrange their gestures in order; he is content to *see* them in their living complexity.

Will this opposition between the French and English minds be recognized by us when we study the philosophical writings? Let us substitute Descartes and Bacon for Corneille and Shakespeare.

What is the preface with which Descartes opens his work? A *Discours de la Méthode.* What is the method of this strong but narrow mind? It consists in "conducting one's thoughts in order, by beginning with the simplest objects, easiest to know, in order to rise gradually, step by step, so to speak, to the knowledge of the more composite ones, and even presupposing an order among those objects which do not follow one another naturally."

And what are these objects "easiest to know" with which "it is necessary to begin"? Descartes repeated the answer on several occasions: These objects are the *simplest* objects, and by these words he understood the notions that are most abstract and most stripped of sensible accidents, the most universal principles, the most general judgments concerning existence and thought, the first truths of geometry.

Starting from these ideas, from these principles, the deductive method will unroll its syllogisms whose long chain of links, all tested, will firmly tie the most minute consequences to the foundations of the system. "These long chains of reasons, all simple and easy, which geometers customarily employ in order to carry through their most difficult demonstrations, had prompted me to suppose that the things which may fall within the province of human knowledge follow one another in the same order, and, provided only that we abstain from accepting any as true which are false and that we always keep the order necessary to deduce them from one another, that there cannot be any so remote as to be inaccessible, or so concealed as to be undiscoverable."

In the use of such a very precise and rigorous method, what is the only cause of error which Descartes feared? It is omission, for he was aware that he had a narrow geometrical mind and that it was hard for him to keep in his mind a complex whole. With respect to the latter alone did he take precautions by preparing a check or test, proposing "to make from time to time such complete enumerations and such general reviews that he is sure nothing has been omitted."

Such is the Cartesian method, exactly applied in the *Principes de Philosophie*, where the strong and restricted mind of the geometer has clearly expounded the mechanism by which it operates.

Let us now open the *Novum Organum*. There is no use in looking for Bacon's method in it, for there is none. The arrangement of his book is based on a childishly simple division. In the *Pars destruens*, he called Aristotle names for having "corrupted natural philosophy with his dialectic and constructed the world with his categories." In the *Pars aedificans*, he praised the true philosophy, whose object is not to construct a clear and well-ordered system of truths logically deduced from warranted principles. Its object is quite practical, I should go so far as to say industrial: "We must see what instructions or directions we may especially desire in order to produce or create in a given body some new property, and explain it in simple terms as clearly as possible.

"For example, if we wish to give the color of gold to silver, or a greater weight (in conformity with the laws of matter) or transparency to a non-diaphanous stone, or infrangibility to glass, or vegetation to some non-vegetating body, we say we must see what instructions or directions it would be most desirable to receive."

Will these instructions teach us to conduct and arrange our experiments in accordance with fixed rules? Will these directions teach us the way to classify our observations? Not in the least. Experiments will be made without any preconceived idea, observations will be made by chance, results will be recorded in crude form, as they happen to present themselves, in tables of "positive facts," "negative facts," "degrees" or "comparisons," and "exclusion" or "rejection," in which a French mind would see only a disordered mass of useless reports. Bacon agreed, it is true, to establish certain categories of prerogatives or privileged facts, but thèse categories were not classified by him, nor enumerated. He did not analyze them in order to bring under the same heading those categories which might well be reducible to one another. He listed twenty-seven of them as kinds and left us in the dark as to why he closed the list after the twenty-seventh kind. He did not seek an exact formula which characterizes and defines each of the categories of prerogatives, but was content to masquerade it under a name suggestive of a sensuous image, such as isolated facts, or facts that are labelled migratory, indicative, clandestine, clustered, boundary-line, hostile, negotiated, crucial, divorced, luminiferous, gateway, fluid, etc. It is this chaos that certain people—who have never read Bacon—call the Baconian method in opposition to the Cartesian method. In no work does the ampleness of the English mind show more transparently the weakness that it conceals.

If the mind of Descartes seems to haunt French philosophy, the imaginative faculty of Bacon, with its taste for the concrete and practical, its ignorance and dislike of abstraction and deduction, seems to have passed into the life-blood of English philosophy. "One by one, Locke, Hume, Bentham, and the two Mills have expounded the philosophy of experience and observation. Utilitarian ethics, inductive logic, associationist psychology, these are the great contributions of English philosophy"[9] to universal thought. All these thinkers proceed not so much by a consecutive line of reasoning as by a piling-up of examples. Instead of linking up syllogisms, they accumulate facts. Darwin and Spencer did not engage their adversaries in the learned fencing of discussion; they crushed them by throwing rocks.

The opposition of the French genius and English genius is observed in every work of the mind. It is likewise noticeable in every manifestation of social life.

What can be more different, for example, than French laws, grouped by codes in which the articles of law are methodically arranged under headings stating clearly defined abstract ideas, and English legislation, a prodigious mass of laws and customs, disparate and often contradictory, juxtaposed since the Magna Carta, one after the other, without any new laws abrogating those that preceded them. English judges do not feel embarrassed by this chaotic state of legislation; they do not boast of a Pothier* or a Portalis;† they are not bothered by the disorderly state of the texts they have to apply; the need for order is a sign of narrowness of mind which, not being able to embrace a whole all at once, needs a guide that can introduce it to each of the elements of that whole, one after the other, without omission or repetition.

The Englishman is essentially conservative; he keeps every tradition, regardless of its source. He is not shocked to see a relic of

[9] André Chevrillon, *Sydney Smith et la renaissance des idées libérales en Angleterre au* XIXe *siècle* (Paris, 1894), p. 90.

* Translator's note: Robert Joseph Pothier (1699-1772), a French jurisconsult born in Orleans, whose works were used in drafting the Civil Code of France.

† Translator's note: Jean-Etienne-Marie Portalis (1745-1807), also a French jurisconsult, one of the editors of the *Code civil.*

Cromwell's time next to one of the era of Charles I; the history of his country appears to him just as it has been: a series of diverse contrasting facts, in which each political party has perchance met with success or failure, and has committed in turn criminal and glorious deeds. Such traditionalism, respectful of the whole past, is incompatible with the strictures of the French mind. The Frenchman wishes to have a clear and simple history which has developed in an orderly and methodic way, a history in which all the events have proceeded strictly from the political principles he boasts of, just as corollaries are deduced from a theorem. And if reality does not furnish him with that history, it will be so much the worse for reality; he will alter the facts, suppress them, invent them, preferring to have to deal with a novel, clear, and methodic history than with a true but confused and complex one.

It is this straitness of mind which makes the Frenchman eager for clarity and method, and it is this love of clarity, order, and method which leads him, in every domain, to throw out or raze to the ground everything bequeathed to him by the past, in order to construct the present on a perfectly coordinated plane. Descartes, who was perhaps the most typical representative of the French mind, took it upon himself to formulate (in his *Discours de la Méthode*) the principles proclaimed by all those who so often have broken the ties of our traditions: "Thus we see that the buildings undertaken and completed by a single architect are generally more beautiful and better arranged than those which several persons have tried to repair by making use of old walls that had been built for other purposes. So with those old sites which, at first having been only little villages, have become in the course of time big cities, and are usually so badly encompassed, compared with those regular placements that an engineer draws on a plain in his imagination. Although, when we consider each of these buildings by itself, we often find as much or more art in one as in any other, yet, on seeing how they are arranged helter-skelter in all sizes and how they make streets crooked and unequal, one would say that chance, rather than the will of a few men using reason, has thus disposed them." In this passage the great philosopher praised in advance the vandalism which in the age of Louis XIV was to raze so many monuments of the past; he was a prophet of the Versailles to come.

The Frenchman conceives the development of social and political life only as a perpetual cycle of new beginnings, an indefinite series of revolutions. The Englishman sees in it a continuous evolution. Taine has shown what a dominant influence the "classical spirit," that is to say, the strong but narrow mind prevalent in most Frenchmen, has had on the history of France. We might just as correctly trace through the course of the history of England the effects of the ample but weak mind of the English people.[10]

Now that we have become acquainted with the diverse manifestations of the power to imagine a multitude of concrete facts accompanied by an inaptitude for abstract ideas, we shall not be astonished to learn that this amplitude and weakness of mind have offered a new type of physical theory—new in contrast to that conceived by the strong but narrow mind—and we shall not be astonished, either, to see this new type attain its highest growth in the works of "that great English school of mathematical physics whose works are one of the glories of the nineteenth century."[11]

5. *English Physics and the Mechanical Model*

In the treatises on physics published in England, there is always one element which greatly astonishes the French student; that element, which nearly invariably accompanies the exposition of a theory, is the model. Nothing helps us better understand how very different from ours is the manner in which the English mind proceeds in the construction of science than this use of the model.

Two electrically charged bodies are before us, and the problem is to give a theory of their mutual attractions or repulsions. The French or German physicist, be he a Poisson or a Gauss, will by an act of thought postulate in the space outside these bodies that abstraction called a material point and, associated with it, that other abstraction called an electric charge. He then tries to calculate a third abstraction: the force to which the material point is subjected. He gives formulas which permit one to determine the magnitude and direction of this force for each possible position of this material point. From these formulas he deduces a series of consequences: he shows clearly that in each point of space the force is directed along the tangent to a certain line called the line of force, and that all the lines of force cross normally (at right angles) certain surfaces—viz., equipotential surfaces—whose equations he gives. In particular, he shows that they are normal to the surfaces

[10] The reader will find a very profound, very subtle, and well-documented analysis of an English mind at once ample and weak in André Chevrillon, *op.cit.*

[11] O. Lodge, *Les Théories modernes de l'Électricité: Essai d'une théorie nouvelle*, tr. E. Meylan (Paris, 1891), p. 3. (Translator's note: Originally published as *Modern Theories of Electricity: Essay in a New Theory* [London, 1890].)

of the two charged conductors which are included among the number of equipotential surfaces. He calculates the force to which each element of these surfaces is subjected. Finally, he integrates all these elementary forces according to the rules of statics; he then knows the laws of the mutual actions of the two charged bodies.

This whole theory of electrostatics constitutes a group of abstract ideas and general propositions, formulated in the clear and precise language of geometry and algebra, and connected with one another by the rules of strict logic. This whole fully satisfies the reason of a French physicist and his taste for clarity, simplicity, and order.

The same does not hold for an Englishman. These abstract notions of material points, force, line of force, and equipotential surface do not satisfy his need to imagine concrete, material, visible, and tangible things. "So long as we cling to this mode of representation," says an English physicist, "we cannot form a mental representation of the phenomena which are really happening."[12] It is to satisfy this need that he goes and creates a model.

The French or German physicist conceives, in the space separating two conductors, abstract lines of force having no thickness or real existence; the English physicist materializes these lines and thickens them to the dimensions of a tube which he will fill with vulcanized rubber. In place of a family of lines of ideal forces, conceivable only by reason, he will have a bundle of elastic strings, visible and tangible, firmly glued at both ends to the surfaces of the two conductors, and, when stretched, trying both to contract and to expand. When the two conductors approach each other, he sees the elastic strings drawing closer together; then he sees each of them bunch up and grow large. Such is the famous model of electrostatic action imagined by Faraday and admired as a work of genius by Maxwell and the whole English school.

The employment of similar mechanical models, recalling by certain more or less rough analogies the particular features of the theory being expounded, is a regular feature of the English treatises on physics. Here is a book (O. Lodge, *op.cit.*) intended to expound the modern theories of electricity and to expound a new theory. In it there are nothing but strings which move around pulleys, which roll around drums, which go through pearl beads, which carry weights; and tubes which pump water while others swell and contract; toothed wheels which are geared to one another and engage

[12] O. Lodge, *Les Théories modernes* . . . , p. 16.

hooks. We thought we were entering the tranquil and neatly ordered abode of reason, but we find ourselves in a factory.

The use of such mechanical models, very far from facilitating the understanding of a theory by a French reader, requires him, in many cases, to make a serious effort to grasp the operation of what is often a very complicated apparatus, as described to him by the English author. Quite an effort is required in order to recognize the analogies between the properties of this apparatus and the propositions of the theory that is being illustrated. This effort is often much greater than the one the Frenchman needs to make in order to understand in its purity the abstract theory which it is claimed the model embodies.

The Englishman, on the other hand, finds the use of the model so necessary to the study of physics that to his mind the sight of the model ends up by being confounded with the very understanding of the theory. It is curious to see this confusion formally accepted and proclaimed by one who is the highest expression of the English scientific genius, one who, famous for a long time under the name of William Thomson, has been raised to the peerage with the title of Lord Kelvin. In his *Lectures on Molecular Dynamics*, he says:

"My object is to show how to make a mechanical model which shall fulfill the conditions required in the physical phenomena that we are considering, whatever they may be. At the time when we are considering the phenomenon of elasticity in solids, I want to show a model of that. At another time, when we have vibrations of light to consider, I want to show a model of the action exhibited in that phenomenon. We want to understand the whole about it; we only understand a part. It seems to me that the test of 'Do we or do we not understand a particular subject in physics?' is 'Can we make a mechanical model of it?' I have an immense admiration for Maxwell's mechanical model of electromagnetic induction. He makes a model that does all the wonderful things that electricity does in inducing currents, etc., and there can be no doubt that a mechanical model of that kind is immensely instructive and is a step towards a definite theory of electro-magnetism."[13]

In another passage, Thomson again says: "I never satisfy myself until I can make a mechanical model of a thing. If I can make a mechanical model, I understand it. As long as I cannot make a

[13] W. Thomson, *Lectures on Molecular Dynamics, and the Wave Theory of Light* (Baltimore: Johns Hopkins University, 1884), pp. 131-32. See also Sir W. Thomson (Lord Kelvin), "Conférences scientifiques et allocutions," tr. L. Lugol, annotated by M. Brillouin, *Constitution de la matière* (Paris, 1893).

mechanical model all the way through I cannot understand, and that is why I cannot get the electromagnetic theory of light. I believe firmly in *an* electromagnetic theory of light, and that when we understand electricity and magnetism and light, we shall see them all together as part of a whole. But I want to understand light as well as I can without introducing things that we understand even less of. That is why I take plain Dynamics. If I can get a model in plain Dynamics, I cannot in Electromagnetics."[14]

Understanding a physical phenomenon is, therefore, for the physicists of the English school, the same thing as designing a model imitating the phenomenon; whence the nature of material things is to be understood by imagining a mechanism whose performance will represent and simulate the properties of bodies. The English school is completely committed to the purely mechanical explanations of physical phenomena.

The purely abstract theory highly regarded by Newton, and studied steadily by us, will appear scarcely intelligible to adepts of this school like Thomson:

"Another class of mathematical theories, based to some extent on experiment, is at present useful, and has even in certain cases pointed to new and important results, which experiment has subsequently verified. Such are the Dynamical Theory of Heat, the Undulatory Theory of Light, etc., etc. In the former, which is based upon the conclusion from experiment that *heat is a form of energy*, many formulae are at present obscure and uninterpretable, because we do not know the mechanism of the motions or distortions of the particles of bodies. Results of the theory in which these are not involved, are of course experimentally verified. The same difficulties exist in the Theory of Light. But before this obscurity can be perfectly cleared up, we must know something of the ultimate, or *molecular*, constitution of the bodies, or groups of molecules, at present known to us only in the aggregate."[15]

This predilection for explanatory and mechanical theories is, of course, not a sufficient basis for distinguishing English doctrines from the scientific traditions thriving in other countries. Mechanical theories have been dressed up in their most complete form by the French genius Descartes; the Dutchman Huygens and the Swiss school of the Bernouilli family have contended for the retention of

[14] W. Thomson, *Lectures on Molecular Dynamics . . .*, p. 270.

[15] W. Thomson and P. G. Tait, *Treatise on Natural Philosophy*, Vol. I, Part I, Sec. 385. (Translator's note: First published in 1867, this work was revised in later editions.)

the principles of atomism in all their rigidity. What distinguishes the English school is not merely that it has tried to reduce matter to mechanism, but the particular form its attempts have taken in obtaining this reduction.

No doubt, wherever mechanical theories have been planted and cultivated, they have owed their birth and progress to a lapse in the faculty of abstracting, that is, to a victory of imagination over reason. When Descartes and his philosophic followers refused to attribute to matter any quality not purely geometrical or kinematic, they did so because such a quality was occult, conceivable only by reason, and thus remaining inaccessible to the imagination. The reduction of matter to geometry by the great thinkers of the seventeenth century clearly indicates that at that time the sense for profound metaphysical abstractions, exhausted by the excesses of a decaying Scholasticism, had been somnolent.

But among the great physicists of France, Holland, Switzerland, and Germany the sense for abstraction may lapse but it never falls asleep completely. It is true that the hypothesis that everything in material nature is reducible to geometry and to kinematics is a triumph of imagination over reason. But after yielding on this essential point, reason recovers its rights when it comes to deducing consequences and constructing the mechanism which is to represent matter. The properties of this mechanism should result logically from the hypotheses taken as the foundations of the cosmological system. Descartes, for example, and Malebranche after him, once having admitted the principle that extension is the essence of matter, took pains to deduce from it that matter has everywhere the same nature, that there cannot be several different material substances, and that shapes and motions alone can distinguish the different parts of matter from one another; also, that the same quantity of matter always occupies the same volume, whence it follows that matter is incompressible; and they aimed at constructing logically a system which explains natural phenomena by permitting only two elements to enter: the shape of the parts moved and the motion with which they are animated.

Not only is the mechanical construction, which will serve to explain the laws of physics, subject to certain logical requirements and bound to respect certain principles, but also the bodies which are used to make up these mechanisms are in no wise similar to the visible and concrete bodies we observe and manipulate every day. These bodies are formed from an abstract and ideal matter defined by the principles of the cosmology favored by the physicist, a matter which never comes to our senses and is visible and accessible only by reason. This is the case with Cartesian matter which is merely extension and motion, as well as with atomistic matter which possesses no property other than shape and hardness.

When an English physicist seeks to construct a model appropriate enough to represent a group of physical laws, he is not embarrassed by any cosmological principle, and is not constrained by any logical necessity. He does not aim to deduce his model from a philosophical system, nor even to put it into accord with such a system. He has only one object: to create a visible and palpable image of the abstract laws that his mind cannot grasp without the aid of this model. Provided that the mechanism is quite concrete and visible to the eyes of the imagination, it matters little to him whether the atomistic cosmology declares itself satisfied with it or whether the principles of Cartesianism condemn it.

The English physicist does not, therefore, ask any metaphysics to furnish the elements with which he can design his mechanisms. He does not aim to know what the irreducible properties of the ultimate elements of matter are. W. Thomson, for example, never asks himself such philosophical questions as: Is matter continuous or is it formed of individual elements? Is the volume of one of the ultimate elements of matter variable or invariable? What is the nature of an atom's actions: are they efficacious at a distance or only by contact? These questions do not even enter his mind, or else, when they are presented to him, he pushes them away as otiose and injurious to scientific progress. For instance, he says: "The idea of an atom has been so constantly associated with incredible assumptions of infinite strength, absolute rigidity, mystical actions at a distance, and indivisibility, that chemists and many other reasonable naturalists of modern times, losing all patience with it, have dismissed it to the realms of metaphysics, and made it smaller than 'anything we can conceive.' But if atoms are inconceivably small, why are not all chemical actions infinitely swift? Chemistry is powerless to deal with this question, and many others of paramount importance, if barred by the hardness of its fundamental assumptions, from contemplating the atom as a real portion of matter occupying a finite space, and forming a not immeasurably small constituent of any palpable body."[16]

The bodies with which the English physicist constructs his models

[16] W. Thomson, "The Size of Atoms," *Nature*, March 1870, reprinted in Thomson and Tait, *op.cit.*, Appendix F.

are not abstract conceptions elaborated by metaphysics. They are concrete bodies, similar to those surrounding us; namely, bodies that are solid or liquid, rigid or flexible, flowing or viscous; and with solidity, fluidity, rigidity, flexibility, and viscosity it is not necessary to understand abstract properties defined in terms of a certain cosmology. These properties are nowhere defined, but imagined by means of observable examples: rigidity calls up the image of a block of steel; flexibility, that of a silk thread; viscosity, that of glycerine. In order to express in a more tangible manner the concrete character of the bodies with which he builds his mechanisms, W. Thomson is not afraid to designate them by the most everyday names: he calls them bell-cranks, cords, jellies. He could not indicate more clearly that what he is concerned with are not combinations intended to be conceived by reason, but mechanical contrivances intended to be seen by the imagination.

Neither could he warn us more clearly that the models he proposes should not be taken as *explanations* of natural laws; anyone who should attribute such a meaning to them would be exposed to strange surprises.

Navier and Poisson have formulated a theory of the elasticity of crystalline bodies; 18 moduli, distinct from one another in general, characterize each of these bodies.[17] Thomson sought to illustrate this theory by means of a mechanical model. "We could not," he says, "be satisfied unless we could see our way to make a model with the 18 independent moduluses."[18] Eight rigid spheres placed at the eight vertices of a parallelepiped, and connected with one another by a sufficient number of spiral springs, constitute the proposed model. One look at it would be enough to disappoint greatly anyone who might have expected an *explanation* of the laws of elasticity; how, indeed, would the elasticity of the spiral springs be explained? Hence, the great English physicist did not offer this model as an explanation. "Although the molecular constitution of solids supposed in these remarks and mechanically illustrated in our model, *is not to be accepted as true in nature*, still the construction of a mechanical model of this kind is undoubtedly very instructive."

[17] At least, according to Thomson, Navier never dealt with any but isotropic bodies. According to Poisson's theory, the elasticity of a body depends on only fifteen moduli; the principles of Navier's theory applied to crystalline bodies lead to a similar result.

[18] W. Thomson, *Lectures on Molecular Dynamics . . .*, p. 131.

6. *The English School and Mathematical Physics*

Pascal very rightly regarded ampleness of mind as the faculty which plays a role in a good many geometrical investigations; more clearly still, it is the characteristic quality of the genius of the pure algebraist. The algebraist is not concerned with analyzing abstract notions and discussing the exact scope of general principles, but simply with combining skillfully, according to fixed rules, signs capable of being drawn as he writes. In order to be a great algebraist, there is scarcely any need for intellectual strength; a great ampleness of mind suffices, for skill in algebraic calculation is not a gift of reason but an ornament of the imaginative faculty.

It is not, therefore, astonishing to note that algebraic skill is very widespread among English mathematicians. This appears not only in the number of very great algebraists among English scientists, but also in the Englishman's predilection for the diverse forms of symbolic calculation.

One word of explanation on this subject.

A man whose mind is not of the ample sort will play a better game of checkers than of chess. In fact, whenever he wants to play a combination of moves at checkers, the elements he has at his disposal will be only of two kinds, the single checkers and the kings, both of which move according to very simple rules. On the other hand, the tactics of chess combine as many distinct elementary moves as there are kinds of pieces, and some of these—the knight's way of moving, for example—are complex enough to disconcert a weak imaginative faculty.

The difference between the games of checkers and chess reappears in the difference between the classical algebra that all French scientists use and the various symbolic algebras created in the nineteenth century. Classical algebra comprises only a few elementary operations, each represented by a special symbol, and each one being a very simple operation; a complicated algebraic calculation is only a long series of these scarcely varied elementary operations or only a lengthy manipulation of these few signs. The object of symbolic algebra is to abbreviate the length of these calculations. Toward this end, it adjoins to the elementary operations of classical algebra other operations which it treats as elementary and represents by a special symbol, and each of which is a combination or condensation effected according to a fixed rule by operations borrowed from the old algebra. In a symbolic algebra you can carry out almost at once a calculation which in the old algebra is composed of a long series of intermediate calculations, but you will have to make use of a very large number of different kinds of signs, each obeying a very complex rule. Instead of playing checkers, you will be playing a sort of chess in which many distinct pieces are to move, each in its own way.

It is clear that the taste for symbolic algebras is an index of that ampleness of mind which we would expect to be particularly widespread among the English.

This predisposition of the English genius for condensed algebraic calculations would not be definitely recognized, perhaps, as so distinctive, if we limited ourselves to even a brief review of the mathematicians who have created such systems of calculation. The English school would cite with pride the calculus of quaternions conceived by Hamilton, but the French could match it with Cauchy's theory of keys, and the Germans with Grassmann's theory of extension (*Ausdehnungslehre*). We must not wonder at that, for ample minds are to be found in every nation.

But only among the English is ampleness of mind found so frequently as an endemic, traditional habit; thus it is only among English men of science that symbolic algebras, the calculus of quaternions, and "vector analysis" are customary, most of the English treatises making use of these complex and shorthand languages. French and German mathematicians do not learn these languages readily; they never succeed in speaking them fluently or, above all, in thinking directly in the forms which constitute these languages. In order to follow a calculation based on the method of quaternions or of "vector analysis" they have to translate it into a version of classical algebra. One of the French mathematicians, Paul Morin, who had studied most profoundly the different kinds of symbolic calculi, once told me: "I am never sure of a result obtained by the method of quaternions until I have checked it by our old Cartesian algebra."

The frequent use the English physicists make of the different sorts of symbolic algebras is therefore evidence of their ampleness of mind, but if this usage imposes a peculiar garb on their mathematical theory, it does not impose any special physiognomy on the very body of the theory itself; by taking off its garb, we could easily dress this theory in the style of classical algebra.

Now in many cases this change of clothing would scarcely suffice to disguise the English origin of a theory of mathematical physics and to have it mistaken for a French or German theory. On the contrary, it would permit one to recognize that in the construction of a physical theory, the English do not always attribute to mathematics the same role the Continental scientists do.

To a Frenchman or a German, a physical theory is essentially a logical system. Perfectly rigorous deductions unite the hypotheses at the base of a theory to the consequences which are derivable from it and are to be compared with experimental laws. If algebraic calculation intervenes, it is only for the sake of making the chain of syllogisms connecting consequences to hypotheses lighter and easier to handle; but in a theory soundly constituted, this purely auxiliary role of algebra should never be forgotten. We must always be aware of the possibility of replacing the calculation by purely logical reasoning of which it is a shorthand expression; and in order that this substitution may be made in as precise and certain a manner as possible, there must be established a very exact and very strict correspondence between the symbols, or letters combined by the symbolic algebra, and the properties that the physicist measures, between the fundamental equations which serve as a point of departure for the analyst and the hypotheses at the base of theory.

Thus the French or German founders of mathematical physics, the Laplaces, the Fouriers, the Cauchys, the Ampères, the Gausses, the Franz Neumanns, have constructed with extreme caution the bridge intended to connect the point of departure of the theory, the definition of the magnitudes it is to deal with, and the justifications for the hypotheses which will bear its deductions to the road on which its algebraic development will proceed. Whence those preambles, models of clarity and method, with which most of their memoirs begin.

These preambles, devoted to *setting up the equations* of a physical theory, are nearly always to be sought in vain among the writings of the English authors. Consider a striking example.

To the electrodynamics of conductors, created by Ampère, Maxwell added a new electrodynamics, that of dielectric bodies. This branch of physics is the outcome of the consideration of an essentially new element, which has been called, and improperly at that, the displacement current.

This displacement current was introduced by Maxwell in order to complete the definition of the properties of a dielectric at a given instant, properties not determined wholly by the known polarity at that instant, just as the conduction current has been added to the electric charge in order to complete the definition of the variable condition of a conductor. The displacement current has some close analogies with the conduction current, but at the same time some profound differences. Thanks to the intervention of this new element, electrodynamics was thrown into disorder; phenomena, never even hinted at by experience and to be discovered only twenty years later by Hertz, were announced as present. We see the germ of a new theory of the propagation of electrical actions in nonconducting media, and this theory leads to an unforeseen interpretation of optical phenomena, namely, to the electromagnetic theory of light.

Of course, we expect that such a new and unforeseen element, which turns out to show such fruitful, surprising, and important consequences, would not have been introduced by Maxwell into his equations until he had defined and analyzed it with the utmost of minute precautions. But open Maxwell's memoir where he expounded his new theory of the electromagnetic field, and you will find only these two lines to justify the introduction of the displacement flux into the equations of electrodynamics:

"The variations of the electrical displacement should be added to the currents in order to obtain the total movement of the electricity."

How can we explain this almost complete absence of definition even when it is a question of the most novel and most important elements, and this indifference to setting up equations for a physical theory? The answer does not seem to us to be doubtful: Whereas the French or German physicist intends the algebraic part of a theory to replace just the series of syllogisms used to develop this theory, the English physicist regards the algebra as playing the part of a *model.* It is an apparatus functioning by signs accessible to the imagination and subject to the rules of algebra; it imitates more or less faithfully the laws of the phenomena under study, as an apparatus of different bodies moving in accordance with the laws of mechanics would imitate the laws of the phenomena.

Hence, when a French or German physicist introduces definitions permitting him to substitute an algebraic calculus for a logical deduction, he must do it with extreme care under penalty of losing the rigor and accuracy he would have required in his syllogism. When, on the other hand, W. Thomson offers a mechanical model for a group of phenomena, he does not impose on himself any very detailed rational argument in order to establish a connection between this apparatus of concrete bodies and the physical laws he is called upon to represent; for imagination, the sole concern of the model, will be the exclusive judge of the resemblance between the drawing and the object drawn. That is what Maxwell did. To the intuitions of the imaginative faculty he left the task of comparing physical laws with the algebraic model which has to imitate them. Without waiting for this comparison, he followed the operation of this model and combined the equations of electrodynamics very often without aiming at a coordination of physical laws in each of his combinations.

The French or German physicist is most often disconcerted by such a conception of mathematical physics. He does not realize that all he has before him is a model mounted to satisfy his imagination rather than his reason. He persists in looking for a series of deductions, in the algebraic transformations, from clearly formulated hypotheses to empirically verifiable consequences. Not finding them, he wonders anxiously what Maxwell's theory really amounts to. To this, one who understands the mind of the English mathematical physicist answers that there is nothing analogous in Maxwell to the physical theory one seeks, but only algebraic formulas which are combined and transformed. Hertz said: "To this question, 'What is Maxwell's theory?' I cannot give any clearer or briefer answer than the following: 'Maxwell's theory is the system of Maxwell's equations.' "[19]

7. *The English School and the Logical Coordination of a Theory*

The theories created by the great Continental mathematicians, whether French or German, Dutch or Swiss, may be classified in two large categories: explanatory and purely representative theories. But these two kinds of theories offer a common feature: they are understood to be systems constructed according to the rules of strict logic. Products of a reason unafraid of profound abstractions or long deductions, but mainly eager for order and clarity, their theories demand that an impeccable method characterize the series of their propositions from beginning to end, from the basic hypotheses to the consequences that can be compared with the facts.

It was this method that brought forth those majestic systems of nature claiming to bestow on physics the formal perfection of Euclid's geometry. These systems take as their foundations a certain number of very clear postulates, and try to erect a perfectly rigid

[19] H. Hertz, *Untersuchungen über die Ausbreitung der elektrischen Kraft, Einleitende Uebersicht* (Leipzig, 1892), p. 23. (Translator's note: Translated into English by D. E. Jones under the title *Electric Waves; Being Researches on the Propagation of Electric Action with Finite Velocity through Space*, Preface by Lord Kelvin [London and New York: Macmillan, 1893, 1900].)

and logical structure in which each experimental law is exactly lodged. From the era in which Descartes constructed his *Principes de Philosophie* to the day when Laplace and Poisson built on the hypothesis of attraction the ample edifice of their mechanics, such an edifice stood as the perpetual ideal of abstract intellects, especially of the French genius. In pursuing this ideal it has raised monuments whose simple lines and grand proportions are still an object of delight and admiration, especially today when these structures are shaky on account of their generally undermined foundations.

This unity of theory and this logical linkage among all the parts of a theory are such natural and necessary consequences of the idea that strength of mind imputes to a physical theory, that to disturb this unity or to break this linkage is to violate the principles of logic or to commit an absurdity, from its viewpoint.

Not at all like that is the case of the ample but weak mind of the English physicist.

Theory is for him neither an explanation nor a rational classification of physical laws, but a model of these laws, a model not built for the satisfying of reason but for the pleasure of the imagination. Hence, it escapes the domination of logic. It is the English physicist's pleasure to construct one model to represent one group of laws, and another quite different model to represent another group of laws, notwithstanding the fact that certain laws might be common to the two groups. To a mathematician of the school of Laplace or Ampère, it would be absurd to give two distinct theoretical explanations for the same law, and to maintain that these two explanations are equally valid. To a physicist of the school of Thomson or Maxwell, there is no contradiction in the fact that the same law can be represented by two different models. Moreover, the complication thus introduced into science does not shock the Englishman at all; for him it adds the extra charm of variety. His imagination, being more powerful than ours, does not know our need for order and simplicity; it finds its way easily where we would lose ours.

Thus, in English theories we find those disparities, those incoherencies, those contradictions which we are driven to judge severely because we seek a rational system where the author has sought to give us only a work of imagination.

For example, here is a series of W. Thomson's lectures devoted to an exposition of molecular dynamics and the wave theory of light.[20] The French reader who reads these notes thinks he is going to find in them a set of neatly formulated hypotheses about the constitution of the ether and of ponderable matter, a series of methodically conducted calculations starting from these hypotheses, and an exact comparison between the consequences of these calculations and the facts of experience, but great will his disappointment be, and brief his illusion! For such a well-ordered theory is not what Thomson intended to construct. He has simply wished to consider diverse kinds of experimental laws and to construct a mechanical model for each of them.[21] There are as many distinct models to represent the role of the material molecule in phenomena as there are categories of these phenomena.

Is the problem to represent the marks of elasticity in a crystal? The material molecule is represented by eight spherical masses occupying the vertices of a parallelepipedon, and these masses are connected to one another by a greater or lesser number of spiral springs.[22]

Is it the theory of the dispersion of light that is to be made clear to the imagination? Then the material molecule is found to be composed of[23] a certain number of rigid, concentric, spherical shells held in that position by springs. A multitude of these little mechanisms is imbedded in the ether. The latter is a homogeneous incompressible body,[24] inelastic for very rapid vibrations, perfectly soft for actions of a certain duration. It resembles a jelly or glycerine.[25]

Is a model suitable to represent rotational polarization desired? Then the material molecules that we scatter by thousands in our "jelly" will no longer be built on the plan we have just described; they will be constructed of little rigid shells in each of which a gyrostat will rotate rapidly around an axis fixed to the shell.[26]

But that is too crude a performance for our "crude gyrostatic molecule,"[27] so that a more perfect mechanism is soon installed to replace it.[28] The rigid shell no longer contains merely one gyrostat, but two of them turning in opposite directions; ball and socket joints and sheaths connect them to each other and to the sides of the spherical shell, allowing a certain play to their axes of rotation.

It would be difficult to choose from among these diverse models,

[20] W. Thomson, *Notes of Lectures on Molecular Dynamics and the Wave Theory of Light* (Baltimore, 1884). The reader should also consult Sir W. Thomson (Lord Kelvin), "Conférences scientifiques. . ."

[21] *Notes of Lectures* . . . , p. 132. [22] *ibid.*, p. 127.

[23] *ibid.*, pp. 10, 105, 118. [24] *ibid.*, p. 9. [25] *ibid.*, p. 118.

[26] *ibid.*, pp. 242, 290. [27] *ibid.*, p. 327. [28] *ibid.*, p. 320.

exhibited in the course of the *Lectures on Molecular Dynamics*, one which best represents the material molecule. But how much more embarrassing will our choice be if we survey the other models imagined by Thomson in the course of his other writings!

In one place, we find an incompressible, homogeneous fluid without viscosity filling all of space. Certain portions of this fluid are animated by persisting vortical motions, and these portions represent the atoms of matter.[29]

In another place, the incompressible liquid is represented by a collection of rigid balls which are connected to one another by conveniently located bars.[30]

Elsewhere, he appeals to Maxwell's and Tait's kinetic theories in order to imagine the properties of solids, liquids, and gases.[31]

Will it be easier to define the constitution attributed to the ether by Thomson?

When Thomson developed his theory of vortex atoms, the ether formed a part of this fluid—homogeneous, incompressible, devoid of all viscosity—which filled space. The ether was represented by that part of this fluid which is exempt from any whirlpool motion. But soon, in order to represent gravitation, which drives molecules of matter toward one another, the great physicist complicated this composition of the ether;[32] by reviving an old hypothesis of Fatio de Duilliers and Lesage, he threw across the homogeneous fluid a whole swarm of small solid corpuscles moving in all directions with extremely high speed.

In another work, the ether became again a homogeneous and incompressible body, but this body was now similar to a very viscous fluid or a jelly.[33] This analogy was abandoned in its turn; in order to represent the properties of the ether, Thomson took up some formulas[34] due to MacCullagh,[35] and in order to make them im-

[29] W. Thomson, "On Vortex Atoms," *Edinburgh Philosophical Society Proceedings*, Feb. 18, 1867.

[30] W. Thomson, *Scientific Papers*, III, 466. Originally in *Comptes rendus de l'Académie des Sciences* (Sept. 16, 1889).

[31] W. Thomson, "Molecular Constitution of Matter," *Proceedings of the Royal Society of Edinburgh, July 1 and 15, 1889*, Secs. 29-44; *Scientific Papers*, III, 404; *Lectures on Molecular Dynamics . . .*, p. 280.

[32] W. Thomson, "On the Ultramundane Corpuscles of Lesage," *Philosophical Magazine*, XLV (1873), 321.

[33] W. Thomson, *Lectures on Molecular Dynamics . . .*, pp. 9, 118.

[34] W. Thomson, "Equilibrium or Motion of an Ideal Substance Called for Brevity Ether," *Scientific Papers*, III, 445.

[35] J. MacCullagh, "An Essay Towards a Dynamical Theory of Crystalline Reflection and Refraction," *Transactions of Royal Irish Academy*, Vol. XXI (Dec. 9, 1839), reprinted in *The Collected Works of James MacCullagh* (1880), p. 145.

aginable he represented them in a mechanical model: rigid boxes, each containing a gyrostat animated by a movement of rapid rotation around an axis fixed to the sidewalls, are attached to one another by strips of flexible but inelastic cloth.[36]

This incomplete enumeration of the diverse models by which Thomson has sought to represent the diverse properties of the ether or of ponderable molecules still gives us only a feeble idea of the crowd of images aroused in his mind by the words: "the constitution of matter." For us to understand, it would be necessary to join all the models created by other physicists but whose use he recommends, for example, to annex that model of electrical actions which Maxwell built[37] and for which Thomson has constantly professed his admiration. There we should see the ether and all poor conductors of electricity fashioned like a cake of honey, the sidewalls of the cells formed not of wax but of an elastic body whose deformations represent electrostatic actions, and the honey replaced by a perfect fluid animated by a rapid vortical motion, the image of magnetic actions.

This collection of engines and mechanisms disconcerts the French reader who was looking for a coordinated sequence of hypotheses on the constitution of matter and a hypothetical explanation of this constitution. But at no time was it Thomson's intention to give such an explanation; at all times the very kind of language he employs prevents the reader from making such an interpretation of his thought. The mechanisms he proposes are "crude models"[38] of "rude representations,"[39] they are "unnatural mechanically."[40] "The molecular constitution of solids supposed in these remarks and mechanically illustrated in our model is not to be accepted as true in nature. . . .";[41] ". . . there is hardly any need to remark that the *ether* we have imagined is a purely ideal substance."[42] The very provisional character of each of these models is noticed in the zigzag way in which the author abandons or takes them up again according to the needs of the phenomena he is studying. "Go back to our spherical

[36] W. Thomson, "On a Gyrostatic Adynamic Constitution of the Ether," *Proceedings of the Edinburgh Royal Society*, March 17, 1890, reprinted in *Scientific Papers*, III, 466. Also "Ether, Electricity and Ponderable Matter," *Scientific Papers*, III, 505.

[37] J. Clerk Maxwell, "On Physical Lines of Force," Part III: "The Theory of Molecular Vortices Applied to Statical Electricity," *Philosophical Magazine*, Jan. and Feb. 1882. See *Scientific Papers*, ed. W. D. Niven (Cambridge: Cambridge University Press, 1890), I, 491.

[38] *Lectures on Molecular Dynamics* . . . , pp. 11, 105.

[39] *ibid.*, p. 11. [40] *ibid.*, p. 105. [41] *ibid.*, p. 131.

[42] W. Thomson, *Scientific Papers*, II, 464.

molecule with its central spherical shells—that is the rude mechanical illustration, remember. I think it is very far from the actual mechanism of the thing, but it will give us a mechanical model."[43] At best he sometimes yields to the hope that these ingeniously imagined models may indicate the road which will lead in the remote future to a physical explanation of the material world.[44] The multiplicity and variety of the models proposed by Thomson to represent the constitution of matter does not astonish the French reader very long, for he very quickly recognizes that the great physicist has not claimed to be furnishing an explanation acceptable to reason, and that he has only wished to produce a work of imagination. The French reader's astonishment is, however, profound and lasting when he again finds the same absence of order and method, the same lack of concern for logic not only in the collection of mechanical models but in a series of algebraic theories. How could he conceive of the possibility of an illogical mathematical development? Hence, the stupefaction he experiences while studying a work like Maxwell's *Treatise on Electricity*:

"The first time a French reader opens Maxwell's book," writes Poincaré, "a feeling of discomfort, and often even of distrust, is at first mingled with his admiration. . . .

"The English scientist does not seek to construct a single, definitive, and well-ordered structure; he seems rather to raise a great number of provisional and independent houses among which communication is difficult and at times impossible.

"Take, for example, the chapter in which electrostatic attractions are explained by pressures and tensions prevailing in the dielectric medium. This chapter might be suppressed without making the rest of the volume less clear and less complete, and, on the other hand, it contains a theory sufficient in itself which we could understand without having read a single line preceding or coming after it. But not only is it independent of the rest of the work; it is difficult to reconcile it[45] with the fundamental ideas of the book, as a more thorough discussion will show later on. Maxwell does not even at-

[43] *Lectures on Molecular Dynamics* . . . , p. 280.

[44] W. Thomson, *Scientific Papers*, III, 510.

[45] In reality this theory of Maxwell's proceeds from a complete miscomprehension of the laws of elasticity; we have given proof of this and developed the correct theory to be substituted for Maxwell's errors in our *Leçons sur l'Electricité et le Magnetisme*, Vol. II, I, XII (Paris, 1892). A term, neglected by mistake in our calculation, has been supplied by Liénard (*La Lumière Électrique*, LII [1894], 7, 67), whose results we have reestablished by a direct analysis (P. Duhem, *American Journal of Mathematics*, XVII [1895], 117).

tempt such a reconciliation. He limits himself to saying: 'I have not been able to make the next step, namely to account by means of mechanical considerations for these stresses in the dielectric.'

"This example will suffice to make my thought understandable. I might mention many other examples; thus, who would doubt while reading the pages devoted to rotational magnetic polarization that there is an identity between optical and magnetic phenomena?"[46]

Maxwell's *Treatise on Electricity and Magnetism* was in vain attired in a mathematical form. It is no more of a logical system than Thomson's *Lectures on Molecular Dynamics.* Like these *Lectures* it consists of a succession of models, each representing a group of laws without concern for the other models representing other laws and, at times representing these very same laws or some of them; except that these models, instead of being constructed out of gyrostats, spiral springs, and glycerine, are an apparatus of algebraic signs. These different partial theories, each developed in isolation, indifferent to the previous one but at times covering part of the field covered by this predecessor, are less properly addressed to our reason than to our imagination. They are paintings, and the artist, in composing each of them, has selected with complete freedom the objects he would represent and the order in which he would group them. It matters little whether one of his clients has already posed in a different attitude for another portrait. The logician would be out of place in being shocked by this; a gallery of paintings is not a chain of syllogisms.

8. *The Diffusion of English Methods*

The English mind is clearly characterized by the ample use it imaginatively makes of concrete collections and by the meagre way in which it makes abstractions and generalizations. This peculiar type of mind produces a peculiar type of physical theory; the laws of the same group of phenomena are not coordinated in a logical system, but are represented by a model. This model, moreover, may be a mechanism built of concrete bodies or an apparatus of algebraic signs; in any case, the English type of theory does not subject itself

[46] H. Poincaré, *Électricité et Optique,* Vol. I: *Les théories de Maxwell et la théorie électro-magnétique de la lumière,* Introduction, p. viii. Poincaré quoted from J. C. Maxwell, *Treatise on Electricity and Magnetism,* I, 174. The reader who desires to know to what degree the lack of concern for all logic and even for any mathematical exactitude went in Maxwell's mind will find numerous examples in P. Duhem, *Les Théories électriques de J. Clerk Maxwell: Étude historique et critique* (Paris, 1902).

in its development to the rules of order and unity demanded by logic.

For a long time these peculiarities were a kind of trade-mark of the physical theories manufactured in England, and there was little use made of them on the Continent. In the last few years there has been a change, and the English manner of dealing with physics has spread everywhere with extreme rapidity. Today it is customarily used in France as well as in Germany. We shall inquire into the causes of this diffusion.

In the first place, it is well to recall that though the type of mind called ample but shallow by Pascal is very widespread among the English, it is nevertheless neither their eminent domain nor their exclusive property.

Surely Newton yielded nothing either to Descartes or to any of the great classical thinkers in the ability to offer very abstract ideas with perfect clarity and very general principles with great accuracy, as well as in the art of carrying out in an irreproachable order a series of experiments or a chain of deductions. His intellectual strength was one of the most powerful known to mankind.

Just as we can find among the English—and Newton's case assures us of this—minds that are strong and accurate, so we can meet outside England minds that are ample but shallow.

Gassendi had such a mind.

The contrast between the two intellectual types so clearly defined by Pascal appears in a very forceful manner in the famous discussion which engaged Gassendi and Descartes in polemics.[47] How ardently Gassendi insisted on the argument "that the mind is not really distinguished from the imaginative faculty"; how forcefully he asserted that "the imagination is not distinguished from the intellect," and that "there is in us only a single faculty by means of which we generally know all things"![48] With what haughtiness Descartes replied to Gassendi: "What I have said about the imagination is clear enough if one wishes to be on one's guard against it, but it is no wonder that my view seems obscure to those who do not reflect on what they imagine!"[49] The two opponents seem to have understood that their debate was assuming a complexion different from that of most of the arguments so frequent among philosophers, that it was not a dispute between two men or two doctrines but

[47] P. Gassendi, *Disquisitio metaphysica, seu dubitationes et instantiae adversus Renati Cartesii Metaphysicam, et responsa.*

[48] P. Gassendi, *Dubitationes in Meditationem* II^am^.

[49] *Cartesii "Responsum ad Dubitationem* V *in Meditationem* II^am^."

a struggle between two types of mentality, the ample but shallow one versus the strong but narrow one. *Oh, soul! Oh, Mind!* cried Gassendi, challenging the champion of abstraction. *Oh, body!* replied Descartes, crushing with haughty disdain the imagination that is limited to concrete objects.

Henceforth, we shall understand Gassendi's predilection for the Epicurean cosmology. Save for their extremely small size, the atoms he imagined strongly resemble the bodies which he daily had occasion to see and touch. This concrete character and accessibility to the imagination of Gassendi's physics appear in a fuller light in the following passage, in which the philosopher explained in his own way the "sympathies" and "antipathies" of the Scholastics: "We must realize that these actions are produced like those which operate in a more observable way among bodies; the only difference is that the mechanisms which are gross in the latter case are very attenuated in the former. Wherever ordinary sight shows us an attraction and union we see hooks and strings, something which catches and something caught; wherever it shows us a repulsion and separation, we see spikes and pikes, a body of some sort or other causing an explosion, etc. In the same way, in order to explain the actions not coming under common observation, we have to imagine little hooks, little strings, little spikes, little pikes, and other instruments of the same kind which are imperceptible and intangible, but we must not infer from this that they do not exist."[50]

In every period of scientific development we are likely to run across, among the French, physicists who are intellectually akin to Gassendi in their desire to give explanations that the imagination can grasp. Among the theorists who do honor to our time, one of the most ingenious and most productive, J. Boussinesq, has expressed with perfect clarity this need felt by certain minds to imagine the objects about which they reason: "The human mind while observing natural phenomena recognizes in them, besides many confused elements which it does not get to unravel, one clear element, which by virtue of its precision is liable to be the object of truly scientific knowledge. That element is the geometric one, pertaining to the localization of objects in space which permits one to represent them, to draw them, or to construct them in a more or less ideal manner. It is made up of the dimensions and shapes of bodies or of systems of bodies, in what is called, in a word, their *configuration* at a given instant. These shapes or configurations,

[50] P. Gassendi, *Syntagma Philosophicum* (Lyons, 1658), Part II, I, VI, Ch. XIV.

whose measurable parts are distances or angles, on the one hand are conserved, at least nearly so for a certain time, and even appear to maintain themselves in the same regions of space and constitute what we call *rest*; on the other hand, they change incessantly but continuously, and their changes of place are called *local motion*, or simply motion."[51]

These diverse configurations of bodies and their changes from one instant to another are the only elements the geometer can draw, and also the only things the imagination can represent clearly to itself. Consequently, they are, according to him, the only proper objects of science. A physical theory will be truly constituted when it will have reduced the study of a group of laws to the description of such local motions. "Until now science, considered with respect to its established part, or what is capable of being part of its structure, has grown in going from Aristotle to Descartes and to Newton, from the ideas of *qualities* or *changes* of state which are not drawn to the idea of *forms* or local *motions* which are drawn or seen."[52]

No more than Gassendi does Boussinesq wish theoretical physics to be a work of reason from which the imagination will be banished. He expresses his thought in this respect in formulas which stand out in a way that recalls certain words of Lord Kelvin.

Let us not be mistaken about it, however; Boussinesq would not follow the great Englishman to the very end. If he wishes the imagination to be able to grasp the constructions of theoretical physics in all their parts, he does not intend to do without the cooperation of logic in outlining the plan of his constructions. He does not by any means allow them, nor would Gassendi for that matter, to be so devoid of all order and unity that they consist of nothing but a labyrinth of independent and incoherent bits of masonry.

At no time have French or German physicists by themselves, of their own free will, reduced physical theory to being nothing but a collection of models. That opinion did not originate spontaneously in Continental science; it is an English importation.

It is due above all to the vogue set by the work of Maxwell and introduced into science by this great physicist's commentators and followers. Thus it was diffused from the very start in apparently one of its most disconcerting forms. Before the French or German physicists came to adopt mechanical models, several of them were

[51] J. Boussinesq, *Leçons synthétiques de Mécanique générale* (Paris, 1889), p. 1.

[52] J. Boussinesq, *Théorie analytique de la Chaleur* (Paris, 1901), Vol. I, p. xv.

already accustomed to treat mathematical physics as a collection of algebraic models.

Among the best of those who have helped promote such a fashion of treating mathematical physics, the distinguished Heinrich Hertz is properly quoted. We heard him declare: "Maxwell's theory is the equations of Maxwell." In conformity with this principle, even before he had formulated it, Hertz had developed a theory of electrodynamics with Maxwell's equations serving as the foundation.[53] These equations were accepted just as they stood without discussion of any kind, without examination of the definitions and hypotheses from which they are derivable. They were treated as self-sufficient without submitting the consequences obtained to experimental test.

Such a method of proceeding would be understandable on the part of an algebraist if he were to study equations drawn from principles accepted by all physicists and completely confirmed by experiments; we should not be surprised to find him quietly indifferent to the setting-up of equations and to an experimental verification concerning neither of which would anybody entertain the least doubt. But such is not the case with the electrodynamic equations studied by Hertz. The reasoning and calculation by which Maxwell tried on several occasions to justify them abound in contradictions, obscurities, and plain mistakes; and the confirmation which experiment may bring to them can only be quite partial and limited. Indeed, we have to face the fact that the simple existence of a piece of magnetized steel is incompatible with such an electrodynamics, and this colossal contradiction was plain to Hertz.[54]

One might perhaps think that the acceptance of so contestable a theory is made necessary by the absence of any other doctrine capable of having a more logical foundation and a more exact agreement with facts. Nothing of the sort is the case. Helmholtz gave an electrodynamic theory which proceeds very logically from the best-established principles of electrical science, and their formulation in equations is exempt from the paradoxes arising too frequently in Maxwell's work. His formulation explains all the facts which the equations of Hertz and Maxwell take into account, without running into the refutations that reality harshly presents in

[53] H. Hertz, "Ueber die Grundgleichungen der Electrodynamik für ruhende Korper," *Göttinger Nachrichten,* March 19, 1890. *Wiedemann's Annalen der Physik und Chemie,* XL, 577. *Gesammelte Werke von H. Hertz,* Vol. II: *Untersuchungen über die Ausbreitung der elektrischen Kraft* (2nd ed.), p. 208.

[54] H. Hertz, *Gesammelte Werke von . . . ,* Vol. II: *Untersuchungen über Ausbreitung der elektrischen Kraft* (2nd ed.), p. 240.

opposition to the latter. It cannot be doubted that reason requires us to prefer this theory, but imagination prefers to play with the elegant algebraic model fashioned by Hertz and at the same time by Heaviside and Cohn. The employment of this model was very quickly diffused among minds too weak to be unafraid of lengthy deductions. We have seen writings multiply in which Maxwell's equations were accepted without discussion, as though they were a revealed dogma whose obscurities are revered like sacred mysteries.

More formally than Hertz, Poincaré has proclaimed the right of mathematical physics to shake off the yoke of too rigorous a logic and to break the connection which joined his diverse theories to one another. He has written:

"We should not flatter ourselves on avoiding all contradiction. But we must take sides. Two contradictory theories may, in fact, provided that we do not mix them and do not seek the bottom of things, both be useful instruments of research. Perhaps the reading of Maxwell would be less suggestive if he had not opened so many new, divergent paths."[55]

These words, which encouraged the practice in France of the methods of English physics and gave free play to the ideas advocated so brilliantly by Lord Kelvin, were not without repercussions which for many reasons were sure to be powerful and prolonged.

I do not have to mention either the high authority of the man who proffered these words or the importance of the discoveries concerning which they were expressed; the reasons I wish to point out are less legitimate though no less powerful.

Among these reasons we must, in the first place, mention the taste for the exotic, the desire to imitate the foreign, the need to dress the mind as well as the body in the fashion of London. Among those who declare the physics of Maxwell and of Thomson preferable to the physics until now classical in France, how many simply invoke the one theme: It is English!

Moreover, the loud admiration for the English method is for too many a means of forgetting how little apt they are in the French method, that is, how difficult it is for them to conceive an abstract idea and to follow a rigorous line of reasoning. Deprived of strength of mind, they try, by taking on the outward ways of the ample mind, to make one believe they possess intellectual amplitude.

[55] H. Poincaré, *Électricité et Optique*, Vol. I: *Les théories de Maxwell et la théorie électro-magnétique de la lumière*, Introduction, p. ix.

These causes, however, would not perhaps suffice to guarantee the vogue which English physics enjoys today, were it not for the industrial needs joined to them.

The industrialist has very often an ample mind; the need to combine machinery, to deal with business matters, and to handle men has early accustomed him to see clearly and rapidly complicated assemblages of concrete facts. On the other hand, his is nearly always a very shallow mind. His daily occupation keeps him removed from abstract ideas and general principles. Gradually the faculties constituting strength of intellect have atrophied in him, as happens with organs no longer functioning. The English model cannot, therefore, fail to appear to him as the form of physical theory most appropriate to his intellectual aptitudes.

Naturally, he desires to have physics expounded in that form to those who will have to direct workshops and factories. Besides, the future engineer requires instruction in a short time; he is in a hurry to make money with his knowledge, and he cannot waste time, which for him is money. Now, abstract physics, preoccupied above all with the absolute solidity of the building which it is raising, does not know such feverish haste. It intends to build on rock and, in order to achieve this, digs as long as is necessary. It requires of those who wish to be its pupils a mind broken in by various exercises of logic, and made supple by the gymnastics of mathematical sciences; it will not welcome as a substitute for them any intermediary or complication. How could those concerned with the useful and not with the true be expected to submit themselves to this rigorous discipline? Why would they not prefer to the latter the more rapid procedures of the theories addressed to the imagination? Those who are commissioned to teach engineering are therefore eager to adopt the English methods and teach this sort of physics, which sees even in mathematical formulas nothing but models.

To this pressure the majority of them offer no resistance, but, on the contrary, even magnify the scorn for order and hatred of logical rigor professed by the English physicists. When they admit a formula in their lectures or treatises, they never ask whether this formula is accurate or exact but only whether it is convenient and appeals to the imagination. To one who has not had the painful obligation of reading closely many writings devoted to the applications of physics, it is hardly credible to what extent this hatred of all rational method and of all exact deduction is carried in these writings. The most glaring paralogisms and the falsest calculations are piled up in broad daylight; under the influence of industrial instruction, theoretical physics has become a perpetual challenge to the integrity of the accurate mind.

For the evil has not only touched the texts and courses intended for future engineers. It has penetrated everywhere, propagated by the hatreds and prejudices of the multitude of people who confuse science with industry, who, seeing the dusty, smoky, and smelly automobile, regard it as the triumphal chariot of the human spirit. Higher education is already contaminated by utilitarianism and secondary education is prey to the epidemic. In the name of this utilitarianism a clean sweep is made of the methods which served until now to expound the physical sciences. Abstract and deductive theories are rejected in favor of offering students concrete and inductive views. We no longer think of putting into young minds ideas and principles, but substitute numbers and facts.

We shall not take time to discuss at great length these inferior and degraded theories of the imagination.

We shall remark to snobs that if it is easy to ape the defects of foreigners it is more difficult to acquire the hereditary qualities which characterize them, that snobs may well be able to give up the strength of the French mind but not its straitness, and that they will easily rival the English in shallowness of mind but not in ampleness. Thus they will condemn themselves to having minds that are both weak and narrow, that is to say, false minds.

We shall remind industrialists, who have no care for the accuracy of a formula provided it is convenient, that the simple but false equation sooner or later becomes, by an unexpected act of revenge of logic, the undertaking which fails, the dike which bursts, the bridge which crashes; it is financial ruin when it is not the sinister reaper of human lives.

Finally, we shall declare to the utilitarians, who think they are making practical men by teaching only concrete things, that their pupils will sooner or later become routine manipulators, mechanically applying formulas they do not understand; for only abstract and general principles can guide the mind in unknown regions and suggest to it the solutions of unforeseen difficulties.

9. *Is the Use of Mechanical Models Fruitful for Discoveries?*

In order to appreciate with justice the imaginative type of physical theory, let us not take it just as it is presented to us by those who claim to make use of it without possessing the ampleness of mind that would be needed to treat it worthily. Let us consider it just as it was presented by those whose powerful imaginations gave birth to it, that is, in particular, the great English physicists.

Concerning the procedures employed by the English in dealing with physics, there is current a commonplace opinion, according to which the abandonment of care for logical unity which was so important with the old theories, and the substitution of models independent of one another for the rigorously linked deductions formerly in use bestow upon the physicist's inquiries a suppleness and freedom which are eminently fruitful for discoveries.

This opinion appears to us to contain a very great share of illusion. Too often those who maintain it attribute to the use of models discoveries which have been made by quite different procedures.

In a great number of cases a model has been constructed of a theory already formed either by the author of the theory himself or by some other physicist; then, gradually, the model has relegated to oblivion the abstract theory that preceded it and without which the model would not have been conceived. It appears as the instrument of discovery whereas it has only been a means of exposition. The reader who is not forewarned and who lacks the leisure to make historical inquiries that go back to origins may be a dupe to this deception.

Take, for example, the report on the sciences (at the Paris Exposition of 1900) in which Emile Picard draws in such broad and sober lines the picture of the state of the sciences in 1900.[56] Read the passages devoted to two important theories of contemporary physics: the theory of the continuity of the liquid state and the theory of osmotic pressure. It will appear as though the part played in the creation and development of these theories by mechanical models and imaginative hypotheses concerning molecules and their motions and collisions has been a very large one. In suggesting such a view to us, Picard's report reflects very exactly the opinions heard every day in courses and laboratories. But these opinions are without foundation. The employment of mechanical models had almost no part in the creation and development of the two doctrines before us for study.

The idea of a continuity between the liquid and gaseous states was presented to the mind of Andrews by experimental induction. It was also induction and generalization that led James Thomson

[56] *Exposition universelle de 1900 à Paris. Rapport du Jury international* (Paris, 1901), General Introduction, Part II: "Sciences," by Emile Picard, pp. 53ff.

to conceive the idea of the theoretical isotherm. From a doctrine which is the paradigm of abstract theories, namely, from thermodynamics, Gibbs deduced a perfectly concatenated exposition of this new branch of physics, while the same thermodynamics furnished Maxwell with an essential relation between the theoretical and practical isotherms.

While abstract thermodynamics thus manifested its fertility, Van der Waals on his side, by means of assumptions about the nature and movements of molecules, touched on the study of the continuity between the liquid and the gaseous states. The contribution of kinetic hypotheses to this study consisted in an equation of the theoretical isotherm, an equation from which was deduced a corollary, the law of corresponding states. But in contact with facts, it had to be recognized that the isothermal equation was too simple and the law of corresponding states too crude for a physics concerned with some degree of exactitude to be able to preserve them.

The history of osmotic pressure is not less clear. Abstract thermodynamics furnished Gibbs from the start with the fundamental equations for it. Thermodynamics has also been the sole guide of J. H. Van't Hoff in the course of his first works, while experimental induction furnished Raoult with the laws necessary for the progress of the new doctrine. The latter had reached maturity and constitutional vigor when the mechanical models and kinetic hypotheses came to bring to it the assistance it did not ask for, with which it had nothing to do, and to which it owed nothing.

Hence, before attributing the discovery of a theory to the mechanical models which encumber it today, it is well to make sure that these models have really presided over or aided its birth, and that they have not come like a parasitic growth and fastened themselves on a tree already robust and full of life.

It is well also, if we wish to appreciate with accuracy the fruitfulness that the use of models may have, not to confuse this use with the use of analogy.

The physicist who seeks to unite and classify in an abstract theory the laws of a certain category of phenomena, lets himself be guided often by the analogy that he sees between these phenomena and those of another category. If the latter are already ordered and organized in a satisfactory theory the physicist will try to group the former in a system of the same type and form.

The history of physics shows us that the search for analogies between two distinct categories of phenomena has perhaps been the surest and most fruitful method of all the procedures put in play in the construction of physical theories.

Thus, it is the analogy seen between the phenomena produced by light and those constituting sound which furnished the notion of light wave from which Huygens drew such a wonderful result. It is this same analogy which later led Malebranche and then Young to represent monochromatic light by a formula similar to the one representing a simple sound.

A similar insight concerning the propagation of heat and that of electricity within conductors permitted Ohm to transport all in one piece the equations Fourier had written for the former to the second category of phenomena.

The history of theories of magnetism and dielectric polarization is simply the development of analogies seen for a long time by physicists between magnets and bodies which insulate electricity. Thanks to this analogy each of the two theories benefits from the progress of the other.

The use of physical analogy often takes a more precise form.

Two categories of very distinct and very dissimilar phenomena having been reduced by abstract theories, it may happen that the equations in which one of the theories is formulated are algebraically identical to the equations expressing the other. Then, although these two theories are essentially heterogeneous by the nature of the laws which they coordinate, algebra establishes an exact correspondence between them. Every proposition of one of the theories has its homologue in the other; every problem solved in the first poses and resolves a similar problem in the second. Each of these two theories can serve to *illustrate* the other, according to the word used by the English: "By physical analogy," Maxwell said, "I mean that partial resemblance between the laws of a science and the laws of another science which makes one of the two sciences serve to illustrate the other."[57]

Of this mutual illustration of two theories, here is an example among many others:

The idea of a warm body and the idea of an electrostatically charged body are two essentially heterogeneous notions. The laws which govern the distribution of stationary temperatures in a group of good conductors of heat and the laws which fix the state of electrical equilibrium in a group of good conductors of electricity pertain to absolutely different physical objects. However, the two theories whose object is to classify these laws are expressed in two

[57] J. Clerk Maxwell, *Scientific Papers*, I, 156.

groups of equations which the algebraist cannot distinguish from each other. Thus, each time that he solves a problem about the distribution of stationary temperatures, he solves by that very fact a problem in electrostatics, and vice versa.

Now, this sort of algebraic correspondence between two theories, this illustration of one by the other, is an infinitely valuable thing: not only does it bring a notable intellectual economy since it permits one to transfer immediately to one of the theories all the algebraic apparatus constructed for the other, but it also constitutes a method of discovery. It may happen, in fact, that in one of these two domains which the same algebraic scheme covers, experimental intuition quite naturally poses a problem and suggests a solution for it, while in the other domain the physicist might not be so easily led to formulating this question or to giving it this response.

These diverse ways of appealing to the analogy between two groups of physical laws or between two distinct theories are therefore fruitful for discoveries, but we should not confuse them with the use of models. Analogies consist in bringing together two abstract systems; either one of them already known serves to help us guess the form of the other not yet known, or both being formulated, they clarify each other. There is nothing here that can astonish the most rigorous logician, but there is nothing either that recalls the procedures dear to ample but shallow minds, nothing which substitutes the use of the imagination for the use of reason, nothing which rejects the logically conducted understanding of abstract notions and general judgments in order to replace it with a vision of concrete collections.

If we avoid attributing to the use of models the discoveries which are due in reality to the use of abstract theories and if we are also careful not to confuse the use of such models with the use of analogy, what will be the exact role of imaginative theories in the progress of physics?

It seems to us that this role will be a very meager one.

The physicist who has most formally identified the understanding of a theory with the vision of a model, Lord Kelvin, has distinguished himself by admirable discoveries, but we do not see that any one of them may have been suggested by imaginative physics. His most beautiful findings, the electrical transfer of heat, the properties of variable currents, the laws of oscillating discharge, and many others too long to recite, have been made by means of the abstract systems of thermodynamics and of classical electrodynamics. Whenever he calls to his aid mechanical models, he limits himself to the task of expounding or representing results already obtained, that is, when he is not making a discovery.

In the same way, it does not appear that the model of electrostatic and electromagnetic actions aided Maxwell to create the electromagnetic theory of light. No doubt he tried to obtain from this model the two essential formulas of this theory; the very manner in which he directed his attempts shows, however, that the results he obtained were known to him through some other means. In his desire to retain these results at any cost he went so far as to falsify one of the fundamental formulas of elasticity.[58] He was not able to create the theory that he envisaged except by giving up altogether the use of any model, and by extending by means of analogy the abstract system of electrodynamics to displacement currents.

Thus, neither in Lord Kelvin's nor in Maxwell's work has the use of mechanical models shown that fruitfulness nowadays attributed so readily to it.

Does this mean that no discovery has ever been suggested to any physicist by this method? Such an assertion would be a ridiculous exaggeration. Discovery is not subject to any fixed rule. There is no doctrine so foolish that it may not some day be able to give birth to a new and happy idea. Judicial astrology has played its part in the development of the principles of celestial mechanics.

Besides, anyone who would deny any fruitfulness to the employment of mechanical models would be seen to be in conflict with some very recent examples. We should cite to him the electro-optical theory of Lorentz, anticipating the doubling of spectral lines in a magnetic field and provoking Zeeman to look for and observe this phenomenon. We should cite the mechanisms imagined by J. J. Thomson to represent the passage of electricity through a gas and the curious experiments connected with it.

Without doubt these same examples would lend themselves to discussion.

We might observe that the electro-optical system of Lorentz, although founded on mechanical hypotheses, is no longer simply a model but an extensive theory the various parts of which are logically connected and coordinated. Besides, Zeeman's phenomenon, far from confirming the theory which suggested the discovery, resulted first of all in proving that Lorentz's theory could not

[58] P. Duhem, *Les Théories électriques de J.-Clerk Maxwell: Étude historique et critique* (Paris, 1902), p. 212.

be maintained just as it was, and in demonstrating that it required at the very least some profound modifications.

We could also remark that the connection is a loose one between the representations which J. J. Thomson offers our imagination and the well-observed facts of the ionization of gases. Perhaps the mechanical models juxtaposed beside these facts obscure the discoveries already made rather than throw light on the discoveries to be made.

But let us not waste time on these fine points. Let us admit frankly that the use of mechanical models has been able to guide certain physicists on the road to discovery and that it is still able to lead to other findings. At least it is certain that it has not brought to the progress of physics that rich contribution boasted for it. The share of booty it has poured into the bulk of our knowledge seems quite meager when we compare it with the opulent conquests of abstract theories.

10. *Should the Use of Mechanical Models Suppress the Search for an Abstract and Logically Ordered Theory?*

We have seen the most distinguished physicists, among those who recommend the use of mechanical models, making use of this form of theory far less as a means of discovery than as a method of exposition. Lord Kelvin did not himself proclaim the divining ability of the mechanisms that he constructed in such large numbers. He limited himself to declaring that the aid of such concrete representations was indispensable to his understanding and that he could not arrive at a clear comprehension of a theory without them.

Strong minds, those that do not need to embody an idea in a concrete image in order to conceive it, cannot reasonably deny to ample but weak minds, which cannot easily conceive of things devoid of shape or color, the right to sketch and paint the objects of physical theories in their visual imagination. The best means of promoting the development of science is to permit each form of intellect to develop itself by following its own laws and realizing fully its type; that is, to allow strong minds to feed on abstract notions and general principles, and ample minds to consume visible and tangible things. In a word, do not compel the English to think in the French manner, or the French in the English style. The principle of this intellectual liberalism, too rarely understood and practiced, was formulated by Helmholtz, who was so highly gifted with an exact and strong mind:

"The English physicists, like Lord Kelvin when he formulated his theory of vortex-atoms or Maxwell when he imagined the hypothesis of a system of cells whose contents are animated by a rotary motion, the hypothesis which serves as a basis for his attempt at a mechanical explanation of electromagnetism, have evidently found in such explanations a more lively satisfaction than if they had contented themselves with the very general representation of facts and their laws by the system of differential equations of physics. As for me, I must confess that I remain attached to this latter mode of representation, and I place more confidence in it than in the other. But I cannot raise any objection in principle against a method pursued by such great physicists."[59]

Besides, it is no longer a question today of knowing whether strong minds will tolerate the use by the imaginative ones of representations and models; the problem is rather to know whether they will retain the right to impose on physical theories unity and logical coordination. The imaginative ones do not in fact limit themselves to the claim that the use of concrete representations is indispensable to them for understanding abstract theories; they assert that by creating for each of the chapters of physics an appropriate mechanical or algebraic model without reference to the model which served to illustrate the preceding chapter, they are satisfying all the legitimate wishes of understanding. They assert that the attempts by which certain physicists try to construct a logically concatenated theory based on the smallest possible number of independent hypotheses is a labor which does not answer any need of a soundly constituted mind, and that, consequently, those whose duty it is to direct studies and orient scientific research ought to divert physicists from this idle labor.

What shall we say against these assertions which we hear repeated every day, in a hundred different forms, by every weak and utilitarian mind, in order for us to maintain the legitimacy, necessity, and preeminent value of abstract theories, logically coordinated? How shall we answer this question which at the present time is put to us in so pressing a manner: Is it permissible to symbolize several distinct groups of experimental laws, or even a single group of laws, by means of several theories each resting on hypotheses incompatible with the hypotheses supporting the others?

To this question we do not hesitate to answer: *If we confine ourselves strictly to considerations of pure logic*, we cannot prevent

[59] H. von Helmholtz, Preface to H. Hertz's work *Die Principien der Mechanik* (Leipzig, 1894), p. 21. (Translator's note: English translation entitled *The Principles of Mechanics* [London, 1899].)

a physicist from representing by several incompatible theories diverse groups of laws, or even a single group of laws; we cannot condemn incoherence in physical theory.

Such a declaration will appear very scandalous to those who regard a physical theory as an explanation of the laws of the inorganic world. It would, indeed, be absurd to explain one group of laws by supposing that matter is constituted in a certain way, and then explain another group of laws by supposing matter constituted in quite a different manner. An explanatory theory should by all necessity avoid even the appearance of a contradiction.

But if we admit what we have aimed at establishing, that a physical theory is simply a system intended to classify a set of experimental laws, how can we draw from the code of logic the right to condemn a physicist who employs, for the sake of ordering laws, different methods of classification, or a physicist who proposes, for the same set of laws, diverse classifications resulting from different methods? Does logical classification forbid naturalists to classify one group of animals according to the structure of the nervous system, and another group according to the structure of the circulatory system? Does a malacologist fall into absurdity when he expounds first Bouvier's system which groups mollusks according to the arrangement of their nerve filaments, and then Rémy Perrier's system which makes comparisons based on the study of the organ of Bojanus? Thus, a physicist will logically have the right first to regard matter as continuous and then to consider it as formed of separate atoms, to explain capillary phenomena by forces of attraction acting between stationary particles, and then to endow these same particles with rapid motion in order to explain heat phenomena. None of these discrepancies will be a violation of logical principles.

Logic evidently imposes on the physicist only one obligation: not to confuse or mix up the various methods of classification he employs. That is, when he establishes a certain relationship between two laws, he is logically obliged to note in a precise manner which of the proposed methods justifies this relationship. Poincaré expressed this when he wrote the words we have already quoted: "Two contradictory theories can, in fact, both be useful instruments of research, *provided that we do not mix them together* and provided that we do not seek the bottom of things in them."[60]

Logic does not, therefore, furnish any unanswerable argument

[60] H. Poincaré, *Électricité et Optique*, Vol. I: *Les théories de Maxwell et la théorie électro-magnétique de la lumière*, Introduction, p. ix.

to anyone who claims we must impose on physical theory an order free from all contradiction. Are there sufficient grounds for imposing such an order if we take as a principle the tendency of science toward the greatest intellectual economy? We do not think so.

At the beginning of this chapter we showed how diverse sorts of minds would judge differently the economy of thought resulting from an intellectual operation. We have seen that where the strong but narrow mind would feel it had made a theory lighter, the other sort, the ample but weak mind, would feel extremely fatigued.

It is clear that minds adapted to the conception of abstract ideas, to the formation of general judgments, and to the construction of rigorous deductions, but easily lost in a somewhat complicated collection of things, will find a theory so much more satisfactory and economical as the order in it is more perfect and less often broken by lacunas or contradictions.

But an imagination ample enough to grasp in a glance a complicated collection of disparate things, without feeling the need for putting such a collection in order, generally accompanies a reason weak enough to fear abstraction, generalization, and deduction. Minds in which these two dispositions are associated will find that the logical labor of coordinating diverse fragments of theory into a single system is considerable, and that it causes them more trouble than the sight of these disjoined fragments. They will not by any means judge the passage from incoherence to unity an economical intellectual operation.

Neither the principle of contradiction nor the law of the economy of thought permits us to prove in an irrefutable manner that a physical theory should be logically coordinated; from what source then can we draw an argument in favor of this opinion?

This opinion is a legitimate one because it results from an innate feeling of ours which we cannot justify by purely logical considerations but which we cannot stifle completely either. Those very physicists who have developed theories whose various parts cannot be fitted together, and whose various chapters describe just so many isolated mechanical or algebraic models, have only done so reluctantly and with regret. It suffices to read the preface written by Maxwell at the beginning of that *Treatise on Electricity and Magnetism* in which insoluble contradictions abound, in order to see that these contradictions were not intended or desired, and that the author wished to obtain a coordinated theory of electromagnetism. Lord Kelvin, while constructing his innumerable and disparate models continued to hope that some day it would be possible to give a mechanical *explanation* of matter. He liked to regard his models as serving to break a path which would lead to the discovery of this explanation.

Every physicist naturally aspires to the unity of science. That is why the employment of disparate and incompatible models has been proposed only for the last few years. Reason, which calls for a theory whose parts are all logically united, and imagination, which desires to embody these diverse parts of the theory in concrete representations, might have seen both their tendencies joined together if it had been possible to arrive at a complete and detailed mechanical explanation of the laws of physics. Hence, the ardor with which theorists have worked for a long time toward such an explanation. When the futility of these efforts had clearly proved that the hope for such an explanation was a chimera,[61] physicists convinced that it was impossible to satisfy at the same time the requirements of reason and the needs of the imagination had to make a choice. The strong and exact minds, subject above all to the empire of reason, ceased asking physical theory for an explanation of natural laws in order to safeguard its unity and rigor; the ample but weak minds, led by an imagination more powerful than reason, gave up constructing a logical system in order to be able to put the fragments of their theory in a visible and tangible form. But the renunciation in the latter persons, at least in those whose thought deserves to be taken into account, was never complete and final; they never offered their isolated and disparate constructions except as provisional shelters, as scaffoldings intended to be removed. They did not despair of someday seeing an architect of genius raise a structure whose parts would all function according to a plan of perfect unity. Only those who affect a hatred of intellectual strength were mistaken to the extent of taking the scaffolding for a completed building.

Thus, all those who are capable of reflecting and of taking cognizance of their own thoughts feel within themselves an aspiration, impossible to stifle, toward the logical unity of physical theory. This aspiration toward a theory whose parts all agree logically with one another is, moreover, the inseparable companion of that other aspiration, whose irresistible power we have previously ascertained,[62]

[61] For more details on this point, the reader is referred to our work *L'Évolution de la Mécanique* (Paris, 1903).

[62] See Ch. II, Sec. 4, above.

toward a theory which is a natural classification of physical laws. We, indeed, feel that if the real relations of things, not capable of being grasped by the methods used by the physicist, are somehow reflected in our physical theories, this reflection cannot be devoid of order or unity. To prove by convincing arguments that this feeling is in conformity with truth would be a task beyond the means afforded by physics; how or to what could we assign the characters that the reflection should present when the objects which are the source of this reflection escape visibility? And yet, this feeling surges within us with indomitable strength; whoever would see in this nothing more than a snare and a delusion cannot be reduced to silence by the principle of contradiction; but he would be excommunicated by common sense.

In this situation, as in all others, science would be impotent to establish the legitimacy of the principles themselves which outline its methods and guide its researches, were it not to go back to common sense. At the bottom of our most clearly formulated and most rigorously deduced doctrines we always find again that confused collection of tendencies, aspirations, and intuitions. No analysis is penetrating enough to separate them or to decompose them into simpler elements. No language is precise enough and flexible enough to define and formulate them; and yet, the truths which this common sense reveals are so clear and so certain that we cannot either mistake them or cast doubt on them; furthermore, all scientific clarity and certainty are a reflection of the clarity and an extension of the certainty of these common-sense truths.

Reason has, therefore, no logical argument to stop a physical theory which would break the chains of logical rigor; but "nature supports reason when impotent and prevents it from talking nonsense even at that point."[63]

[63] B. Pascal, *Pensées*, ed. Havet, Art. 8.

PART II

The Structure of Physical Theory

CHAPTER I

QUANTITY AND QUALITY

1. *Theoretical Physics Is Mathematical Physics*

The discussions developed in the first part of this book have informed us exactly about the aim the physicist should have when he constructs a theory.

A physical theory will then be a system of logically linked propositions and not an incoherent series of mechanical or algebraic models. This system will have for its object not the furnishing of an explanation but the representation and natural classification of experimental laws, taken in a group.

To require that a great number of propositions be linked in a perfect logical order is not a slight or easy condition to satisfy. The experience of centuries testifies how easily a fallacy slips into what appears to be the most irreproachable series of syllogisms.

There is, however, one science in which logic attains a degree of perfection which makes it easy to avoid error and easy to recognize it when it has been committed, namely, the science of numbers, arithmetic, with its extension in algebra. It owes this perfection to an extremely abbreviated symbolic language in which each idea is represented by an unambiguously defined sign, and in which each sentence of the deductive reasoning is replaced by an operation combining the signs in accord with strictly fixed rules and by a calculation whose accuracy is always easy to test. This rapid and precise symbolism of algebra guarantees progress which disregards almost entirely the opposing doctrines of competing schools.

One of the claims to fame of the geniuses who made the sixteenth and seventeenth centuries distinguished was the recognition of the truth that physics would not become a clear and precise science, exempt from the perpetual, sterile disputes characterizing its history till then, and would not be capable of demanding universal assent for its doctrines so long as it would not speak the language of geometers. They created a true theoretical physics by their understanding that it had to be mathematical physics.

Created in the sixteenth century, mathematical physics proved it was the sound method of physics by the wonderful, steady progress it made in the study of nature. Today it would be impossible, without shocking the plainest good sense, to deny that physical theories should be expressed in mathematical language.

In order for a physical theory to be able to present itself in the form of a chain of algebraic calculations, all the ideas employed in the theory must be capable of being represented by numbers. This leads us to ask ourselves the following question: Under what conditions may a physical attribute be signified by a numerical symbol?

2. *Quantity and Measurement*

The first answer to this question appears at once to the mind to be as follows: In order that an attribute found in a body may be expressed by a numerical symbol, it is necessary and sufficient that this attribute belong, in Aristotle's language, to the category of quantity and not to the category of quality. In the more readily accepted language of modern geometry, it is necessary and sufficient that this attribute be a magnitude.

What are the essential characteristics of a magnitude? By what mark do we recognize that the length of a line, for example, is a magnitude?

By comparing different lengths with one another we come across the notions of equal and unequal lengths which present the following characteristics:

Two lengths equal to the same length are equal to each other.

If the first length is greater than the second and the second greater than a third, the first is greater than the third.

These two characteristics already permit us to express the fact that two lengths are equal to each other by making use of the arithmetical symbol =, and by writing $A = B$. They permit us to express the fact that A is greater than B in length by writing $A > B$ or $B < A$. In fact, the only properties of the signs of equality or inequality invoked in arithmetic or in algebra are the following:

1. The two equalities $A = B$ and $B = C$ imply the equality $A = C$.

2. The two inequalities $A > B$ and $B > C$ imply the inequality $A > C$.

These properties still belong to the signs of equality and inequality when we make use of them in the study of lengths.

Let us place several lengths end to end; we obtain a new length S which is greater than each of the component lengths A, B, and C. S does not change if we change the order in which we put the components end to end; neither does it change if we replace some of the component lengths (e.g., B and C) by the length obtained by putting them end to end.

These several characteristics authorize us to employ the arithmetical sign of addition to represent the operation which consists in putting several lengths end to end, and to write $S = A + B + C + \ldots$.

In fact, from what we have just said, we can write

$$A + B > A, \quad A + B > B$$
$$A + B = B + A$$
$$A + (B + C) = (A + B) + C$$

Now these equalities and inequalities represent the only fundamental postulates of arithmetic. All the rules of calculation conceived in arithmetic to combine numbers are going to be extended to lengths.

The most immediate of these extensions is that of multiplication; the length obtained by placing end to end n lengths, equal to one another and to A, may be represented by the symbol $n \times A$. This extension is the starting point for the measurement of lengths and will permit us to represent each length by a number accompanied by the name of a certain standard or unit of length chosen once for all lengths.

Let us choose such a standard of length, for example, the meter, which is the length given to us under very specific conditions by a certain metal bar deposited in the International Bureau of Weights and Measures.

Certain lengths may be reproduced by placing n lengths equal to a meter end to end; the number n followed by the name meter will adequately represent such a length; we say that it is a length of n meters.

Other lengths cannot be represented in that way, but they can be reproduced by placing end to end p equal segments when q of these same segments subsequently placed one after the other would reproduce the length of a meter. Such a length will be entirely known when we state the fraction p/q followed by the name meter; it will be a length of p/q meters.

An incommensurable number, still followed by the name of the standard, will permit us to represent as well any length not belonging to either of the two categories we have just defined. In short, any length whatsoever will be perfectly known when we say it is a length of x meters, whether x is an integer, fraction, or incommensurable number.

Then the symbolic addition of $A + B + C + \ldots$, by which we represent the operation of bringing several lengths end to end, will be replaceable by a true arithmetic sum. It will suffice to measure each of the lengths $A, B, C, \ldots$ with the same unit, the meter, for example; we thus obtain numbers of meters $a, b, c, \ldots$. The length S which is formed by placing $A, B, C, \ldots$ end to end, measured also in meters, will be represented by a number s equal to the arithmetic sum of the numbers $a, b, c, \ldots$, which measure the lengths $A, B, C, \ldots$. For the symbolic equality

$$A + B + C + \ldots = S$$

between the component lengths and the resultant length, we substitute

$$a + b + c + \ldots = s$$

the arithmetic equality of the numbers of meters representing these lengths.

Thus, through the choice of a standard length and through measurement, we give to the signs of arithmetic and algebra, set up to represent operations done with numbers, the power to represent operations performed with lengths.

What we have just said about lengths could be repeated concerning surfaces, volumes, angles, and times; all the physical attributes which are magnitudes would show analogous characteristics. In every case we should see the different states of a magnitude show relations of equality or inequality susceptible of representation by the signs $=$, $>$, and $<$; we should always be able to submit this magnitude to an operation having the double property of being commutative and associative, and consequently, capable of being represented by the arithmetic symbol of addition, the sign $+$. Through this operation, measurement would be introduced into the study of this magnitude, and would enable one to define it fully by means of the union of an integer, fraction, or surd, and a unit of measurement; such a union is known by the name of a concrete number.

3. *Quantity and Quality*

The essential character of any attribute belonging to the category of quantity is therefore the following: Each state of a quantity's magnitude may always be formed through addition by means of other smaller states of the same quantity; each quantity is the union through a commutative and associative operation of quantities smaller than the first but of the same kind as it is, and they are parts of it.

The Aristotelian philosophy expressed this in a formula, too concise to give in full all the details of the thought, by saying: Quantity is that which has parts external to one another.

Every attribute that is not *quantity* is *quality*.

"Quality," said Aristotle, "is one of those words which are taken in many senses." The shape of a geometrical figure which makes a circle or a triangle of it is a quality; the observable properties of bodies, such as being warm or cold, light or dark, red or blue, are qualities; to be in good health is a quality; to be virtuous is a quality; to be a grammarian, mathematician, or musician—all are qualities.

"There are qualities," added the Stagirite, "which are not susceptible of more or less: a circle is not more or less circular; a triangle is not more or less triangular. But most qualities are susceptible of more or less; they are capable of intensity; a white thing can become whiter."

At first blush, we are tempted to establish a correlation between the various intensities of the same quality and the various states of the same quantity's magnitude, that is, to compare the heightening of intensity (*intensio*) or the lowering of intensity (*remissio*) to the increase or diminution of length, surface, or volume.

$A, B, C, \ldots$ are different mathematicians. A may be as good a mathematician as B, or a better, or not so good a mathematician. If A is as good a mathematician as B, and B is as good as C, then A is as good as C. If A is a better mathematician than B and B is better than C, then A is a better mathematician than C.

$A, B, C, \ldots$ are red materials whose shades we are comparing. Material A may be as brilliant a red as B, or less, or more brilliant than the material B. If the shade of A is as brilliant as the shade of B and that of B as brilliant as the shade of C, then the shade of A is as brilliant as the shade of C. If the material A is a deeper red than the material B, and the latter is a deeper hue of red than the material C, then the material A is a deeper red than the material C.

Thus, in order to express the fact that two qualities of the same kind do or do not have the same intensity, we can employ the signs $=$, $>$, and $<$, which will preserve the same properties they have in arithmetic.

The analogy between quantities and qualities stops there.

A large quantity, we have seen, may always be formed by the addition of a certain number of small quantities of the same kind. The large number of grains inside a sack of wheat may always be obtained by the summation of piles of wheat each containing a smaller number of grains. A century is a succession of years; a year is a succession of days, hours, and minutes. A road several miles long is traveled by putting end to end the short segments which the hiker crosses with each step. A field which has a large surface may be broken up into pieces of smaller surface.

Nothing like this applies to the category of quality. Bring together in a vast meeting as many mediocre mathematicians as you can find, and you will not have the equal of an Archimedes or of a Lagrange. Sew together the cloth remnants of a dark red hue, and the piece obtained will not be a brilliant red.

No quality of a certain kind and intensity results in any manner from several qualities of the same kind but of lesser intensity. Each intensity of quality has its own individual characteristics which make it absolutely unlike lesser and greater intensities. A quality of a certain intensity does not include as an integral part of itself the same quality made more intense. Boiling water is hotter than boiling alcohol and the latter hotter than boiling ether, but neither the boiling point of alcohol nor that of ether is part of the boiling point of water. Whoever would say that the heat of boiling water is the sum of the heat of boiling alcohol and the heat of boiling ether would be talking nonsense.[1] Diderot used to ask jokingly how many snowballs would be required to heat an oven; the question is embarrassing only for one who confuses quantity with quality.

Thus, in the category of quality we find nothing which resembles the formation of a large quantity by means of the small quantities which are its parts. We find no operation both commutative and associative which merits the name "addition" and may be represented by the + sign. Measurement stemming from the idea of addition cannot capture quality.

4. *Purely Quantitative Physics*

Every time an attribute is capable of being measured, or is a quantity, algebraic language becomes apt for expressing different states of this attribute. Is this aptitude for algebraic expression peculiar to quantities, and are qualities entirely deprived of it? The philosophers who in the seventeenth century created mathematical physics certainly thought so. Hence, in order to realize the mathematical physics to which they aspired, they had to require their

[1] It is, of course, understood that we are taking the word "heat" in its everyday meaning, which has nothing in common with what physicists attribute to "quantity of heat."

theories to deal exclusively with quantities and to rigorously banish any qualitative notion.

Moreover, these same philosophers all saw in physical theory not the representation but the explanation of empirical laws. The ideas combined in the propositions of physical theory were not, for them, the signs and symbols of observable properties, but the very expression of the reality hidden under appearances. The physical universe, which our senses present to us as an immense assemblage of qualities, had therefore to be offered to the mind as a system of quantities.

These common aspirations of the great scientific reformers who ushered in the seventeenth century culminated in the creation of the Cartesian philosophy.

To eliminate qualities completely from the study of material things is the aim and virtually the defining character of Cartesian physics.

Among the sciences only arithmetic, with its extension to algebra, is free from any notion borrowed from the category of quality, and it alone conforms to the ideal which Descartes proposed for the complete science of nature.

When it comes to geometry the mind runs into a qualitative element, for this science remains "so confined to the consideration of diagrams that it cannot exercise the understanding without fatiguing the imagination a great deal." "The scruples the ancients had against using arithmetical terms in geometry, which could only come from their not seeing clearly their relationships, caused much obscurity and difficulty in the manner with which they explained themselves." This obscurity and difficulty are to disappear when we get rid of the qualitative notion of geometrical form and shape, and keep only the quantitative notion of distance and the equations which connect the mutual distances of the different points studied. Although their objects are of different natures, the various branches of mathematics do not consider in these objects "anything else than the various relations or proportions found in them," so that it suffices to deal with these proportions in general with the methods of algebra, without being concerned about the objects in which they are encountered or the diagrams in which they are embodied; consequently, "everything which mathematicians have to consider is reduced to problems of one and the same kind, namely, to finding the values of the roots of some equation." All mathematics is reduced to the science of numbers in which only quantities are dealt with; qualities no longer have any place in it.

Qualities having been eliminated from geometry, they must now be banned from physics. In order to succeed in this, it suffices to reduce physics to mathematics, which has become the science of quantity alone. That is the task Descartes set out to accomplish:

"I admit no principles in physics which are not also accepted in mathematics." "For I profess plainly not to recognize any other substance in material things than the matter capable of all sorts of divisions, configurations, and motions which the geometers call quantity and which they take as the object of their demonstrations; and in this matter I consider absolutely nothing but these divisions, configurations, and motions. Concerning them, I admit nothing as true which cannot be deduced from axioms impossible for us to doubt and deduced in so evident a manner that the deduction amounts to a mathematical demonstration. And as all the phenomena of nature may be explained in that way, as we shall see in the sequel, I think we should admit no other principles in physics, nor wish for any other sort."[2]

What, then, is matter, first of all? "Its nature does not consist in hardness, nor in weight, heat, or other qualities of this kind," but only in "extension in length, breadth, and depth," in what "the geometers call quantity"[3] or volume. Matter is therefore quantity; the quantity of a certain portion of matter is the volume it occupies. A vessel contains as much matter as its volume, whether it is filled with mercury or filled with air. "Those who claim to distinguish material substance from extension or from quantity either have no idea of what comes under the name substance or else have a confused idea of immaterial substance."[4]

What is motion? Also a quantity. Multiply the quantity of matter that each of the bodies of a system contains by the speed with which it is set in motion, add together all the products, and you will have the quantity of motion of the system. So long as the system will not collide with any external body which may give motion to it or take motion away from it, it will conserve an invariable quantity of motion.

Thus, there is spread throughout the universe a single, homogeneous, incompressible, and inelastic matter about which we know nothing except that it is extended. This matter is divisible into parts of various shapes, and these parts can be moved into different relations with one another. Such are the only genuine properties of what constitutes bodies, and all the apparent qualities affecting our senses

[2] R. Descartes, *Principia Philosophiae,* Part II, Art. LXIV.
[3] *ibid.,* Part II, Art. IV. [4] *ibid.,* Part II, Art. IX.

reduce to these properties. The object of Cartesian physics is to explain how this reduction is made.

What is gravitation? The effect produced on bodies by vortices of ethereal matter. What is a hot body? A body "composed of small parts which agitate one another with a very sudden and violent motion." What is light? A pressure exerted on the ether by the motion of fiery bodies and transmitted instantaneously through the greatest distances. All the qualities of bodies without a single exception are explained by a theory in which we consider only geometric extension, the different configurations that can be traced in it, and the different motions which these can have. "The universe is a machine in which there is nothing at all to consider except the shapes and motions of its parts." Thus the entire science of material nature is reduced to a sort of universal arithmetic from which the category of quality radically is banned.

5. *The Various Intensities of the Same Quality Are Expressible in Numbers*

Theoretical physics, as we conceive it, does not have the power to grasp the real properties of bodies underneath the observable appearances; it cannot, therefore, without going beyond the legitimate scope of its methods, decide whether these properties are qualitative or quantitative. By insisting that it could decide for the quantitative, Cartesianism was making claims which do not appear tenable to us.

Theoretical physics does not grasp the reality of things; it is limited to representing observable appearances by signs and symbols. Now, we wish our theoretical physics to be a mathematical physics starting with symbols that are algebraic symbols or numerical combinations. If, therefore, only magnitudes can be expressed in numbers, we ought not to introduce into our theories any notion which is not a magnitude. Without asserting that everything at the very bottom of material things is merely quantity, we should admit nothing but what is quantitative in the picture we make of the totality of physical laws; quality would have no place in our system.

Now, there is no good ground on which to subscribe to this conclusion; the purely qualitative character of a notion is not opposed to the use of numbers to symbolize its various states. The same quality may appear with an infinity of different intensities. We can affix a label and number, so to speak, to each of these various intensities, registering the same number in two circumstances where the same quality is found with the same intensity, and identifying a second case, where the quality considered is more intense than in the first case, by a second number greater than the first.

Take, for example, the quality of being a mathematician. When a certain number of young mathematicians take a competitive examination, the examiner who is to judge gives a mark to each of them, assigning the same mark to two candidates who seem to him to be equally good mathematicians, and giving a better mark to one or the other if one appears to him to be a better mathematician than the other.

These pieces of material are red in varying degrees of intensity; the merchant who arranges them on his racks assigns numbers to them; to each number a very definite shade of red corresponds, and the higher the number, the more intense the brightness of red.

Here are some heated bodies. This first one is as hot as the second, hotter or colder than it; that body is hotter or colder at this instant than this one. Each part of a body, small as we suppose it to be, seems to us endowed with a certain quality which we call heat, and the intensity of this quality is not the same at a given instant when we compare one part of the body to another; at the same point of the body, it varies from one instant to the next.

We might in our reasoning speak of this quality of heat and of its various intensities, but wishing to employ the language of algebra as much as possible, we proceed to substitute for this quality of heat that of a numerical symbol, the temperature.

Temperature will then be a number assigned to each point of a body at each instant; it will be correlated to the heat prevailing at that point and in that instant. To two equally intense heats will be correlated two numerically equal temperatures. If it is hotter at one point than at another, the temperature at the first point will be a greater number than the temperature at the second point.

If, therefore, M, M', M'' are different points, and if T, T', T'' are the numbers expressing the temperatures at those points, the equality $T = T'$ has the same meaning as the following sentence: It is as warm at point M' as at point M. The inequality $T' > T''$ is equivalent to the sentence: It is warmer at point M' than at point M''.

The use of a number, the temperature, to represent an intensity of heat as a quality rests entirely on the following two propositions:

If body A is as warm as body B and body B is as warm as body C, then body A is as warm as body C.

If body *A* is warmer than body *B* and body *B* is warmer than body *C*, then body *A* is warmer than body *C*.

These two propositions, in fact, suffice to enable the signs $=$, $>$, and $<$ to represent the possible relations of different intensities of heat, as they permit representation of either the mutual relations of numbers or the mutual relations of different states of magnitude of the same quantity.

If I am told that two lengths are respectively measured by the numbers 5 and 10, without any further indication, I am being given certain information about these lengths: I know that the second is longer than the first, and even that it is double the first. This information is, however, very incomplete; it will not permit me to reproduce one of these lengths, or even to know whether it is large or small.

This information will be more complete if, not content with being given the numbers 5 and 10 as measuring two lengths, I am told that these lengths are measured in meters, and if I am shown the standard meter or a copy of it. Then I shall be able to reproduce and bring into existence these two lengths whenever I wish.

Thus, the numbers measuring magnitudes of the same kind inform us fully about these magnitudes only when we join to them concrete knowledge of the standard which represents the unit.

Some mathematicians have been examined in a competition; I am told they have earned the marks 5, 10, and 15, and that furnishes me certain information about them which will allow me, for example, to classify them. But this information is not complete, and does not allow me to form an idea of each one's talent. I do not know the absolute value of the marks which have been given to them; I lack knowledge of the scale to which these marks refer.

Similarly, if I am told simply that the temperatures of different bodies are represented by the numbers 10, 20, and 100, I learn that the first body is not as hot as the second and the second not as hot as the third. But is the first warm or cold? Can it melt ice or not? Would the last one burn me? Would it cook an egg? I do not know these things so long as I am not given the thermometric scale to which these temperatures 10, 20, and 100 refer, that is to say, a procedure allowing me to realize in a concrete manner the intensities of heat indicated by the numbers 10, 20, and 100. If I am given a graduated glass tube containing mercury and if I am taught that the temperature of a mass of water should be taken as equal to 10, 20, and 100 every time I see the mercury rise to these calibrations when the thermometer is plunged into the water, my doubts will be completely dissipated. Every time the numerical value of a temperature is indicated to me, I shall be able, if I wish, to realize in fact that a mass of water will have that temperature, since I possess the thermometer on which it is read.

So, just as a magnitude is not defined simply by an abstract number but by a number joined to concrete knowledge of a standard, in the same way the intensity of a quality is not entirely represented by a numerical symbol, but to this symbol must be joined a concrete procedure suitable for obtaining the scale of these intensities. Only the knowledge of this scale allows one to give a physical meaning to the algebraic propositions which we state concerning the numbers representing the different intensities of the quality studied.

Naturally, the scale which serves to calibrate the different intensities of a quality is always some quantitative effect having this quality as its cause. We choose this effect in such a way that its magnitude increases in time as the quality which causes it becomes more intense. Thus, in a glass vessel surrounded by a warm body, the mercury undergoes an apparent expansion which becomes greater as the body becomes warmer; this is the quantitative phenomenon provided by a thermometer which allows us to construct a scale of temperatures suitable for calibrating numerically different intensities of heat.

In the domain of quality, there is no room for addition; the latter does apply, however, when we study the quantitative phenomenon which provides a suitable scale on which to calibrate the different intensities of a quality. The various intensities of heat are not additive, but the apparent expansions of a liquid in a solid vessel are additive; we can get the sum of several numbers representing the temperatures.

Thus, the choice of a scale allows us to substitute for the study of the various intensities of a quality the consideration of numbers subject to the rules of algebraic calculation. The advantages sought by past physicists when they substituted a hypothetical quantity for the qualitative property revealed to the senses, and measured the magnitude of that quantity, can very often be obtained without employing that hypothetical quantity, simply by the choice of a suitable scale.

The electric charge will furnish us with an example of this.

What experiment shows us at first in very small bodies electrically charged is something qualitative. Soon, this quality of being charged electrically ceases to appear simple; it is capable of two forms which oppose and destroy each other: the resinous (negative) and the vitreous (positive).

Whether it is resinous or vitreous, the charged state of a small body may be more or less powerful; it is capable of different intensities.

Franklin, Oepinus, Coulomb, Laplace, Poisson—all the creators of the science of electricity thought that qualities could not be admitted into the constitution of a physical theory and that only quantities have the right of entry. Hence, underneath this quality of electric charge manifest to their senses, their reason sought a quantity, "the quantity of electricity." In order to arrive at an understanding of this quantity, they imagined that each of the two charges was due to the presence within the charged body of a certain "electrical fluid"; that the charged body showed an intensity of charge that varied with the mass of the electrical fluid; and that the magnitude of this mass then yielded the quantity of electricity.

The study of this quantity enjoyed a central role in the theory, a role which proceeded from these two laws:

The algebraic sum of the quantities of electricity spread over a group of bodies (a sum in which the quantities of vitreous electricity are prefixed by the + sign, and the quantities of resinous electricity have the — sign) does not change so long as this group is isolated from other bodies.

At a given distance two small charged bodies repel each other with a force proportional to the product of the quantities of electricity they carry.

Well now, these two propositions can be preserved intact without appealing to hypothetical and very improbable electrical fluids, and without depriving the electrical charge of the qualitative character our immediate observations confer on it. All we have to do is choose a suitable scale to which we refer the intensities of the electrical quality.

Let us take a small body charged vitreously (positively) in a manner that is always the same; at a fixed distance, we cause to act on it each one of the small bodies whose electrical state we wish to study. Each one of them will exert on the first body a force whose magnitude we shall be able to measure and to which we shall attach the + sign when there is a repulsion and a — sign when there is an attraction. Then each small body charged vitreously will exert on the first body a positive force whose magnitude will be greater as its charge is greater in intensity; each small body charged resinously will exert a negative force whose absolute value will increase in proportion as the charge on it is more powerful.

It is this force, a quantitalive elcment which is measurable and additive, which we shall choose as an electrometric scale and which will supply different positive numbers to represent the diverse intensities of vitreous electricity, and different negative numbers to calibrate the diverse degrees of resinous electrical charge. To these numbers or readings, furnished by this electrometric method, we can, if we wish, give the name "quantities of electricity"; and then the two essential propositions that the doctrine of electrical fluids formulated will become meaningful and true again.

No better example seems to us to make evident the following truth: In order to make a universal arithmetic out of physics, as Descartes desired to do, it is not at all necessary to imitate the great philosopher and to reject all quality, for the language of algebra allows us to reason as well about the various intensities of a quality as about the various magnitudes of a quantity.

CHAPTER II

PRIMARY QUALITIES

1. *On the Excessive Multiplication of Primary Qualities*

FROM THE MIDST of the empirically given physical world we shall detach the qualities which appear to us the ones that should be regarded as primary. We shall not try to explain these qualities or to reduce them to other more hidden attributes. We shall accept them just as our means of observation make us acquainted with them, whether they appear to us in the form of quantities or are given to us as perceived qualities; in either case we shall regard them as irreducible notions, as the very *elements* which are to constitute our theories. But we shall correlate these properties, qualitative or quantitative, with corresponding mathematical symbols which will allow us in reasoning about them to borrow the language of algebra.

Will this manner of proceeding commit us to the abuse for which the promoters of Renaissance science harshly scolded Scholastic physics and for which they rigorously and definitively brought it to justice?

Undoubtedly the scientists or scholars to whom we owe modern physics could not pardon the Scholastic philosophers for being averse to discussion of natural laws in mathematical language: "If we know anything," cried Gassendi, "we know it by mathematics; but those people have no concern for the true and legitimate science of things! They cling to trivialities!"[1]

But this was not the grievance most often and most sharply brought against the Scholastic doctors by the reformers of physics. Above all, their charge was that the Scholastics invented a new quality every time they looked at a new phenomenon, attributing to a special virtue each effect they had neither studied nor analyzed, and imagining they had given an explanation when they had only given a name. They had thus transformed science into a vain and pretentious jargon.

"This manner of philosophizing," Galileo used to say, "has to my

[1] P. Gassendi, *Exercitationes paradoxicae adversus Aristotelicos* (Grenoble, 1624), Problem I.

mind a great analogy with the manner one of my friends had in painting; he would write on the canvas with chalk: 'Here, I want a fountain with Diana and her nymphs, as well as some hunting dogs; there, a hunter with a stag's head; in the distance, a little woods, a field, a hill'; then he left to the artist the trouble of painting all these things and went away convinced that he had painted the metamorphosis of Acteon when he had only given some names."[2] And Leibniz compared the method followed in physics by the philosophers who on every occasion introduced new forms and new qualities with one "who would be content to say that a clock has the clocklike quality derived from its form without considering in what the latter consists."[3]

A laziness of the mind, which finds it convenient to be paid with words, and an intellectual dishonesty, which finds it profitable to pay others with words, are vices that are widespread in mankind. Certainly the Scholastic physicists, so prompt in endowing the form of each body with all the virtues their vague and superficial systems proclaimed, were often deeply tainted with these vices. But the philosophy that admits qualitative properties does not have a sad monopoly of these faults, for we find them as well among the followers of schools that pride themselves on reducing everything to quantity.

Gassendi, for example, was a convinced atomist; for him every observable quality was but an appearance; in reality there was nothing but the atoms, their shapes, groupings, and motions. But if we had asked him to explain essential physical qualities according to these principles—if we had asked him, "What is taste? What is odor? What is sound? What is light?"—how would he have answered us?

"In the very thing we call tasteful, taste does not seem to consist in anything else than in corpuscles of such a configuration that by penetrating the tongue or palate they affect the contexture of this organ and set it in motion in a manner that gives rise to the sensation we call taste.

"In reality, odor seems to be nothing else than certain corpuscles of such a configuration that when they are exhaled and penetrate the nostrils they conform to the contexture of these organs so as to give rise to the sensation we call olfaction or odor.

"Sound does not seem to be anything else than certain corpuscles

[2] Galileo, *Dialogo sopra i due massimi sistemi del mundo* (Florence, 1632), "Giornata terza."

[3] G. W. Leibniz, *Die philosophischen Schriften*, 7 vols., ed. C. I. Gerhardt (Berlin, 1875-1890), IV, 434.

which, configurated in a certain fashion and transmitted rapidly far from the sounding body, penetrate the ear, set it in motion, and cause the sensation called hearing.

"In a luminous body light does not seem to be anything else than very tenuous corpuscles configurated in a certain fashion and emitted by the luminous body with incredible velocity; they penetrate the organ of sight, and are apt to set it in motion and to create the sensation of vision."[4]

It was an Aristotelian, a *doctus bachelieurus* (learned doctor), who when asked:

> Demandabo causam et rationem quare
> Opium facit dormire?*

answered:

> Quiat est in eo
> Virtus dormitiva
> Cujus est natura
> Sensus assoupire.†

If this bachelor of science had given up Aristotle and had made himself an atomist, Molière would have undoubtedly met him at the philosophical lectures given at Gassendi's home, where the great writer of comedies often visited.

Moreover, the Cartesians would have been mistaken in shouting so triumphantly at the common ridicule into which they saw the Aristotelians and atomists fall. Pascal must have been thinking of one of these Cartesians when he wrote: "There are some who go to the absurd extreme of explaining a word by the same word. One of them, I know, defined light as follows: 'Light is a luminary motion of a luminous body,' as if we could understand the words 'luminary' and 'luminous' without understanding 'light.' "[5]

The allusion, in fact, was to Father Noël, a former teacher of Descartes at the school in La Flèche, who later became one of his fervent disciples, and who in a letter to Pascal on the vacuum had written: "Light, or rather illumination, is a luminary motion of the rays constituting lucid bodies which fill transparent bodies and are not moved luminarily except by other lucid bodies."

When one attributes light to a virtue of brightening, to luminous

[4] P. Gassendi, *Syntagma philosophicum* (Florence, 1727), I, V, Chs. IX, X, XI.

* Translator's note: "What is the cause and reason why opium causes one to sleep?"

† Translator's note: "Because there is in it a dormitive virtue whose nature is to cause the senses to become drowsy."

[5] B. Pascal, *De l'esprit géométrique.*

corpuscles, or to a luminary motion, he is an Aristotelian, an atomist, or a Cartesian, respectively; but if one boasts of having in that way added a particle to our knowledge concerning light, he does not have a sound mind. In all the schools we find people with false minds who imagine themselves to be filling a flask with a precious liqueur when they simply stick a fancy label on it; but all physical doctrines soundly interpreted agree in condemning this illusion. We should bend our efforts, therefore, to avoiding it.

2. *A Primary Quality Is a Quality Irreducible in Fact, Not by Law*

Moreover, our principles put us on guard against that travesty of thought which consists in putting into bodies as many distinct qualities, or almost as many, as there are diverse effects to be explained. We propose to give as simplified and as summary a representation of a group of physical laws as is possible; we aspire to achieve the most complete economy of thought realizable. It is therefore clear that for the construction of our theory we shall have to employ the least number of notions regarded as primitive and of qualities taken as simple. We shall push as far as it will go the method of analysis and reduction, a method which dissociates complex properties, especially those grasped by the senses, and reduces them to a small number of elementary properties.

How shall we know that our dissection has been pushed to the very end, and that the qualities at the end of our analysis cannot in turn be resolved into simpler qualities?

Physicists who tried to construct explanatory theories drew upon the philosophical precepts to which they had subjected themselves for touchstones and reagents to enable them to recognize whether the analysis of a quality had penetrated to the elements. For example, so long as an atomist had not reduced a physical effect to the size, configuration, and action of atoms and to the laws of impact, he knew that his task was not accomplished; so long as a Cartesian found something in a quality other than "bare extension and its modification," he was certain its true nature had not been reached.

If we, on our part, do not claim to explain the properties of bodies but only to give a condensed algebraic representation of them, if we do not proclaim in the construction of our theories any metaphysical principle but intend to make physics an autonomous doctrine, where shall we go for a criterion allowing us to declare that such and such a quality is truly simple and irreducible or that such and such a complex is destined for a more penetrating dissection?

When we regard a property as primary and elementary, we shall not in any way assert that this quality is by its nature simple and indecomposable; we shall declare that all our efforts to reduce this quality to others have failed and that it has been impossible for us to decompose it.

Every time, therefore, that a physicist ascertains a set of phenomena hitherto unobserved or discovers a group of laws apparently showing a new property, he will first investigate whether this property is not a combination, formerly unsuspected, of already known qualities accepted in prevailing theories. Only after he has failed in his many varied efforts, will he decide to regard this property as a new primary quality and introduce into his theories a new mathematical symbol.

"Every time an *exceptional* fact has been discovered," wrote H. Sainte-Claire Deville, describing the hesitations of his thought when he recognized the first phenomena of dissociation, "the first job, I shall say the first duty, practically imposed on the man of science has been to make every effort to cause the fact to come under the common rule by means of an explanation which sometimes requires more work and reflection than the discovery itself. When we succeed, we experience a very keen satisfaction in extending, so to speak, the domain of a physical law, and in increasing the simplicity and generality of a great classification. . . . But when an exceptional fact escapes every explanation, or at least resists every effort conscientiously made to subject it to common law, we must look for other facts which are analogous to it; when they are found they must be classified *provisionally* by means of the theory that has been formed."[6]

When Ampère discovered the mechanical action between two electrical wires, each connected to one of the two poles of a battery, the attractions and repulsions between electrical conductors had been known for a long time. The quality manifested in these attractions and repulsions had been analyzed; it had been represented by an appropriate mathematical symbol, the positive or negative charge of each material element. The use of this symbol had led Poisson to build a mathematical theory which represented most felicitously the experimental laws established by Coulomb.

[6] H. Sainte-Claire Deville, "Recherches sur la décomposition des corps par la chaleur et la dissociation," *Archives des Sciences physiques et naturelles* of the Bibliothèque Universelle, new period, IX (1860), 59.

Might not newly discovered laws be reduced to this quality, whose introduction into physics was an accomplished fact? Might not one explain the attractions and repulsions exerted between two wires in a closed circuit by admitting that certain charges are suitably distributed on the surface of these wires or within them, and that these charges attract or repel each other inversely with the square of the distance, according to the fundamental thesis underlying the theory of Coulomb and Poisson? It was legitimate for this question to have been asked and investigated by physicists; if some one of them had succeeded in giving an affirmative answer to it and reduced the laws of the actions observed by Ampère to the laws of electrostatics established by Coulomb, he would have given us an electrical theory free from the consideration of any primary quality other than the electric charge.

Attempts to reduce the laws of the forces Ampère had put in evidence to electrostatic actions were first of all multiplied. But Faraday cut short these attempts by showing that these forces could give rise to movements of continuous rotation; indeed, as soon as Ampère learned of the phenomenon discovered by the great English physicist, he understood its whole import. This phenomenon, he said, "proves that the action emanating from two conductors of electricity cannot be due to a special distribution of certain fluids at rest in these conductors, as are ordinary electrical attractions and repulsions."[7] "In fact, from the principle of the conservation of living force, which is a necessary consequence of the laws of motion, it follows necessarily that when the elementary forces, here the attractions and repulsions in inverse ratio to the squares of the distances, are expressed by simple functions of the mutual distances of the points between which they act, and if some of these points are constantly connected to one another and move only by virtue of these forces, the others remaining fixed, the first points cannot return to the same position relative to the second points with velocities greater than they had when they started from that position. Now, in the continuous motion impressed on a moving conductor by the action of a fixed one, all points of the former return to the same position with velocities increasing with each revolution, until the friction and resistance of the battery acid in which the end of the conductor is immersed puts an end to the increase

[7] A. M. Ampère, "Exposé sommaire des nouvelles expériences électrodynamiques," read before the Academy, April 8, 1822, *Journal de Physique*, xciv, 65.

of the speed of this conductor's rotation, which then becomes constant despite the friction and resistance.

"It is therefore completely proven that we cannot account for the phenomena produced by the action of two voltaic conductors by supposing that electrical molecules acting inversely with the square of the distance are distributed over the conducting wires."[8]

Strict necessity demanded that there be attributed to the various parts of a voltaic conductor a property not reducible to static electricity; it was necessary to recognize a new primary quality whose existence was to be expressed by saying that the wire is "traversed by a current." This electrical current appears to be bound in a certain direction or affected with a certain sense of direction. It shows a lesser or greater intensity which can by a choice of a scale be correlated with a smaller or larger number, a number to which we assign the name "intensity of electrical current." This intensity of electrical current, a mathematical symbol of a primary quality, allowed Ampère to develop his theory of electrodynamic phenomena, a theory which dispenses with the Frenchman's need to envy the Englishman's pride in the glory of Newton.

The physicist who asks a metaphysical doctrine for the principles with which to develop his theories acquires from that doctrine the marks by which he will recognize whether a quality is simple or complex, and these two words have an absolute sense for him. The physicist who seeks to make his theories autonomous and independent of any philosophical system attributes an entirely relative sense to the words "simple quality" or "primary property"; they designate for him simply a property that it has been impossible for him to resolve into other qualities.

The meaning that the chemists attribute to the words "simple body" or "element" has undergone an analogous transformation.

For an Aristotelian only the four elements fire, air, water, and earth deserved the name of a simple body; all other bodies were complex, and so long as they had not been dissociated to the point of separating out the four elements which could enter into their composition, analysis had not reached its end. Similarly, an alchemist knew that his spagyric art of decomposition had not attained the ultimate goal of his operations until he had separated out the salt, sulphur, and mercury whose union made up all mixtures. The alchemist and the Aristotelian both claimed to know the

[8] A. M. Ampère, *Théorie mathématique des phénomènes électrodynamiques uniquement déduite de l'expérience* (Paris, 1826). Reprinted in the edition published by Hermann (Paris, 1883), p. 96.

marks which characterize the truly simple body in an absolute manner.

Lavoisier and his school led chemists to adopt an entirely different idea of a simple body; it is not a body that a certain philosophical doctrine declares indecomposable, but a body that we have not been able to decompose, a body which has resisted every means of analysis used in laboratories.[9]

When the alchemist and the Aristotelian pronounced the word element, they were proudly asserting their claim to know the very nature of the materials which have gone into the construction of every body in the universe. In the mouth of the modern chemist the same word is a gesture of modesty, an admission of impotence; he is confessing that a body has victoriously resisted every effort made to reduce it.

Chemistry has been compensated for this modesty by its enormous fertility. Is it not legitimate to hope that a similar modesty will procure for theoretical physics the same gains?

3. *A Quality Is Never Primary, except Provisionally*

"We can, therefore, never be sure," Lavoisier said, "that what we regard as simple today will be so in fact. All that we can say is that such a substance is the present end-term at which chemical analysis has arrived, and that the substance cannot be subdivided further in the present state of our knowledge. It is presumable that earth-substances will soon cease to be counted among the simple substances. . . ."[10]

Indeed, in 1807 Humphry Davy transformed Lavoisier's guess into a demonstrated truth, and proved that potash and soda are the oxides of two metals which he called potassium and sodium. Since that time a great many bodies which had long resisted every attempt at analysis have been decomposed and are now excluded from the number of elements.

The title "element" which certain bodies bear is a quite provisional one; it is at the mercy of a more ingenious or more powerful analysis than those in use up to date, a means of analysis which will perhaps dissociate the substance regarded as simple into several distinct substances.

No less provisional is the title "primary quality." The quality

[9] The reader who desires to know the phases through which the idea of a simple body has passed may consult our book *Le Mixte et la Combinaison chimique. Essai sur l'évolution d'une idée.* (Paris, 1902), Part II, Ch. 1.

[10] A. L. Lavoisier, *Traité élémentaire de Chimie* (3rd ed.), I, 194.

which today cannot be reduced to any other physical property will cease tomorrow to be independent; tomorrow, perhaps, the progress of physics will make us recognize in the primary quality a combination of properties which some apparently very different effects have revealed to us for a long time.

The study of the phenomena of light leads to the consideration of a primary quality, light. A direction is given to this quality; its intensity, far from being fixed, varies periodically with enormous rapidity, repeating itself several hundred trillion times a second. A line whose length varies periodically with this extraordinary frequency furnishes a geometrical symbol appropriate for imagining light; the symbol, light vibration, will serve to deal with this quality by mathematical reasoning. Light vibration will be the essential element by means of which the theory of light will be built; its components will serve in writing some equations with partial derivatives and some boundary conditions, condensing and classifying with admirable order and brevity all the laws of the propagation of light, its partial or total reflection, its refraction, and its diffraction.

In another quarter, the analysis of the phenomena that insulating substances like sulphur, vulcanized rubber, and wax show in the presence of electrically charged bodies have led physicists to attribute to these dielectric bodies a certain property. After trying in vain to reduce this property to the electric charge, they had to decide to treat it as a primary quality with the name dielectric polarization. The latter has at each point of the insulating substance and at each instant not only a certain intensity but also a certain direction and a certain sense so that a line segment furnishes the mathematical symbol allowing one to speak about dielectric polarization in the language of mathematicians.

A bold extension of the electrodynamics formulated by Ampère furnished Maxwell with a theory of the variable state of dielectrics. This theory condenses and orders the laws of all the phenomena produced inside insulators where the dielectric polarization varies from one instant to the next. All these laws are summarized in a small number of equations, some of which are satisfied at every point of the same insulating body and the others at every point of the surface separating two distinct dielectrics.

The equations governing light vibration have all been established as though the dielectric polarization did not exist; the equations on which dielectric polarization depends have been discovered by a theory in which the word light is not even mentioned.

Now, see how a surprising convergence between these equations is established.

A dielectric polarization which varies periodically has to verify equations all of which are similar to the equations governing light vibration.

And not only do these equations have the same form but also the coefficients figuring in them have the same numerical value. Thus in a vacuum or in air, at first without any electric action polarizing a certain region, electric polarization once begun is propagated with a certain velocity; the equations of Maxwell allow one to determine this velocity by purely electrical procedures wherein nothing is borrowed from optics; numerous measurements agree that the value of this velocity is around 300,000 kilometers per second; this number is precisely equal to the velocity of light in air or in a vacuum, a velocity that four purely optical methods, distinct from one another, have taught us.

The conclusion imposed by this unexpected convergence is: Light is not a primary quality; light vibration is nothing else than a periodically variable dielectric polarization; the electromagnetic theory of light created by Maxwell has resolved a property we thought irreducible; it has derived it from a quality with which for many years there appeared to be no connection.

Thus the progress of theories may itself lead physicists to reduce the number of qualities that they had at first considered as primary, and to prove that two properties regarded as distinct are but two diverse aspects of the same property.

Must we conclude that the number of qualities admitted into our theories will diminish from day to day, that matter which is the subject of our theorizing will be less and less rich in essential attributes, and that it will tend towards a simplicity comparable to that of atomistic or Cartesian matter? I think that would be a rash conclusion. Undoubtedly, the very development of theory may from time to time produce the fusion of two distinct qualities, similar to that fusion of light and dielectric polarization established by the electromagnetic theory of light. But on the other hand, the constant progress of experimental physics frequently brings on the discovery of new categories of phenomena, and in order to classify these phenomena and group their laws, it is necessary to endow matter with new properties.

Which of these two contrary movements will prevail—the one which reduces qualities to other qualities and tends to simplify matter, or the one which discovers new properties and tends to complicate? It would be imprudent to formulate any long-term prediction about this subject. At least, it seems certain that in our time the second trend is much more powerful than the first and is leading our theories toward a more and more complex conception of matter, richer in attributes.

Besides, the analogy between the primary qualities of physics and the simple bodies of chemistry is here again a marked one. If perhaps the day will come when powerful methods of analysis will resolve the numerous bodies we today call simple into a small number of elements, there is no certain or probable sign allowing us to announce the dawn of that day yet. In our own day* chemistry is making progress in constantly discovering new simple bodies. For half a century, the rare earths have continued to furnish new recruits to an already long list of metals; gallium, germanium, scandium, etc. show us chemists proud to inscribe the name of their country on this list. In the air we breathe, a mixture of nitrogen and oxygen apparently so well known since Lavoisier, we see revealed a whole family of new gases: argon, helium, xenon, crypton. Finally, the study of new radiations, which will surely compel physics to enlarge the circle of its primary qualities, furnishes chemistry with hitherto unknown bodies: radium and perhaps polonium and actinium.

Most assuredly we have gone a long way from the beautifully simple bodies which Descartes dreamed up, those bodies which were reduced "simply to bare extension and its modification." Chemistry piles up a collection of about a hundred material bodies irreducible to one another, and to each of these physics associates a form capable of a multitude of diverse properties. Each of these two sciences strives to reduce the number of its elements as much as it can, and yet, in proportion to the progress each science makes, it sees this number grow.

* Translator's note: About 1900.

CHAPTER III

MATHEMATICAL DEDUCTION AND PHYSICAL THEORY

1. *Physical Approximation and Mathematical Precision*

When we set out to construct a physical theory, at first we have to choose among those properties given to observation the ones which we shall take as primary qualities, and represent them by algebraic or geometric symbols.

Having completed this first operation, to the study of which we have devoted the two preceding chapters, we must accomplish a second: Among the algebraic or geometric symbols representing the primary properties we must establish relations; these relations will serve as principles for the deductions through which the theory will be developed.

It would seem natural, therefore, to analyze now this second operation, the statement of hypotheses. But before drawing the plan of the foundations to support a house and before selecting the materials with which to build them, it is indispensable to know what the structure will be and what stresses it will exert on its base. Hence, only at the end of our study shall we be able to state precisely what conditions are imposed on the choice of hypotheses.

Consequently, we are going to take up immediately the third operation constituting any theory, the mathematical development.

Mathematical deduction is an intermediary process; its object is to teach us that on the strength of the fundamental hypotheses of the theory the coming together of such and such circumstances will entail such and such consequences; if such and such facts are produced, another fact will be produced. For example, it will tell us that on the strength of the hypotheses of thermodynamics, when we submit a block of ice to a certain pressure, the block will melt when the thermometer reads a certain number.

Does mathematical deduction introduce directly into its calculations the facts we call circumstances in the concrete form in which we observe them? Does it draw from them the facts we call consequences in the concrete form in which we ascertain them? Certainly not. The apparatus used for compression, a block of ice, and a thermometer are things the physicist manipulates in the laboratory; they are not elements belonging to the domain of algebraic calculation. Hence, in order to enable the mathematician to introduce in his formulas the concrete circumstances of an experiment, it is necessary to translate these circumstances into numbers by the intermediary of measurements. For example, the words "a certain pressure" must be replaced by a certain number of atmospheres which he will substitute for the letter P in his equation. Similarly, what the mathematician will obtain at the end of his calculation is a certain number. It will be necessary to refer back to the method of measurement in order to make this number correspond to a concrete and observable fact; for example, in order to make the numerical value taken by the letter T in the algebraic equation correspond to a certain thermometer reading.

Thus at both its starting and terminal points, the mathematical development of a physical theory cannot be welded to observable facts except by a translation. In order to introduce the circumstances of an experiment into the calculations, we must make a version which replaces the language of concrete observation by the language of numbers; in order to verify the result that a theory predicts for that experiment, a translation exercise must transform a numerical value into a reading formulated in experimental language. As we have already indicated, the method of measurement is the dictionary which makes possible the rendering of these two translations in either direction.

But translation is treacherous: *traduttore, traditore* (to translate is to betray). There is never a complete equivalence between two texts when one is a translated version of the other. Between the concrete facts, as the physicist observes them, and the numerical symbols by which these facts are represented in the calculations of the theorist, there is an extremely great difference. We shall later have an opportunity to analyze and take note of the principal characteristics of this difference. Right now only one of these will occupy our attention.

First of all, let us consider what we shall call a *theoretical* fact, that is to say, that set of mathematical data through which a concrete fact is replaced in the reasoning and calculations of the theorist. For example, let us take this fact: The temperature is distributed in a certain manner over a certain body.

In such a theoretical fact there is nothing vague or indecisive. Everything is determined in a precise manner: the body studied is geometrically defined; its sides are true lines without thickness, its points true points without dimensions; the different lengths and angles determining its shape are exactly known; to each point of this body there is a corresponding temperature, and this temperature is for each point a number not to be confused with any other number.

Opposite this *theoretical* fact let us place the *practical* fact translated by it. Here we no longer see anything of the precision we have just ascertained. The body is no longer a geometrical solid; it is a concrete block. However sharp its edges, none is a geometrical intersection of two surfaces; instead, these edges are more or less rounded and dented spines. Its points are more or less worn down and blunt. The thermometer no longer gives us the temperature at each point but a sort of mean temperature relative to a certain volume whose very extent cannot be too exactly fixed. Besides, we cannot assert that this temperature is a certain number to the exclusion of any other number; we cannot declare, for example, that this temperature is strictly equal to 10°; we can only assert that the difference between this temperature and 10° does not exceed a certain fraction of a degree depending on the precision of our thermometric methods.

Thus, whereas the contours of the drawing are fixed by a line of precise hardness, the contours of the object are misty, fringed, and shadowy. It is impossible to describe the practical fact without attenuating by the use of the word "approximately" or "nearly" whatever is determined too well by each proposition; on the other hand, all the elements constituting the theoretical fact are defined with rigorous exactness.

Whence we have this consequence: An infinity of different theoretical facts may be taken for the translation of the same practical fact.

For example, to say in a proposition of theoretical fact that a certain line has a length of 1 centimeter, or 0.999 cm., or 0.993 cm., or 1.002 cm., or 1.003 cm. is to formulate propositions which are for the mathematician essentially different; but we change nothing of the practical fact translated by the theoretical fact if our means of measurement do not allow us to evaluate lengths of less than 0.001 cm. To say that the temperature of a body is 10°, or 9.99° or 10.01° is to formulate three incompatible theoretical facts, but these three incompatible facts correspond to one and the same practical fact when our thermometer is accurate only to a fifth of a degree.

A practical fact is not translated therefore by a single theoretical fact but by a kind of bundle including an infinity of different theoretical facts. Each of the mathematical elements brought together in order to constitute one of these facts may vary from one fact to another; but the variation to which it is susceptible cannot exceed a certain limit, namely, the limit of error within which the measurement of this element is blotted. The more perfect the methods of measurement are, the closer is the approximation and the narrower the limits but they never become so narrow that they vanish.

2. *Mathematical Deductions Physically Useful and Those Not*

These remarks we have made are very simple, and are commonplaces to the physicist; nevertheless, they imply serious consequences for the mathematical development of a theory.

When the numerical data of a calculation are fixed in a precise manner, this calculation, no matter how long and complicated it is, likewise yields knowledge of the exact numerical value of the result. If we change the value of the data, we generally change the value of the result. Consequently, when we have represented the conditions of an experiment by a clearly defined theoretical fact, the mathematical development will represent by another clearly defined theoretical fact the result that this experiment should provide; if we change the theoretical fact which translates the conditions of the experiment, the theoretical fact which translates the result will change likewise. If, for example, in the formula deduced from thermodynamic hypotheses connecting the melting point of ice with the pressure, we replace the letter P representing the pressure by a certain number, we shall know the number that must be substituted for the letter T, symbol of the temperature of the melting point; if we change the numerical value attributed to the pressure, we also change the numerical value of the melting point.

Now, according to what we have seen in Section 1 of this chapter, if the conditions of an experiment are concretely given, we shall not be able to translate them by a definite theoretical fact without ambiguity; we have to correlate them with a whole bundle of theoretical facts, infinite in number. Consequently, the calculations of the theorist will not forecast the experimental result in the form of a unique theoretical fact but in the form of an infinity of different theoretical facts.

In order to translate, for example, the conditions of our experiment on the melting point of ice, we shall not be able to substitute a single and unique numerical value, say 10 atmospheres, for the symbol P of the pressure; if the limit of error of the manometer we use is 0.10 atmospheres, we shall have to assume that P may take all the values included between 9.95 and 10.05 atmospheres. Naturally, to each of these values of the pressure our formula will correlate a different value of the melting point of ice.

Thus, experimental conditions given in a concrete manner are translated by a bundle of theoretical facts; the mathematical development of the theory correlates this first bundle of theoretical facts with a second, intended to stand for the result of the experiment.

These latter theoretical facts will not be able to serve us in the same form in which we obtain them. We shall have to translate them and put them in the form of practical facts; only then shall we know truly the result assigned to our experiment by our theory. We shall not, for instance, have to stop with the diverse numerical values of the letter T derived from our thermodynamic formula, but it will be necessary to find out to what really observable readings on the graduated scale of our thermometer the indicated values correspond.

Now, when we have made this new translation intended to transform theoretical into practical facts, the inverse of the one with which we first concerned ourselves, what have we obtained?

It may turn out that the bundle of infinitely numerous theoretical facts by which mathematical deduction assigns to our experiment the result that should be produced will not furnish us after the translation with several different practical facts, but only with a single practical fact. It may happen, for instance, that two of the numerical values found for the letter T never differ by even a hundredth of a degree, and that the limit of sensitivity of our thermometer is a hundredth of a degree, so that all these different theoretical values of T correspond practically to one and the same reading on the scale of the thermometer.

In such a case mathematical deduction will have attained its end: it will have allowed us to assert that on the strength of the hypotheses on which our theory rests, a certain experiment done under certain practically given conditions should yield a certain concrete and observable result; it will have made possible the comparison of the consequences of the theory with the facts.

But it will not always be thus. As a result of mathematical deduction an infinity of theoretical facts present themselves as possible consequences of our experiment; by translating these theoretical facts into concrete language it may happen that we obtain not a single practical fact but several practical facts which the sensitivity of our instruments will allow us to distinguish. It may happen, for instance, that the different numerical values given by our thermodynamic formula for the melting point of ice present deviations of a tenth of a degree, or even one degree, whereas our thermometer allows us to evaluate a hundredth of a degree. In that case the mathematical deduction will have lost its usefulness; the conditions of an experiment being practically given, we shall no longer be able to state in a practically definite way the result that should be observed.

A mathematical deduction, stemming from the hypotheses on which a theory rests, may therefore be useful or otiose, according to whether or not it permits us to derive a *practically definite* prediction of the result of an experiment whose conditions are *practically given.*

This evaluation of the utility of a mathematical deduction is not always absolute; it depends on the degree of the sensitivity of the apparatus used in observing the result of the experiment. Let us suppose, for example, that a practically given pressure is correlated with a bundle of melting points of ice, and that between two of the melting points there is sometimes a difference greater than a hundredth of a degree but never one of more than a tenth of a degree. The mathematical deduction that yielded this formula will be called useful by the physicist whose thermometer measures only tenths of a degree, and useless by the physicist whose instrument accurately detects a difference of a hundredth of a degree. In that way we see how much the judgment concerning the utility of a mathematical development will vary from time to time, from one laboratory to another, and from one physicist to another, according to the skill of the designers, the perfection of the equipment, and the intended application of the results of the experiment.

This evaluation may also depend on the sensitivity of the means of measurement used to translate into numbers the practically given conditions of experiment.

Let us take up again the thermodynamic formula which has served us constantly as an example. We are in possession of a thermometer which discriminates accurately a difference of a hundredth of a degree; in order that our formula may state without practical ambiguity the melting point of ice under a given pressure, it will be necessary and sufficient that the formula should yield us the numerical value of the letter T correct to the hundredth of a degree.

Now, if we employ a crude manometer, incapable of distinguishing two pressures when their difference is less than ten atmospheres, it may happen that a practically given pressure corresponds to melting points differing by more than a hundredth of a degree in the formula; whereas, if we determine the pressure with a more sensitive manometer, accurately distinguishing two pressures which differ by one atmosphere, the formula will correlate with a given pressure a melting point known with an approximation higher than a hundredth of a degree. The formula which was useless when we employed the first manometer became useful when we employed the second.

3. *An Example of Mathematical Deduction That Can Never Be Utilized*

In the case of the example we have just taken up, we have increased the precision of the methods of measurement that were used to translate the practically given conditions of an experiment into theoretical facts; in that way, we have tightened more and more the bundle of theoretical facts which this translation correlates with a single practical fact. At the same time we have also tightened the bundle of theoretical facts with which our mathematical deduction represents the result predicted for the experiment; it has become narrow enough for our method of measurement to correlate it with a single practical fact, and at that moment our mathematical deduction has become useful.

It seems as if it should always be so. If we take as a datum a single theoretical fact, mathematical deduction correlates it with another single theoretical fact; as a result, we are naturally led to formulate the following conclusion: Whatever narrowness is needed for the bundle of theoretical facts which we wish to obtain as a result, mathematical deduction will always be able to guarantee it that narrowness, provided that we tighten sufficiently the bundle of theoretical facts representing the data given.

If this intuition encompassed the truth, a mathematical deduction stemming from the hypotheses on which a physical theory rests could never be useless except in a relative and provisional manner; however delicate the methods intended to measure the experimental results, we might always, by making the means of translating the experimental conditions into numbers precise and minute enough, manage to make our deduction draw a practically unique result from practically determined conditions. A deduction which is useless today would become useful the day on which we noticeably increased the sensitivity of the instruments serving to measure the experimental conditions.

The modern mathematician is very much on his guard against these appearances of evidence which so often are only tricks of sleight of hand. What we have just invoked is nothing but a deception. We can cite cases where it is in plain contradiction with the truth. A certain deduction correlates a single theoretical fact, taken as given, with a single theoretical fact, as a result. If the given is a bundle of theoretical facts, the result is another bundle of theoretical facts. But in vain do we tighten indefinitely the first bundle and make it as thin as possible; we are not authorized to diminish as much as we please the deviation of the second bundle; although the first bundle is infinitely narrow, the blades forming the second bundle diverge and separate out without our being able to reduce their mutual deviations below a certain limit. Such a mathematical deduction is and always will remain useless to the physicist; however precise and minute are the instruments by which the experimental conditions will be translated into numbers, this deduction will still correlate an infinity of different practical results with practically determined experimental conditions, and will not permit us to predict what should happen in the given circumstances.

The researches of J. Hadamard provide us with a very striking example of such a deduction that can never be useful. It is borrowed from one of the simplest problems that the least complicated of physical theories, mechanics, has to deal with.

A material mass slides on a surface; no weight and no force act on it; no friction interferes with its motion. If the surface on which it is to remain is a plane, it describes a straight line with uniform velocity; if the surface is a sphere, it describes the arc of a great circle, also with uniform velocity. No matter what surface our material point moves on, it describes a line that geometers call a "geodesic line" of the surface considered. When the initial position of our material point and the direction of its initial velocity are given, the geodesic it should describe is well determined.

Hadamard's researches have dealt especially with geodesics of surfaces of negative curvature, with multiple connections, and with infinite folds.[1] Without stopping here to define such surfaces geometrically, let us restrict ourselves to giving an illustration of one of them.

Imagine the forehead of a bull, with the protuberances from which the horns and ears start, and with the collars hollowed out

[1] J. Hadamard, "Les surfaces à courbures opposées et leurs lignes géodésiques," *Journal de Mathématiques pures et appliquées*, 5th series, Vol. IV (1898), p. 27.

between these protuberances; but elongate these horns and ears without limit so that they extend to infinity; then you will have one of the surfaces we wish to study.

On such a surface geodesics may show many different aspects.

There are, first of all, geodesics which close on themselves. There are some also which are never infinitely distant from their starting point even though they never exactly pass through it again; some turn continually around the right horn, others around the left horn, or right ear, or left ear; others, more complicated, alternate, in accordance with certain rules, the turns they describe around one horn with the turns they describe around the other horn, or around one of the ears. Finally, on the forehead of our bull with his unlimited horns and ears there will be geodesics going to infinity, some mounting the right horn, others mounting the left horn, and still others following the right or left ear.

Despite this complication, if we know with complete accuracy the initial position of a material point on this bull's forehead and the direction of the initial velocity, the geodesic line that this point will follow in its motion will be determined without any ambiguity. In particular, we shall know whether the moving point will always remain at a finite distance from its starting point or whether it will move away indefinitely so as never to return.

It will be quite a different matter if the initial conditions are not mathematically but practically given: the initial position of our material point will no longer be a determinate point on the surface, but some point taken inside a small spot; the direction of the initial velocity will no longer be a straight line defined without ambiguity, but some one of the lines included in a narrow bundle connected by the contour of the small spot; and our practically determined initial conditions will, for the geometer, correspond to an infinite multiplicity of different initial conditions.

Let us imagine certain of these geometrical data corresponding to a geodesic line that does not go to infinity, for example, a geodesic line that turns continually around the right horn. Geometry permits us to assert the following: Among the innumerable mathematical data corresponding to the same practical data, there are some which determine a geodesic moving indefinitely away from its starting point; after turning a certain number of times around the right horn, this geodesic will go to infinity on the right horn, or on the left horn, or on the right or left ear. More than that: despite the narrow limits which restrict the geometrical data capable of representing the given practical data, we can always take these geometrical data in such a way that the geodesic will go off on that one of the infinite folds which we have chosen in advance.

It will do no good to increase the precision with which the practical data are determined, to diminish the spot where the initial position of the material point is, to tighten the bundle which includes the initial direction of the velocity, for the geodesic which remains at a finite distance while turning continually around the right horn will not be able to get rid of those unfaithful companions who, after turning like itself around the right horn, will go off indefinitely. The only effect of this greater precision in the fixing of the initial data will be to oblige these geodesics to describe a greater number of turns embracing the right horn before producing their infinite branch; but this infinite branch will never be suppressed.

If, therefore, a material point is thrown on the surface studied starting from a geometrically given position with a geometrically given velocity, mathematical deduction can determine the trajectory of this point and tell whether this path goes to infinity or not. But, for the physicist, this deduction is forever unutilizable. When, indeed, the data are no longer known geometrically, but are determined by physical procedures as precise as we may suppose, the question put remains and will always remain unanswered.

4. *The Mathematics of Approximation*

The example we have just analyzed came to us, we said, through one of the simplest problems one has to deal with in mechanics, that is, the least complex of physical theories. This extreme simplicity has allowed Hadamard to penetrate far enough into the study of the problem to expose fully the absolutely irremediable physical uselessness of certain mathematical deductions. Should we not meet that ensnaring conclusion in a host of other, more complicated problems, if it were possible to analyze the solutions closely enough? The answer to this question scarcely seems doubtful; the progress of mathematical sciences will undoubtedly prove to us that a great many problems well defined for the mathematician lose all their meaning for the physicist.

Here is one of them whose relationship to the one discussed by Hadamard is apparent; it is very famous.[2]

In order to study the motions of the heavenly bodies that make up the solar system, mathematicians replace all these bodies—sun, planets, asteroids, satellites—by material points; they assume that

[2] *ibid.*, p. 71.

pairs of these points attract each other as the product of the masses and in inverse ratio to the square of the distance separating the two elements. The study of the motion of a system like that is a much more complicated problem than the one we have discussed in the foregoing pages. It is famous in the history of science under the heading "the problem of *n* bodies"; even when the number of bodies subjected to mutual interactions is reduced to three, "the problem of three bodies" remains a formidable puzzle for mathematicians.

Nevertheless, if we know with mathematical precision the position and velocity at a given time of each of the bodies forming the solar system, we may assert that each of the bodies follows a perfectly definite trajectory starting from that instant; the effective determination of this trajectory may oppose to the efforts of mathematicians obstacles which are far from being removed, but we may be allowed to suppose that some day they will be overthrown.

Consequently, the mathematician may ask himself the following question: The positions and velocities of the bodies forming the solar system being what they are today, will they all continue indefinitely to turn around the sun? Will it not, on the contrary, probably come about that one of these bodies will finally escape from the swarm of its companions and get lost in the immensity of space? This question constitutes the problem of the stability of the solar system which Laplace thought he had solved, but whose extreme difficulty has been shown especially by the efforts of modern mathematicians, in particular, Henri Poincaré.

The problem of the stability of the solar system certainly has a meaning for the mathematician, for the initial positions and velocities of the bodies are for him elements known with mathematical precision. But for the astronomer these elements are determined only by physical procedures involving errors which will gradually be reduced by improvements in the instruments and methods of observation, but will never be eliminated. It might be the case, consequently, that the problem of the stability of the solar system should be for the astronomer a question devoid of all meaning; the practical data that he furnishes to the mathematician are equivalent for the latter to an infinity of theoretical data, neighboring on one another but yet distinct. Perhaps among these data there are some that would eternally maintain all heavenly bodies at a finite distance from one another, whereas others would throw some one of these bodies into the vastness of space. If such a circumstance analogous to the one offered by Hadamard's problem should turn up here, any mathematical deduction relative to the stability of the solar system would be for the physicist a deduction that he could never use.

One cannot go through the numerous and difficult deductions of celestial mechanics and mathematical physics without suspecting that many of these deductions are condemned to eternal sterility.

Indeed, a mathematical deduction is of no use to the physicist so long as it is limited to asserting that a given *rigorously* true proposition has for its consequence the *rigorous* accuracy of some such other proposition. To be useful to the physicist, it must still be proved that the second proposition remains *approximately* exact when the first is only *approximately* true. And even that does not suffice. The range of these two approximations must be delimited; it is necessary to fix the limits of error which can be made in the result when the degree of precision of the methods of measuring the data is known; it is necessary to define the probable error that can be granted the data when we wish to know the result within a definite degree of approximation.

Such are the rigorous conditions that we are bound to impose on mathematical deduction if we wish this absolutely precise language to be able to translate without betraying the physicist's idiom, for the terms of this latter idiom are and always will be vague and inexact like the perceptions which they are to express. On these conditions, but only on these conditions, shall we have a mathematical representation of the *approximate.*

But let us not be deceived about it; this "mathematics of approximation" is not a simpler and cruder form of mathematics. On the contrary, it is a more thorough and more refined form of mathematics, requiring the solution of problems at times enormously difficult, sometimes even transcending the methods at the disposal of algebra today.

CHAPTER IV

EXPERIMENT IN PHYSICS[1]

1. *An Experiment in Physics Is Not Simply the Observation of a Phenomenon; It Is, Besides, the Theoretical Interpretation of This Phenomenon*

THE AIM of all physical theory is the representation of experimental laws. The words "truth" and "certainty" have only one signification with respect to such a theory; they express concordance between the conclusions of the theory and the rules established by the observers. We could not, therefore, push our critical examination of physical theory further if we did not analyze the exact nature of the laws stated by experimenters, and if we did not note precisely what sort of certainty they can yield. Moreover, a law of physics is but the summary of an infinity of experiments that have been made or will be performable. Hence we are naturally led to raise the question: What exactly is an experiment in physics?

This question will undoubtedly astonish more than one reader. Is there any need to raise it, and is not the answer self-evident? What more does "doing an experiment in physics" mean to anybody than producing a physical phenomenon under conditions such that

[1] This chapter and the two following it are devoted to the analysis of the experimental method used by the physicist in particular. In this regard, we ask the reader's permission to take note of a few dates. We think we were the first to formulate this analysis in an article entitled "Quelques réflexions au sujet de la Physique expérimentale," *Revue des Questions scientifiques*, 2nd Series, Vol. III (1894). G. Milhaud took as the subject of his course in 1895-96 an exposition of a part of these ideas; he published a summary of his lectures (in which, besides, he quoted us) under the title: "La Science rationelle," *Revue de Métaphysique et de Morale*, 4th year (1896), p. 290; also in book form in *Le Rationnel* (Paris, 1898). The same analysis of the experimental method was adopted by Edouard Le Roy in the second part of his article "Science et Philosophie," *Revue de Métaphysique et de Morale*, 7th year (1899), p. 503, and in another essay entitled "La Science positive et les philosophies de la liberté," *Congrès international de Philosophie* (held in Paris in 1900), Sec. I: "Philosophie générale et Métaphysique," p. 313. E. Wilbois also admits an analogous doctrine in his article "La Méthode des Sciences physiques," *Revue de Métaphysique et de Morale*, 7th year (1899), p. 579. The several authors we have just cited often draw from this analysis of the experimental method used in physics conclusions which go beyond the boundaries of physics; we shall not follow them that far, but shall stay always within the limits of physical science.

it may be observed exactly and minutely by means of appropriate instruments?

Go into this laboratory; draw near this table crowded with so much apparatus: an electric battery, copper wire wrapped in silk, vessels filled with mercury, coils, a small iron bar carrying a mirror. An observer plunges the metallic stem of a rod, mounted with rubber, into small holes; the iron oscillates and, by means of the mirror tied to it, sends a beam of light over to a celluloid ruler, and the observer follows the movement of the light beam on it. There, no doubt, you have an experiment; by means of the vibration of this spot of light, this physicist minutely observes the oscillations of the piece of iron. Ask him now what he is doing. Is he going to answer: "I am studying the oscillations of the piece of iron carrying this mirror?" No, he will tell you that he is measuring the electrical resistance of a coil. If you are astonished and ask him what meaning these words have, and what relation they have to the phenomena he has perceived and which you have at the same time perceived, he will reply that your question would require some very long explanations, and he will recommend that you take a course in electricity.

It is indeed the case that the experiment you have seen done, like any experiment in physics, involves two parts. In the first place, it consists in the observation of certain facts; in order to make this observation it suffices for you to be attentive and alert enough with your senses. It is not necessary to know physics; the director of the laboratory may be less skillful in this matter of observation than the assistant. In the second place, it consists in the interpretation of the observed facts; in order to make this interpretation it does not suffice to have an alert attention and practiced eye; it is necessary to know the accepted theories and to know how to apply them, in short, to be a physicist. Any man can, if he sees straight, follow the motions of a spot of light on a transparent ruler, and see if it goes to the right or to the left or stops at such and such a point; for that he does not have to be a great cleric. But if he does not know electrodynamics, he will not be able to finish the experiment, he will not be able to measure the resistance of the coil.

Let us take another example. Regnault is studying the compressibility of gases; he takes a certain quantity of gas, encloses it in a glass tube, keeps the temperature constant, and measures the pressure the gas supports and the volume it occupies.

There you have, it will be said, the minute and exact observation of certain phenomena and certain facts. Certainly, in the hands and under the eyes of Regnault, in the hands and under the eyes of his assistants, concrete facts were produced; was the recording of these facts that Regnault reported his intended contribution to the advancement of physics? No. In a sighting device Regnault saw the image of a certain surface of mercury become level with a certain line; is that what he recorded in the report of his experiments? No, he recorded that the gas occupied a volume having such and such a value. An assistant raised and lowered the lens of a cathetometer until the image of another height of mercury became level with the hairline of the lens; he then observed the disposition of certain lines on the scale and on the vernier of the cathetometer; is that what we find in Regnault's memoir? No, we read there that the pressure supported by the gas had such and such a value. Another assistant saw the thermometer's liquid oscillate between two line-marks; is that what he reported? No, it was recorded that the temperature of the gas had varied between such and such degrees.

Now, what is the value of the volume occupied by the gas, what is the value of the pressure it supports, what is the degree of temperature to which it is brought? Are they three concrete objects? No, they are three abstract symbols which only physical theory connects to the facts really observed.

In order to form the first of these abstractions, the value of the volume of the enclosed gas, and to make it correspond with the observed fact, namely, the mercury becoming level with a certain line-mark, it was necessary to calibrate the tube, that is to say, to appeal not only to the abstract ideas of arithmetic and geometry and the abstract principles on which they rest, but also to the abstract idea of mass and to the hypotheses of general mechanics as well as of celestial mechanics which justify the use of the balance for the comparison of masses; it was necessary to know the specific weight of mercury at the temperature when the calibration was made, and for that its specific weight at 0° had to be known, which cannot be done without invoking the laws of hydrostatics; to know the law of the expansion of mercury, which is determined by means of an apparatus where a lens is used, certain laws of optics are assumed; so that the knowledge of a good many chapters of physics necessarily precedes the formation of that abstract idea, the volume occupied by a certain gas.

More complex by far and more intimately tied up with the most profound theories of physics is the genesis of that other abstract idea, the value of the pressure supported by the gas. In order to define and measure it, it has been necessary to use ideas of pressure and of force of cohesion that are so delicate and so difficult to acquire; it has been necessary to call for the help of Laplace's formula for the level of a barometer, a formula drawn from the laws of hydrostatics; it has been necessary to bring in the law of the compressibility of mercury whose determination is related to the most delicate and controversial questions of the theory of elasticity.

Thus, when Regnault did an experiment he had facts before his eyes and he observed phenomena, but what he transmitted to us of that experiment is not a recital of observed facts; what he gave us are abstract symbols which accepted theories permitted him to substitute for the concrete evidence he had gathered.

What Regnault did is what every experimental physicist necessarily does; that is why we can state the following principle whose consequences will be developed in the remainder of this book: An experiment in physics is the precise observation of phenomena accompanied by an *interpretation* of these phenomena; this interpretation substitutes for the concrete data really gathered by observation abstract and symbolic representations which correspond to them by virtue of the theories admitted by the observer.

2. *The Result of an Experiment in Physics Is an Abstract and Symbolic Judgment*

The characteristics which so clearly distinguish the experiment in physics from common experience, by introducing into the former, as an essential element, a theoretical interpretation excluded from the latter, also mark the results arrived at by these two sorts of experiences.

The result of common experience is the perception of a relation between diverse concrete facts. Such a fact having been artificially produced some other fact has resulted from it. For instance, a frog has been decapitated, and the left leg has been pricked with a needle; the right leg has been set into motion and has tried to move away from the needle: there you have the result of an experiment in physiology. It is a recital of concrete and obvious facts, and in order to understand it, not a word of physiology need be known.

The result of the operations in which an experimental physicist is engaged is by no means the perception of a group of concrete facts; it is the formulation of a judgment interrelating certain abstract and symbolic ideas which theories alone correlate with the facts really observed. This truth is immediately evident to anyone who thinks at all. Open any report at all of an experiment in physics and read its conclusions; in no way are they purely and simply an exposition of certain phenomena; they are abstract propositions to which you can attach no meaning if you do not know the physical theories admitted by the author. When you read, for example, that the electromotive force of a certain gas battery increases by so many volts when the pressure is increased by so many atmospheres, what does this proposition mean? We cannot attribute any meaning to it without recourse to the most varied and advanced theories of physics. We have already said that pressure is a quantitative symbol introduced by theoretical mechanics and one of the most subtle notions which that science has to deal with. In order to understand the words "electromotive force" we must appeal to the electrokinetic theory founded by Ohm and by Kirchhoff. The volt is the unit of electromotive force in the practical electromagnetic system of units; the definition of this unit is drawn from the equations of electromagnetism and induction established by Ampère, F.-E. Neumann, and W. Weber. Not one of the words serving to state the result of such an experiment directly represents a visible and tangible object; each of them has an abstract and symbolic meaning related to concrete realities only by long and complicated theoretical intermediaries.

In the statement of an experimental result, similar to the one we have just recalled, a person ignorant of physics and for whom such a statement remains a dead letter might be tempted to see simply an exposition in a technical language, not understandable by the profane but clear to the initiated, of facts observed by the experimenter. That would be a mistake.

I am on a sailing ship. I hear the officer on watch shout out the order: "All hands, tackle the halyard and bowlines everywhere!" A stranger to things of the sea, I do not understand these words, but I see the men on ship run to posts assigned in advance, grab hold of specific ropes, and pull on them in regular order. The words uttered by the officer indicate to them very specific and concrete objects, arousing in their mind the idea of a known manipulation to be performed. Such, for the initiated, is the effect of technical language.

Quite different is the language of the physicist. Suppose the following sentence is pronounced to a physicist: "If we increase the pressure by so many atmospheres, we increase the electromotive force of a battery by so many volts." It is indeed true that the initiated person who knows the theories of physics can translate this statement into facts and can do the experiment whose result is thus expressed, but the noteworthy point is that he can do it in an infinity of different ways. He may exert the pressure by pouring mercury into a tube, by raising a reservoir full of liquid, by manipulating a hydraulic press, or by plunging the piston of a screw pump into water. He may measure this pressure with an open-arm manometer, with a closed-arm manometer, or with a metallic manometer. In order to gauge the variation of the electromotive force, he may employ successively all the known types of electrometers, galvanometers, electrodynamometers, and voltmeters. Each new arrangement of the apparatus will furnish him with new facts to observe; he will be able to employ arrangements of apparatus which the first author of the experiment did not suspect, and see phenomena which this author will never have seen. However, all these diverse manipulations, among which the uninitiated would fail to see any analogy, are not really different experiments; they are only different forms of the same experiment; the facts which have been really produced have been as dissimilar as possible, yet the perception of these facts is expressed by a single proposition: The electromotive force of a certain battery increases by so many volts when the pressure is increased by so many atmospheres.

It is therefore clear that the language in which a physicist expresses the results of his experiments is not a technical language similar to that employed in the diverse arts and trades. It resembles a technical language in that the initiated can translate it into facts, but differs in that a given sentence of a technical language expresses a specific operation performed on very specific objects whereas a sentence in the physicist's language may be translated into facts in an infinity of different ways.

Henri Poincaré has offered the very opinion we are now combatting,[2] in opposition to those who, like us, insist with Edouard Le Roy on the considerable part played by theoretical interpretation in the statement of an experimental fact. According to Poincaré, physical theory should be simply a vocabulary permitting one to translate concrete facts into a simple and convenient conventional language. "A scientific fact," he says, "is nothing but a brute fact stated in a convenient language."[3] And again, "All that the scientist creates in a fact is the language in which he states it."[4]

"When I observe a galvanometer and if I ask an ignorant visitor: 'Is a current passing through it?' he goes and looks at the wire in

[2] H. Poincaré, "Sur la valeur objective des théories physiques," *Revue de Métaphysique et de Morale*, 10th year (1902), p. 263.

[3] *ibid.*, p. 272. [4] *ibid.*, p. 273.

order to see something passing through it. But if I put the same question to my assistant who understands my language, he will realize that that means 'Is the spot[5] displaced?' and he will look at the scale.

"What difference is there between the statement of a brute fact and the statement of a scientific fact? It is the same difference that exists between the statement of a brute fact in the French language and the statement of the same fact in the German language. The scientific statement is the translation of the brute statement into a language which is distinguished from everyday French or everyday German primarily because it is spoken by many fewer persons."[6]

It is not correct to say that the words "the current is on" are simply a conventional manner of expressing the fact that the magnetized little bar of the galvanometer has deviated. Indeed, to the question, "Is the current on?" my assistant may very well answer: "The current is on, and yet the magnet has not deviated; the galvanometer shows some defect." Why does he say that the current is on despite the absence of the galvanometer reading? Because he has observed that in a voltameter, placed in the same circuit as the galvanometer, bubbles of gas were being released; or else, that an incandescent lamp inserted on the same wire was glowing; or else, that a coil around which this wire is wrapped was becoming warm; or else, that a break in the conductor was accompanied by sparks; and because, in virtue of accepted theories, each of these facts as well as the deviation of the galvanometer may be translated by the words "the current is on." This group of words does not therefore express in a technical and conventional language a certain concrete fact; as a symbolic formula it has no meaning for one who is ignorant of physical theories; but for one who knows these theories, it can be translated into concrete facts in an infinity of different ways, *because all these disparate facts admit the same theoretical interpretation.*

M. Henri Poincaré knows that this objection can be made to the doctrine he maintains;[7] here is how he expounds it and replies to it:

[5] That is what we call the small patch of light which a mirror attached to the magnet of the galvanometer sends back to a transparent divided ruler.

[6] *ibid.*, p. 270.

[7] There is nothing astonishing about this if we observe that the foregoing doctrine has been published by us in practically identical terms since 1894, whereas M. Poincaré's article appeared in 1902. By comparing our two articles one will be able to see that in this passage M. Poincaré is combatting our way of looking at things as well as M. Le Roy's.

"Let us not go too fast, however. In order to measure a current I may use a very large number of types of galvanometers or an electrodynamometer. And then when I say, 'There is in this circuit a current of so many amperes,' that will mean, 'If I connect such an electrodynamometer to this circuit, I shall see the spot come to the division *b*.' And that will mean still many other things, for the current may manifest itself not only in mechanical effects but in effects that are chemical, thermal, luminous, etc.

"Therefore, you have a statement which agrees with a very large number of absolutely different brute facts. Why? Because I accept a law according to which every time a certain mechanical effect is produced a certain chemical effect is produced on its side. Very many previous experiments have never showed me anything wrong with this law, and then I realized that I might express by means of the same proposition two facts so invariably connected to one another."[8]

M. Poincaré therefore recognizes that the words "a certain wire carries a current of so many amperes" do not express a single fact but an infinity of possible facts, and that in virtue of constant relations among diverse experimental laws. But are not these relations precisely what everybody calls "the theory of the electric current"? It is because this theory is assumed constructed that the words "there is a current of so many amperes in this wire" may condense so many distinct significations. The role of the scientist is not limited to creating a clear and precise language in which to express concrete facts; rather, it is the case that the creation of this language presupposes the creation of a physical theory.

Between an abstract symbol and a concrete fact there may be a correspondence, but there cannot be complete parity; the abstract symbol cannot be the adequate representation of the concrete fact, the concrete fact cannot be the exact realization of the abstract symbol; the abstract and symbolic formula by which a physicist expresses the concrete facts he has observed in the course of an experiment cannot be the exact equivalent or the faithful story of these observations.

This disparity between the *practical* fact, really observed, and the *theoretical* fact, the symbolic, abstract formula stated by the physicist, is revealed to us when very different concrete facts interpreted by a theory fuse into one another to constitute but one and the same experiment, and are expressed by a single symbolic proposition:

[8] *op.cit.*, p. 270.

The same theoretical fact may correspond to an infinity of distinct practical facts.

This same disparity is also plainly translatable into this other consequence: The same practical fact may correspond to an infinity of logically incompatible theoretical facts; the same group of concrete facts may be made to correspond in general not with a single symbolic judgment but with an infinity of judgments different from one another and logically in contradiction with one another.

An experimenter has made certain observations; he has translated them in the statement: An increased pressure of 100 atmospheres causes the electromotive force of a given gas battery to increase by 0.0845 volts. He might just as well have said that this increase of pressure causes an increase of electromotive force of 0.0844 volts, or that it increased it 0.0846 volts. How can these diverse propositions be equivalent for the physicist? For the mathematician they contradict one another; if a number is 845, it is not, and cannot be, 844 or 846.

What the physicist means when he declares that these three judgments are identical in his eyes is this: Accepting the value 0.0845 volts for the e.m.f. drop, he calculates with the aid of accepted theories the deviation that the galvanometer needle will undergo when he throws into the instrument the current supplied by the battery. That, indeed, is the phenomenon his senses will properly observe, and he finds that this deviation will take on a certain value. If he repeats the same calculation by giving the e.m.f. drop of the battery a value of 0.0844 volts or the value of 0.0846 volts, he will find other values for the deviation of the magnet; but the three deviations thus calculated will differ too little to be visibly discernible on the scale. That is why the physicist will mingle together as one measurement of the e.m.f. drop these three evaluations, 0.0845 volts, 0.0844 volts, and 0.0846 volts, whereas the mathematician would regard them as incompatible.

There can be no adequation between the precise and rigorous theoretical fact and the practical fact with vague and uncertain contours such as our preceptions reveal in everything. That is why the same practical fact can correspond to an infinity of theoretical facts. We have in the preceding chapter insisted on this disparity and its consequences enough to make it unnecessary to return to this point in the present chapter.

A single theoretical fact may then be translated into an infinity of disparate practical facts; a single practical fact corresponds to an infinity of incompatible theoretical facts. This double observation presents in a very striking manner the truth we wished to put in evidence: Between the phenomena really observed in the course of an experiment and the result formulated by the physicist, there is interpolated a very complex intellectual elaboration which substitutes for the recital of concrete facts an abstract and symbolic judgment.

3. *The Theoretical Interpretation of Phenomena Alone Makes Possible the Use of Instruments*

The importance of this intellectual operation, by means of which the phenomena really observed by the physicist are interpreted according to admitted theories, is noticed not only in the form taken by the result of an experiment; it is just as clearly shown in the means used by the experimenter.

It would really be impossible to use the instruments we have in physics laboratories if we did not substitute for the concrete objects composing these instruments an abstract and schematic representation which mathematical reasoning takes over, and if we did not submit this combination of abstractions to deductions and calculations implying the assimilation of theories.

At first blush, this assertion will probably astonish the reader.

A great many people employ a magnifying glass, which is an instrument of physics. Yet, in order to make use of it they do not need to replace this piece of convex, polished, shiny, and heavy glass, mounted in copper or horn, with the pair of spherical surfaces bounding a medium having a certain index of refraction, although only this configuration is accessible to mathematical reasoning in dioptrics; they do not need to have studied dioptrics or to know the theory of the magnifying glass. All they have had to do is to look at the same object at first with the naked eye and then with the magnifying glass in order to perceive that this object keeps the same aspect in both cases but that it appears larger in the second than in the first; hence, if the magnifying glass makes them see an object that the naked eye does not perceive, a quite spontaneous generalization, emerging from common sense, permits them to assert that this object has been enlarged by the glass to the point of becoming visible, but that it was not created or deformed by the glass lens. The spontaneous judgments of common sense thus suffice to justify the use people make of the magnifying glass in the course of their observations; the results of these observations will depend in no way on theories of dioptrics.

The example chosen was borrowed from one of the simplest and crudest of the instruments of physics; nevertheless, is it true that we may use this instrument without making any appeal to theories of dioptrics? The objects seen through the magnifying glass appear circled by colors of the rainbow; is it not the theory of dispersion which teaches us to regard these colors as created by the instrument, and to disregard them when we describe the object observed? And how much more important this remark is when it is no longer a matter of a simple magnifying glass but of a powerful microscope! To what strange errors we should be exposed at times, if we naïvely attributed to the observed objects the shape and color revealed by the instrument, or if a discussion drawn from optical theories did not allow us to distinguish the role of appearances from that of realities!

Yet, even with this microscope intended for the purely qualitative description of very small concrete objects, we are still very far from the instruments employed by the physicist; the experiments combined with the aid of these instruments are not to terminate in a recital of real facts or in a description of concrete objects, but in a numerical evaluation of certain symbols created by theories.

Here, for instance, is an instrument called a tangent galvanometer. On a circular frame is wrapped a copper wire covered by silk insulation; in the center of the frame a very small bar of magnetized steel is suspended by a silk thread; an aluminum needle carried by this small bar moves over a circle divided into degrees. This permits one to report with precision the direction in which the small bar is oriented. When the two ends of the copper wire are connected to the poles of a battery, the magnet is deflected with a deviation that can be read on the divided circle; the deviation is, for instance, 30°.

The mere perception of this fact does not imply any commitment to physical theories, but neither does it suffice to constitute an experiment in physics. The physicist, in fact, does not aim to know the deviation experienced by the magnet, but rather to measure the intensity of the current going through the copper wire.

Now, in order to calculate the value of this intensity agreeing with the value, 30°, of the observed deviation, he must bring this latter value into a certain formula. This formula is a consequence of the laws of electromagnetism; to anyone who would not regard the electromagnetic theory of Laplace and Ampère as correct, the use of this formula and the calculation which is to make known the current's intensity would be veritable nonsense.

This formula applies to all possible tangent galvanometers, to all deviations, and to all current intensities. In order to derive the value of the particular intensity we propose to measure, we must restrict the formula not only by introducing into it the particular value of the deviation, 30°, which has just been observed, but also by applying it not to any sort of tangent galvanometer but to the particular one used. How do we make this special application? Certain letters in the formula represent the characteristic constants of the instrument: the radius of the circular wire through which the current goes, the magnetic moment of the magnet, the magnitude and direction of the magnetic field at the place where the instrument is. These letters are replaced by the numerical values suitable for the instrument used and for the laboratory in which it is.

Now, what is presupposed by this way of expressing the fact that we have used a certain instrument in a certain laboratory? It assumes that for the copper wire of a certain thickness, in which we have introduced a current, we have substituted the circumference of a circle or a geometric line wholly defined by its radius; that for the piece of magnetized steel of a certain size and shape and hung by a silk thread, we have substituted an infinitely small horizontal axis, moveable without friction around a vertical axis, and having a certain magnetic moment; that for the laboratory where the experiment was done, we have substituted a certain space entirely defined by a magnetic field having a certain direction and intensity.

Thus, so long as it was a question simply of reading the deviation of the magnet, what we did was touch and look at a collection of copper, steel, aluminum, glass, and silk, lying beside three calibrating screws on a table of a certain laboratory situated in the building of the faculty of sciences of Bordeaux, on the ground floor. But when it comes to finishing the experiment by interpreting the readings made and by applying the formula of the tangent galvanometer, we have left behind us this laboratory, where the visitor ignorant of physics may enter, and this instrument which can be examined without knowing a word of electromagnetism; we have substituted for them the assemblage of a magnetic field, a magnetic axis, a magnetic moment, a circular current of a certain intensity, that is to say, a group of symbols given a meaning only by physical theories inconceivable to those not knowing electromagnetism.

Hence, when a physicist does an experiment, two very distinct representations of the instrument on which he is working fill his mind: one is the image of the concrete instrument that he manipulates in reality; the other is a schematic model of the same instrument, constructed with the aid of symbols supplied by theories; and it is on this ideal and symbolic instrument that he does his reasoning, and it is to it that he applies the laws and formulas of physics.

These principles permit us to define what we agree to understand when we say that we increase the precision of an experiment by eliminating causes of error by appropriate corrections. We shall see, in fact, that these corrections are nothing else than improvements brought in with the theoretical interpretation of the experiment.

As physics gradually progresses, we see a narrowing of the indetermination of the group of abstract judgments that the physicist correlates with the same concrete fact; the degree of approximation of experimental results continues to grow better not only because manufacturers supply instruments that are increasingly more precise, but also because physical theories yield more and more satisfactory rules to establish the correspondence of facts with the schematic ideas serving to represent them. This increasing precision is purchased, it is true, by an increasing complication, by the obligation of observing, at the same time as we observe the main fact, a series of accessory facts, and by the necessity of subjecting the raw data of experience to more and more numerous and delicate transformations and combinations; and these transformations that we make on the immediate data of the experiment are the corrections.

If an experiment in physics were merely the observation of a fact, it would be absurd to bring in corrections, for it would be ridiculous to tell an observer who had looked attentively, carefully, and minutely: "What you have seen is not what you should have seen; permit me to make some calculations which will teach you what you should have observed."

The logical role of corrections, on the other hand, is very well understood when it is remembered that a physical experiment is not simply the observation of a group of facts but also the translation of these facts into a symbolic language with the aid of rules borrowed from physical theories. Indeed, a result of this is that the physicist constantly compares two instruments, the real one that he manipulates and the ideal, symbolic one on which he reasons; for example, the word manometer designated two essentially distinct but inseparable things for Regnault: on the one hand, a series of glass tubes, solidly connected to one another, supported on the walls of the tower of the Lycée Henri IV, and filled with a very heavy metallic liquid called mercury by the chemists; on the other hand, a column of that creature of reason called a perfect fluid in mechanics, and having at each point a certain density and temperature defined by a certain equation of compressibility and expansion. It was on the first of these two manometers that Regnault's laboratory assistant directed the eyepiece of his cathetometer, but it was to the second that the great physicist applied the laws of hydrostatics.

The schematic instrument is not and cannot be the exact equivalent of the real instrument, but we conceive it possible for him to have given a more or less perfect picture of it; we conceive that the physicist, after reasoning on a schematic instrument that is too simple and too remote from reality, will seek to substitute for it a more complicated scheme that resembles reality more. This passage from a certain schematic instrument to another which better symbolizes the concrete instrument is essentially the operation that the word correction designates in physics.

An assistant of Regnault's gives him the height of the column of mercury in a manometer; Regnault corrects it; does he suspect that his assistant has looked poorly and been mistaken in his readings? No, he has full confidence in the observations which have been made; if he did not, he could not correct the experiment, but could only begin it over again. If, therefore, Regnault substitutes for the height determined by his assistant another number, it is on the strength of intellectual operations intended to diminish the disparity between the ideal, symbolic manometer which exists only in his reason and to which he applies his calculations, and the real manometer of glass and mercury which faces his gaze and from which his assistant makes his readings. Regnault could represent this real manometer by an ideal one, formed of an incompressible fluid having the same temperature everywhere and subjected at every point of its free surface to an atmospheric pressure independent of the height; between this oversimplified scheme and reality there would be too great a discrepancy and consequently, the experiment would be insufficiently precise. Then he conceives a new ideal manometer, more complicated than the first, but representing better the real and concrete one; he forms this new manometer with a compressible fluid and allows the temperature to vary from one point to another; he also allows the barometric pressure to change when one goes higher up in the atmosphere. All these retouchings of the primitive scheme constitute so many corrections: a correction relative to the compressibility of mercury, a correction relative to the unequal warming of the mercurial column, a

Laplacean correction relative to the barometric height; and all these corrections go to increase the precision of the experiment.

The physicist who complicates the theoretical representation of the observed facts by corrections, in order to permit this representation to come to closer grips with reality, is similar to the artist who, after finishing the line sketch of a drawing, adds shading in order to express better on a plane surface the profile of the model.

Whoever sees in physical experiments only the observation of facts would not understand the role played by corrections in these experiments; he would not understand, furthermore, what is meant in speaking of "systematic errors" that an experiment may involve.

To allow a cause of systematic error to remain in an experiment is to omit making a possible correction which would increase the precision of an experiment; it means being content with a very simple theoretical picture when we might substitute for it a more complicated one which would better represent reality; it means being content with a line sketch when we could make a shaded drawing.

In his experiments on the compressibility of gases Regnault let exist a cause of systematic error which he did not perceive and which has since been pointed out: he neglected the action of weight on the gas under pressure. What do we mean when we criticize Regnault for not having taken this action into account and for having omitted this correction? Do we mean that his senses deceived him while he was observing the phenomena produced before him? By no means. We are criticizing him for having oversimplified the theoretical picture of these facts by representing the gas under pressure as a homogeneous fluid, whereas by regarding it as a fluid whose pressure varies with the height according to a certain law, he would have obtained a new abstract picture, more complicated than the first but a more faithful reproduction of the truth.

4. *On Criticism of an Experiment in Physics; in What Respects It Differs from the Examination of Ordinary Testimony*

An experiment in physics being quite another matter than the mere observation of a fact, there is no difficulty in conceiving the certainty of an experimental result to be of quite another order than that of a fact merely observed by the senses. It is similarly understandable that these certainties of such different sorts should become known by entirely distinct methods.

When a sincere witness, sound enough in mind not to confuse the play of his imagination with perceptions, and knowing the language he uses well enough to express his thought clearly, says he has observed a fact, the fact is certain: if I declare to you that on such and such a day at such and such an hour I saw a white horse in a certain street, unless you have reasons to consider me a liar or subject to hallucinations, you ought to believe that on that day, at that hour, and in that street there was a white horse.

The confidence which ought to be accorded to a proposition stated by a physicist as the result of an experiment is not the same kind of thing; if the physicist restricts himself to narrating the facts he has seen, in the strict sense of seeing with his own eyes, his testimony should be investigated in accordance with the usual rules for determining the degree of credibility of the testimony of a man; if the physicist were recognized as trustworthy—and this would generally be the case, I think—his testimony ought to be received as the expression of a truth.

But, once again, what the physicist states as the result of an experiment is not the recital of observed facts, but the interpretation and the transposing of these facts into the ideal, abstract, symbolic world created by the theories he regards as established.

Therefore, after submitting the physicist's testimony to the rules determining the credibility of a witness's story, we shall have done only a part, the easiest part at that, of the criticism which should determine the value of his experiment.

In the first place, we must inquire very carefully into the theories which the physicist regards as established and which he used in interpreting the facts he has observed. Without knowing these theories it is impossible for us to understand the meaning he gives to his own statements; this physicist would confront us as a witness confronts a judge who does not understand the witness's language.

If the theories admitted by this physicist are those we accept, and if we have agreed to follow the same rules in the interpretation of the same phenomena, we speak the same language and can understand each other. But that is not always the case. It is not so when we discuss the experiments of a physicist who does not belong to our school; and it is especially not so when we discuss the experiments of a physicist separated from us by fifty years, a century, or two centuries. We must then seek to establish a correspondence between the theoretical ideas of the author we are studying and ours, and to interpret anew with the aid of the symbols we use what he interpreted with the aid of the symbols he used. If we succeed in doing this, the discussion of his experiment will be possible; this experiment will be a piece of testimony given in a language foreign to ours, but one whose vocabulary we possess; we shall be able to translate it and investigate it.

Newton, for example, had done certain experiments concerning the colors of rings; he had interpreted these observations in the optical theory he had created, namely, the theory of emission; he had interpreted them as giving for light corpuscles of each color the distance between a "fit of easy reflection" and a "fit of easy transmission." When Young and Fresnel later brought in the wave theory to replace the emission theory it was possible for them to make certain elements of the new theory correspond with certain elements of the old one; in particular, they saw that the distance between a fit of easy reflection and one of easy transmission corresponded to a quarter of what they called a wave length. Thanks to this observation of theirs the results of Newton's experiments could be translated into the language of waves; the numbers that Newton had obtained multiplied by four gave the wave lengths of the diverse colors.

In like manner, Biot had done a great many detailed experiments on the polarization of light, and had interpreted them in the system of emissions; Fresnel was able to translate them into the language of wave theory and use them as a check on this theory.

If, on the contrary, we cannot obtain sufficient information about the theoretical ideas of the physicist whose experiment we are discussing, and if we fail to establish a correspondence between the symbols he has adopted and the symbols furnished by the theories that we accept, the propositions through which that physicist translated the results of his experiments will be neither true nor false for us; they will be devoid of meaning, a dead letter; to our eyes they will be what Etruscan or Ligurian inscriptions are to the epigrapher's eyes: documents written in an undecipherable language. How many observations accumulated by physicists of former times are thus lost forever! Their authors have neglected to inform us about the methods they used to interpret the facts, and it is impossible to transpose their interpretations into our theories. They have sealed their ideas in signs to which we lack a key.

These first principles will seem naïve perhaps, and one will wonder at our insisting on maintaining them; however, if these rules are commonplace, the lack of them is still more commonplace. How many scientific discussions there are in which each of the contenders claims to have crushed his adversary under the overwhelming testimony of facts! Contradictory observations are offered by each to the other's arguments. The contradiction does not exist in reality, which is always in accord with itself, but lies in the theories through which each of the two champions expresses this reality. How many propositions are regarded as monstrous errors in the writings of those who have preceded us! We should perhaps commemorate them as great truths if we really wished to inquire into the theories which give these propositions their true meaning, and if we took the trouble to translate them into the language of theories praised today.

Suppose that we have perceived the agreement between the theories admitted by an experimenter and those we regard as accurate. There is still much lacking before we can accept offhand the judgments in which he states the results of his experiments: we must now investigate whether in his interpretation of the observed facts he has correctly applied the rules outlined by the theories common to us; at times we shall notice that the experimenter has not satisfied all of the legitimate requirements; he may have committed a fallacy in reasoning or in calculation while applying his theories; then, the reasoning should be resumed or the calculation done over; the result of the experiment will have to be modified, the number obtained replaced by another.

The experiment done has been a continual juxtaposition of two sorts of apparatus, the real apparatus manipulated by the investigator and the ideal and schematic one on which he did his reasoning. The comparison of these two sets of apparatus must be resumed by us, and in order to do that we must know both exactly. We can have an adequate knowledge of the second, for it is defined by mathematical symbols and formulas. But that is not the case with the first apparatus; we have to form as exact an idea of it as possible from the description given to us by the experimenter. Is this description sufficient? Does it supply us with all the information that may be useful to us? The state of the bodies studied, their degree of chemical purity, the circumstances in which they were placed, the perturbations they could have been experiencing, the thousand and one accidents which could have had an influence on the result of the experiment—have all these been determined in a careful and minute way, leaving nothing to be desired?

Once we have answered all these questions, we shall be able to investigate to what extent the schematic apparatus offered a picture resembling the concrete apparatus; we shall be able to find out whether we might not have gained a closer resemblance by complicating the definition of the ideal apparatus; we shall be in a position to ask if all the important causes of systematic error have been eliminated and all the desirable corrections made.

Even assuming that the experimenter has employed, in order to interpret his observations, theories which we accept with him, that he has correctly applied in the course of this interpretation the rules that these theories prescribe, that he has minutely studied and described the apparatus he used, and that he has eliminated the causes of systematic error or corrected their effects—that would still not be sufficient reason to accept the result of his experiment. The abstract and mathematical propositions which theories correlate with observed facts are not, we have said, completely determined; an infinity of different propositions may correspond with the same facts, and an infinity of different numerical evaluations may correspond with the same measurements. The degree of indetermination of the abstract mathematical proposition through which the experimental result is expressed is what we call the degree of approximation of this experiment. We must know the degree of approximation of the experiment we are investigating; if the experimenter has indicated it, we must check the procedures by which he has evaluated it; if he has not indicated it, we must determine it by our own analysis. A complex and infinitely delicate operation! The estimation of the degree of exactness of an experiment requires, in the first place, that we judge the acuteness of the observer's senses. Astronomers try to determine this information in the mathematical form of a personal equation, but this equation partakes very little of the serene constancy of geometry, for it is at the mercy of a splitting headache or painful indigestion. This estimate requires, in the second place, that we evaluate the systematic errors that could not be corrected; but, after making as complete an enumeration as possible of the causes of these errors, we are sure to omit infinitely more than have been enumerated, for the complexity of concrete reality is beyond us. Under the label of accidental errors are lumped together all those systematic errors with unsuspected causes—ignorance of the circumstances which determine them does not allow us to correct them. Mathematicians have taken advantage of the latitude allowed by this ignorance to fabricate hypotheses about these errors, permitting them to attenuate their effect by certain mathematical operations, but the theory of probable errors is worth no more than the validity of these hypotheses; and how shall we know what these hypotheses are worth, since we know nothing of the errors they deal with except that we do not know the sources of these errors?

The estimation of the degree of approximation of an experiment is, therefore, an extremely complex task. It is often difficult to hold to any logical order in this task; reasoning should then make way for that rare and subtle quality, that sort of instinct or flair called the experimental sense, a pennant worn by the penetrating mind (*esprit de finesse*) rather than by the geometrical mind.

The mere description of the rules governing the investigation of a physical experiment and its adoption or rejection suffices to put in evidence the following essential truth: An experimental result in physics does not have the same order of certainty as a fact ascertained by non-scientific methods through mere seeing or touching by a man of sound body and mind; less self-evident and subject to arguments from which everyday testimony escapes, this certainty of physical experiment remains constantly subordinated to the confidence inspired by a whole group of theories.

5. *Experiment in Physics Is Less Certain but More Precise and Detailed than the Non-scientific Establishment of a Fact*

The uninitiated believe that the result of a scientific experiment is distinguished from ordinary observation by a higher degree of certainty. They are mistaken, for the account of an experiment in physics does not have the immediate certainty, relatively easy to check, that ordinary, non-scientific testimony has. Though less certain than the latter, physical experiment is ahead of it in the number and precision of the details it causes us to know: therein lies its true and essential superiority.

Ordinary testimony, which reports a fact established by the procedures of common sense and not by scientific methods, can be certain only at the expense of not being detailed or minute, and by taking the fact as gross or in its most salient aspect. In a certain street of the city and near a certain hour I saw a white horse: that is what I affirm with certainty. Perhaps, I shall be able to add to that general statement some peculiarity which struck my attention to the exclusion of other details: the strange posture of the horse or a colorful piece of his harness, but do not press me with any more questions; my memory would be disturbed and my answers vague; soon I should be reduced to telling you, "I do not know." With rare exceptions, ordinary testimony offers assurance only to the extent that it is less precise, less analytic, and sticks to the grossest and most obvious considerations.

Quite different is the account of a physical experiment: it is not content with letting us know a gross phenomenon; it claims to analyze it, to inform us about the least detail and the most minute particularity, and to take exact note of the rank and relative importance of each detail and peculiarity; it claims to give us information in such a form that we can reproduce, whenever we please, the phenomenon exactly as reported or, at least, a theoretically equivalent phenomenon. This claim would exceed the power of scientific experimentation to fulfill it, as it exceeds that of ordinary observation, were not the former better armed than the latter. The number and minutiae of details constituting and surrounding each phenomenon would rout the imagination, be too much for memory, and defy description, if the physicist had not at his service a wonderful means of classification and expression, an admirably clear and concise symbolic means of representation, viz., mathematical theory, and if he did not have, in order to note the relative importance of each particular, the exact and brief method of judging supplied by numerical evaluation, viz., measurement. If someone, on a bet, undertook to describe a physical experiment today by excluding all theoretical language, for example, if he tried to expound Regnault's experiments on the compressibility of gases by ridding his account of all the abstract and symbolic expressions introduced by physical theories, that is, the words pressure, temperature, density, intensity of weight, optical axis of a lens, etc., he would perceive that the account of these experiments alone would fill a whole volume with the most confused, the most involved, and the least comprehensible recital imaginable.

Therefore, if theoretical interpretation removes from the results of physical experiment the immediate certainty that the data of ordinary observation possess, on the other hand it is theoretical interpretation which permits scientific experiment to penetrate much further than common sense into the detailed analysis of phenomena, and to give a description of them whose precision exceeds by far the accuracy of current language.

CHAPTER V

PHYSICAL LAW

1. *The Laws of Physics Are Symbolic Relations*

JUST AS the laws of common sense are based on the observation of facts by means natural to man, so the laws of physics are based on the results of physical experiments. Of course, the profound differences which separate the non-scientific ascertainment of a fact from the result of a physical experiment will also separate the laws of common sense from the laws of physics; thus, nearly everything we have said about the experiments of physics will extend to the laws that science states.

Let us consider one of the simplest and most certain of common-sense laws: All men are mortal. This law surely relates two abstract concepts, the abstract idea of man in general, rather than the concrete idea of this or that man in particular, and the abstract idea of death, rather than the concrete idea of this or that form of death; indeed, it is only on this condition, viz., that the concepts related are abstract, that the law can be general. But these abstractions are in no way theoretical symbols, for they merely extract what is universal in each of the particular cases to which the law applies. Thus, in each of the particular cases where we apply the law, we shall find concrete objects in which these abstract ideas are realized; each time we might wish to ascertain that all men are mortal we shall find ourselves aware of a certain individual man embodying the general idea of man, and of a certain particular death implying the general idea of death.

Let us take another law, quoted as an example by Milhaud when he expounded these ideas[1] expressed by us a little earlier. It is a law about an object belonging to the domain of physics, but it retains the form that the laws of physics had when this branch of knowledge existed only as a dependency of common sense without yet having acquired the dignity of a rational science.

Here is the law: We see the flash of lightning before we hear thunder. The ideas of lightning and thunder which this statement ties together are abstract and general ideas, but these abstractions

[1] G. Milhaud, "La Science rationnelle," *Revue de Métaphysique et de Morale*, IV (1896), 280. Reprinted in *Le Rationnel* (Paris, 1898), p. 44.

are drawn so instinctively and naturally from particular data that with each bolt of lightning we perceive a glare and a rumbling in which we recognize immediately the concrete form of our ideas of lightning and thunder.

This is not, however, true of the laws of physics. Let us take one of these laws, Mariotte's law,* and examine its formulation without caring for the moment about the accuracy of this law. At a constant temperature, the volumes occupied by a constant mass of gas are in inverse ratio to the pressures they support; such is the statement of the law of Mariotte. The terms it introduces, the ideas of mass, temperature, pressure, are still abstract ideas. But these ideas are not only abstract; they are, in addition, symbolic, and the symbols assume meaning only by grace of physical theories. Let us put ourselves in front of a real, concrete gas to which we wish to apply Mariotte's law; we shall not be dealing with a certain concrete temperature embodying the general idea of temperature, but with some more or less warm gas; we shall not be facing a certain particular pressure embodying the general idea of pressure, but a certain pump on which a weight is brought to bear in a certain manner. No doubt, a certain temperature corresponds to this more or less warm gas, and a certain pressure corresponds to this effort exerted on the pump, but this correspondence is that of a sign to the thing signified and replaced by it, or of a reality to the symbol representing it. This correspondence is by no means immediately given; it is established with the aid of instruments and measurements, and this is often a very long and very complicated process. In order to assign a definite temperature to this more or less warm gas, we must have recourse to a thermometer; in order to evaluate in the form of a pressure the effort exerted by the pump, we must use a manometer, and the use of the thermometer and manometer imply, as we have seen in the preceding chapter, the use of physical theories.

The abstract terms referred to in a common-sense law being no more than whatever is general in the concretely observed objects, the transition from the concrete to the abstract is made in such a necessary and spontaneous operation that it remains unconscious; placed in the presence of a certain man or of a certain case of death, I associate them immediately with the general idea of man and with the general idea of death. This instinctive and unreflective operation yields unanalyzed general ideas, abstractions

* Translator's note: Boyle's law.

taken grossly, so to speak. No doubt, the thinker may analyze these general and abstract ideas, he may wonder what man is, what death is, and seek to penetrate the deep and full sense of these words. This inquiry will lead him to a better understanding of the reasons for the law, but it is not necessary to do that in order to understand the law; it is sufficient to take the terms related in their obvious sense in order to understand this law, which is clear to us whether we are philosophers or not.

The symbolic terms connected by a law of physics are, on the other hand, not the sort of abstractions that emerge spontaneously from concrete reality; they are abstractions produced by slow, complicated, and conscious work, i.e., the secular labor which has elaborated physical theories. If we have not done this work or if we do not know physical theories, we cannot understand the law or apply it.

According to whether we adopt one theory or another, the very words which figure in a physical law change their meaning, so that the law may be accepted by one physicist who admits a certain theory and rejected by another physicist who admits some other theory.

Take a peasant who has never analyzed the notions of man or of death and a metaphysician who has spent his life analyzing them; take two philosophers who have analyzed and adopted different, irreconcilable notions of man and of death; for all, the law "All men are mortal" will be equally clear and true. In the same way, the law "We see the flash of lightning before we hear thunder" has for the physicist who knows thoroughly the laws of disruptive electrical discharge the same clarity and certainty as it had for the Roman plebeian who saw in a stroke of lightning the anger of Capitoline Jupiter.

On the other hand, let us consider the following physical law: "All gases contract and expand in the same manner;" and let us ask different physicists whether this law is or is not violated by iodine vapor. The first physicist professes theories according to which iodine vapor is a single gas, and draws from the foregoing law the consequence that the density of iodine vapor relative to air is a constant. Now, experiment shows that this density depends on the temperature and pressure; therefore, our physicist concludes that iodine vapor is not subject to the law stated. A second physicist will have it that iodine vapor is not a single gas but a mixture of two gases which are polymers of each other and capable of being transformed into each other; consequently, the law mentioned does not require the iodine-vapor density relative to air to be constant, but claims this density varies with the temperature and pressure according to a certain formula established by J. Willard Gibbs. This formula represents, indeed, the results of experimental determinations; our second physicist concludes that iodine vapor is not an exception to the rule which states that all gases contract and expand in the same manner. Thus our two physicists have entirely different opinions concerning a law which both enunciate in the same form: one finds fault with it because of a certain fact, the other finds that it is confirmed by that very fact. That is because the different theories they hold do not determine uniquely the meaning suited to the words "a single gas," so that though they both pronounce the same sentence, they mean two different propositions; in order to compare his proposition with reality each makes different calculations, so that it is possible for one to verify this law which the other finds contradicted by the same facts. This is plain proof of the following truth: A physical law is a symbolic relation whose application to concrete reality requires that a whole group of laws be known and accepted.

2. *A Law of Physics Is, Properly Speaking, neither True nor False but Approximate*

A common-sense law is merely a general judgment; this judgment is either true or false. Take, for instance, the law that everyday observation reveals: In Paris, the sun rises every day in the east, goes up in the heavens, then comes down and sets in the west. There you have a true law without conditions or restrictions. On the other hand, take this statement: The moon is always full. That is a false law. If the truth of a common-sense law is questioned, we can answer this question by yes or no.

Such is not the case with the laws that a physical science, come to full maturity, states in the form of mathematical propositions; such laws are always symbolic. Now, a symbol is not, properly speaking, either true or false; it is, rather, something more or less well selected to stand for the reality it represents, and pictures that reality in a more or less precise, a more or less detailed manner. But applied to a symbol the words "truth" and "error" no longer have any meaning; so, the logician who is concerned about the strict meaning of words will have to answer anyone who asks whether physics is true or false, "I do not understand your question." Let us comment on this answer which may seem paradoxical but the understanding of which is necessary for anyone who claims to know what physics is.

The experimental method, as practiced in physics, does not make a given fact correspond to only one symbolic judgment, but to an infinity of different symbolic judgments; the degree of symbolic indetermination is the degree of approximation of the experiment in question. Let us take a sequence of analogous facts; finding the law for these facts means to the physicist finding a formula which contains the symbolic representation of each of these facts. The symbolic indetermination corresponding to each fact consequently entails the indetermination of the formula which is to unite these symbols; we can make an infinity of different formulas or distinct physical laws correspond to the same group of facts. In order for each of these laws to be accepted, there should correspond to each fact not *the* symbol of this fact, but some one of the symbols, infinite in number, which can represent the fact; that is what is meant when the laws of physics are said to be only approximate.

Let us imagine, for example, that we refuse to be satisfied with the information supplied by the common-sense law about the sun's rising in the east, climbing the sky, descending, and setting in the west every day in Paris; we address ourselves to the physical sciences in order to have a precise law of the motion of the sun seen from Paris, a law indicating to the observer in Paris what place the sun occupies in the sky at each moment. In order to solve the problem, the physical sciences are not going to use sensed realities, say of the sun just as we see it shining in the sky, but will use symbols through which theories represent these realities: the real sun, despite the irregularities of its surface, despite the enormous protuberances it has, will be replaced in their theories by a geometrically perfect sphere, and it is the position of the center of this ideal sphere that these theories will try to determine; or rather, they will seek to determine the position that this point would occupy if astronomical refraction did not deviate the rays, and if the annual aberration did not modify the apparent position of the heavenly bodies. It is, therefore, a symbol that is substituted for the sole sensible reality offered to our observation, for the shiny disk that our lens may sight. In order to make the symbol correspond to the reality, we must effect complicated measurements, we must make the edges of the sun coincide with the hairlines of a lens equipped with a micrometer, we must make many readings on divided circles, and subject these readings to diverse corrections; we must also develop long and complex calculations whose legitimacy depends on admitted theories, on the theory of aberration, and on the theory of atmospheric refraction.

The point symbolically called the center of the sun is not yet obtained by our formulas; they tell us only the coordinates of this point, for instance, its longitude and latitude, coordinates whose meaning cannot be understood without knowing the laws of cosmography, and whose values do not designate a point in the sky that you can indicate with your finger or that a telescope can sight except by virtue of a group of preliminary determinations: the determination of the meridian of the place, its geographical coordinates, etc.

Now, can we not make a single value for the longitude and a single value for the latitude of the sun's center correspond to a definite position of the solar disk, assuming the corrections for aberration and refraction to have been made? Indeed not. The optical power of the instrument used to sight the sun is limited; the diverse operations and readings required of our experiment are of a limited sensitivity. Let the solar disk be in such a position that its distance from the next position is small enough, and we shall not be able to perceive the deviation. Admitting that we cannot know the coordinates of a fixed point on the celestial sphere with a precision greater than 1′, it will suffice, in order to determine the position of the sun at a given instant, to know the longitude and latitude of the sun's center to approximately 1′. Hence, to represent the path of the sun, despite the fact that it occupies only one position at each instant, we shall be able to give for each instant not one value alone for the longitude and only one value for the latitude, but an infinity of values for each, except that for a given instant two acceptable values of the longitude or two acceptable values of the latitude will not differ by more than 1′.

We now proceed to seek the law of the sun's motion, that is to say, two formulas permitting us to calculate at each instant of a period the value of the longitude and latitude, respectively, of the center of the sun. Is it not evident that, in order to represent the path of the longitude as a function of the time, we shall be able to adopt not a single formula, but an infinity of different formulas, provided that for a given instant all these formulas give us values for the longitude differing by less than 1′? And is not the same evident for the latitude? We shall then be able to represent equally well our observations on the path of the sun by an infinity of different laws; these diverse laws will be expressed by equations which algebra regards as incompatible, by equations such that if one of them is verified, no other is. They will each trace a different curve on the celestial sphere, and it would be absurd to say that the same point describes two of these curves at the same time; yet, to the physicist all these laws are equally acceptable, for all determine the position of the sun with a closer approximation than can be observed with our instruments. The physicist does not have the right to say that any of these laws is true to the exclusion of the others.

No doubt the physicist has the right to choose between these laws, and generally he will choose; but the motives which will guide his choice will not be of the same kind or be imposed with the same imperious necessity as those which compel him to prefer truth to error.

He will choose a certain formula because it is simpler than the others; the weakness of our minds constrains us to attach great importance to considerations of this sort. There was a time when physicists supposed the intelligence of the Creator to be tainted with the same debility, when the simplicity of these laws of nature was imposed as an indisputable dogma in the name of which any experimental law expressing too complicated an algebraic equation was rejected, when simplicity seemed to confer on a law a certainty and scope transcending those of the experimental method which supplied it. It was than that Laplace, speaking of the law of double refraction discovered by Huygens, said: "Until now this law has been only the result of observation, approximating the truth within the limits of error to which the most exact experiments are subject. Now the simplicity of the law of action on which it depends should make us consider it a rigorous law."[2] That time no longer exists. We are no longer dupes of the charm which simple formulas exert on us; we no longer take that charm as the evidence of a greater certainty.

The physicist will especially prefer one law to another when the first follows from the theories he admits; he will, for example, ask the theory of universal attraction to decide which formulas he should prefer among all those which could represent the motion of the sun. But physical theories are only a means of classifying and bringing together the approximate laws to which experiments are subject; theories, therefore, cannot modify the nature of these experimental laws and cannot confer absolute truth on them.

Thus, every physical law is an approximate law. Consequently, it cannot be, for the strict logician, either true or false; any other

[2] P. S. Laplace, *Exposition du système du monde* I, IV, Ch. XVIII: "De l'attraction moléculaire."

law representing the same experiments with the same approximation may lay as just a claim as the first to the title of a true law or, to speak more precisely, of an acceptable law.

3. *Every Law of Physics Is Provisional and Relative because It Is Approximate*

What is characteristic of a law is that it is fixed and absolute. A proposition is a law only because once true, always true, and if true for this person, then also for that one. Would it not be contradictory to say that a law is provisional, that it may be accepted by one person and rejected by another? Yes and no. Yes, certainly, if we mean by "laws" those that common sense reveals, those we can call true in the proper sense of the word; such laws cannot be true today and false tomorrow, and cannot be true for you and false for me. No, if we mean by "laws" the laws that physics states in mathematical form. Such laws are always provisional; not that we must understand this to mean that a physical law is true for a certain time and then false, but at no time is it either true or false. It is provisional because it represents the facts to which it applies with an approximation that physicists today judge to be sufficient but will some day cease to judge satisfactory. Such a law is always relative; not because it is true for one physicist and false for another, but because the approximation it involves suffices for the use the first physicist wishes to make of it and does not suffice for the use the second wishes to make of it.

We have already noticed that the degree of approximation is not something fixed; it increases gradually as instruments are perfected, and as the causes of error are more rigorously avoided or more precise corrections permit us to evaluate them better. As experimental methods gradually improve, we lessen the indetermination of the abstract symbol brought into correspondence with the concrete fact by physical experiment; many symbolic judgments which might have been regarded at one time as adequately representing a definite, concrete fact will no longer be accepted at another time as signifying this fact with sufficient precision. For example, the astronomers of one century will, in order to represent the position of the sun's center at a given instant, accept all the values of the longitude which do not differ from each other by more than 1′ and all the values of the latitude confined within the same interval. The astronomers of the next century will have telescopes with greater optical power, more perfectly divided circles, more minute and precise methods of observation; they will require then that the diverse determinations of the longitude and latitude, respectively, of the sun's center at a given instant agree within about 10″; an infinity of determinations which their predecessors were willing to permit would be rejected by them.

As the indetermination of experimental results becomes narrower, the indetermination of the formulas used to condense these results becomes more restricted. One century would accept as the law of the sun's motion any group of formulas which gave for each instant the coordinates of the center of this star within approximately 1′; the next century will impose on any law of the sun's motion the condition that the coordinates of the sun's center be known within approximately 10″; an infinity of laws accepted by the first century will thus be rejected by the second.

This provisional character of the laws of physics is made plain every time we read the history of this science. For Dulong and Arago, Mariotte's [Boyle's] law was an acceptable form of the law of the compressibility of gases because it represented the experimental facts with deviations that remained less than the possible errors of the methods of observation used by them. When Regnault had improved the apparatus and experimental method, this law had to be rejected; the deviations of Mariotte's law from the results of observation were much greater than the uncertainties affecting the new apparatus.

Now, given two contemporary physicists, the first may be in the circumstances Regnault was in, whereas the second may still be working under conditions under which Dulong and Arago worked. The first possesses very precise apparatus and plans to make very exact observations; the second possesses only crude instruments and, in addition, the investigations he is making do not demand close approximation. Mariotte's law will be accepted by the latter and rejected by the former.

More than that, we can see the same physical law simultaneously adopted and rejected by the same physicist in the course of the same work. If a law of physics could be said to be true or false, that would be a strange paradox; the same proposition would be affirmed and denied at the same time, and this would constitute a formal contradiction.

Regnault, for example, is making inquiries about the compressibility of gases for the purpose of finding a more approximate formula to substitute for Mariotte's law. In the course of his experiments he needs to know the atmospheric pressure at the level reached by the mercury in his manometer; he uses Laplace's formula to obtain this pressure, and Laplace's formula rests on the use of Mariotte's law. There is no paradox or contradiction here. Regnault knows that the error introduced by this particular employment of Mariotte's law is much smaller than the uncertainties of the experimental method he is using.

Any physical law, being approximate, is at the mercy of the progress which, by increasing the precision of experiments, will make the degree of approximation of this law insufficient: the law is essentially provisional. The estimation of its value varies from one physicist to the next, depending on the means of observation at their disposal and the accuracy demanded by their investigations: the law is essentially relative.

4. *Every Physical Law Is Provisional because It Is Symbolic*

Physical law is provisional not only because it is approximate, but also because it is symbolic: there are always cases in which the symbols related by a law are no longer capable of representing reality in a satisfactory manner.

In order to study a certain gas, for example, oxygen, the physicist has created a schematic representation of it which can be grasped in mathematical reasoning and algebraic calculation. He has pictured this gas as one of the perfect fluids that mechanics studies: it has a certain density, is brought to a certain temperature, and is subject to a certain pressure. Among these three elements, density, temperature, and pressure, he has established a certain relation that a certain equation expresses: that is the law of the compressibility and expansion of oxygen. Is this law definitive?

Let the physicist place some oxygen between the plates of a strongly charged electrical condenser; let him determine the density, temperature, and pressure of the gas; the values of these three elements will no longer verify the law of the compressibility and expansion of oxygen. Is the physicist astonished to find his law at fault? Not at all. He realizes that the faulty relation is merely a symbolic one, that it did not bear on the real, concrete gas he manipulates but on a certain logical creature, on a certain schematic gas characterized by its density, temperature, and pressure, and that this schematism is undoubtedly too simple and too incomplete to represent the properties of the real gas placed in the conditions given now. He then seeks to complete this schematism and to make it more representative of reality: he is no longer content to represent oxygen by means of its density, its temperature, and the pressure it supports; he introduces into the construction of the new schematism the intensity of the electrical field in which the gas is placed; he subjects this more complete symbol to new studies and obtains the law of the compressibility of oxygen endowed with dielectric polarization. This is a more complicated law; it includes the former as a special case, but it is more comprehensive and will be verified in cases where the original law would fail.

Is this new law definitive?

Take the gas to which it applies and place it between the poles of an electromagnet; you will see the new law falsified in its turn by the experiment. Do not think that this new falsity upsets the physicist; he knows that he has to deal with a symbolic relation and that the symbol he has created, though a faithful picture of reality in certain cases, cannot resemble it in all circumstances. Hence, without being discouraged, he again takes up the schematism by which he pictures the gas on which he is experimenting. In order to have this sketch represent the facts he burdens it with new features: it is not enough for the gas to have a certain density, a certain temperature, and a certain dielectric power, to support a certain pressure, and to be placed in an electrical field of a given intensity; in addition, he assigns to it a certain coefficient of magnetization; he takes into account the magnetic field in which the gas is and, connecting all these elements by a group of formulas, he obtains the law of the compressibility and expansion of the polarized and magnetized gas, a more complicated and more comprehensive law than those he had at first obtained, a law which will be verified in an infinity of cases where the former would be falsified; and yet it is a provisional law. Some day the physicist expects to find conditions in which this law will in its turn be faulty; on that day, he will have to take up again the symbolic representation of the gas studied, add new elements to it and enounce a more comprehensive law. The mathematical symbol forged by theory applies to reality as armor to the body of a knight clad in iron: the more complicated the armor, the more supple will the rigid metal seem to be; the multiplication of the pieces that are overlaid like shells assures more perfect contact between the steel and the limbs it protects; but no matter how numerous the fragments composing it, the armor will never be exactly wedded to the human body being modelled.

I know what is going to be said in objection to this. I shall be told that the law of compressibility and expansion formulated at the very first has not in any way been upset by the later experiments; that it remains the law according to which oxygen is compressed and dilated when all electrical and magnetic actions are eliminated; that the physicist's later inquiries have taught us only that it was suitable to join to this law, whose validity was unaffected, the law of the compressibility of an ionized gas and the law of the compressibility of a magnetized gas.

These same persons who take things so obliquely ought to recognize that the original law could lead to serious mistakes if taken without caution, for the domain it governs has to be delimited by the following double restriction: the gas studied is removed from all electrical action as well as magnetic action. Now the necessity for this restriction did not appear at first but was imposed by the experiments we have mentioned. Are such restrictions the only ones which should be imposed on the law's statement? Will not experiments done in the future indicate other restrictions as essential as the former? What physicist would dare to pronounce judgment on this and assert that the present statement is not provisional but final?

The laws of physics are therefore provisional in that the symbols they relate are too simple to represent reality completely. There are always circumstances in which the symbol ceases to picture concrete things and to announce phenomena exactly; the statement of the law must then be accompanied by restrictions which permit one to eliminate these circumstances. It is the progress of physics which brings knowledge of these restrictions; never is it permissible to affirm that we possess a complete enumeration of them or that the list drawn up will not undergo some addition or modification.

This task of continual modification by which the laws of physics avoid more and more adequately the refutations provided by experiment plays such an essential role in the development of the science that we may be permitted to insist somewhat further on its importance and to study its course in a second example.

Here is some water in a vessel. The law of universal attraction teaches us what force acts on each of the particles of this water: this force is the weight of the particle. Mechanics indicates to us what shape the water should assume: whatever the nature and shape of the vessel are, the water should be bounded by a horizontal plane. Look closely at the surface bounding the water: horizontal at a distance from the edge of the vessel, it stops being so in the vicinity of the walls of glass, and rises along these walls; in a narrow tube the water rises very high and becomes altogether concave. There you have the law of universal attraction failing. In order to prevent capillary phenomena from refuting the law of gravitation, it will be necessary to modify it: we shall no longer have to regard the formula of the inverse ratio of the square of the distance as an exact formula but as an approximate one; we shall have to suppose that this formula shows with sufficient precision the attraction of two distant material particles but that it becomes very incorrect when the problem is to express the mutual action of two elements very close to each other; we shall have to introduce into the equations a complementary term which, while complicating them, will make them capable of representing a wider class of phenomena and will permit them to include the motions of heavenly bodies and capillary effects under the same law.

This law will be more comprehensive than Newton's law, but will not be, for all that, safe from all contradiction. At two different points of a liquid mass, let us insert the metallic wires coming from two poles of a battery: there you see the laws of capillarity in disagreement with observation. In order to remove this disagreement, we must again take up the formula for capillary action, and modify and complete it by taking into account the electrical charges carried by the fluid's particles and the forces acting among these ionized particles. Thus, this struggle between reality and the laws of physics will go on indefinitely: to any law that physics formulates, reality will oppose sooner or later the harsh refutation of a fact, but indefatigable physics will touch up, modify, and complicate the refuted law in order to replace it with a more comprehensive law in which the exception raised by the experiment will have found its rule in turn.

Physics makes progress through this unceasing struggle and the work of continually supplementing laws in order to include the exceptions. It was because the laws of weight were contradicted by a piece of amber rubbed by wool that physics created the laws of electrostatics, and because a magnet lifted iron despite these same laws of weight that physics formulated the laws of magnetism; it was because Oersted had found an exception to the laws of electrostatics and of magnetism that Ampère invented the laws of electrodynamics and electromagnetism. Physics does not progress as does geometry, which adds new final and indisputable propositions to the final and indisputable propositions it already possessed; physics makes progress because experiment constantly causes new disagreements to break out between laws and facts, and because physicists constantly touch up and modify laws in order that they may more faithfully represent facts.

5. *The Laws of Physics Are More Detailed than the Laws of Common Sense*

The laws that ordinary non-scientific experience allows us to formulate are general judgments whose meaning is immediate. In the presence of one of these judgments we may ask, "Is it true?" Often the answer is easy; in any case the answer is a definite yes or no. The law recognized as true is so for all time and for all men; it is fixed and absolute.

Scientific laws based on the experiments of physics are symbolic relations whose meaning would remain unintelligible to anyone who did not know physical theories. Since they are symbolic, they are never true or false; like the experiments on which they rest, they are approximate. The degree of approximation of a law, though sufficient today, will become insufficient in the future through the progress of experimental methods; sufficient for the needs of the physicist, it would not satisfy somebody else, so that a law of physics is always provisional and relative. It is provisional also in that it does not connect realities but symbols, and that is because there are always cases where the symbol no longer corresponds to reality; the laws of physics cannot be maintained except by continual retouching and modification.

The problem of the validity of the laws of physics hence poses itself in an entirely different manner, infinitely more complicated and delicate than the problem of the certainty of the laws of common sense. One might be tempted to draw the strange conclusion that the knowledge of the laws of physics constitutes a degree of knowledge inferior to the simple knowledge of the laws of common sense. We are content to reply to those who would deduce this paradoxical conclusion from the foregoing considerations by repeating for the laws of physics what we have said about scientific experiments: A law of physics possesses a certainty much less immediate and much more difficult to estimate than a law of common sense, but it surpasses the latter by the minute and detailed precision of its predictions.

Take the common-sense law "In Paris the sun rises every day in the east, climbs the sky, then comes down and sets in the west" and compare it with the formulas telling us the coordinates of the sun's center at each instant within about a second, and you will be convinced of the accuracy of this proposition.

The laws of physics can acquire this minuteness of detail only by sacrificing something of the fixed and absolute certainty of common-sense laws. There is a sort of balance between precision and certainty: one cannot be increased except to the detriment of the other. The miner who presents me with a stone can tell me without hesitation or qualification that it contains gold; but the chemist who shows me a shiny ingot, telling me, "It is pure gold," has to add the qualification "or nearly pure"; he cannot affirm that the ingot does not retain minute traces of impurities.

A man may swear to tell the truth, but it is not in his power to tell the whole truth and nothing but the truth. "Truth is so subtle a point that our instruments are too blunt to touch it exactly. When they do reach it, they crush the point and bear down around it, more on the false than on the true."[3]

[3] B. Pascal, *Pensées*, ed. Havet, Art. III, No. 3.

CHAPTER VI

PHYSICAL THEORY AND EXPERIMENT

1. *The Experimental Testing of a Theory Does Not Have the Same Logical Simplicity in Physics as in Physiology*

The sole purpose of physical theory is to provide a representation and classification of experimental laws; the only test permitting us to judge a physical theory and pronounce it good or bad is the comparison between the consequences of this theory and the experimental laws it has to represent and classify. Now that we have minutely analyzed the characteristics of a physical experiment and of a physical law, we can establish the principles that should govern the comparison between experiment and theory; we can tell how we shall recognize whether a theory is confirmed or weakened by facts.

When many philosophers talk about experimental sciences, they think only of sciences still close to their origins, e.g., physiology or certain branches of chemistry where the experimenter reasons directly on the facts by a method which is only common sense brought to greater attentiveness but where mathematical theory has not yet introduced its symbolic representations. In such sciences the comparison between the deductions of a theory and the facts of experiment is subject to very simple rules. These rules were formulated in a particularly forceful manner by Claude Bernard, who would condense them into a single principle, as follows:

"The experimenter should suspect and stay away from fixed ideas, and always preserve his freedom of mind.

"The first condition that has to be fulfilled by a scientist who is devoted to the investigation of natural phenomena is to preserve a complete freedom of mind based on philosophical doubt."[1]

If a theory suggests experiments to be done, so much the better: ". . . we can follow our judgment and our thought, give free rein to our imagination provided that all our ideas are only pretexts for instituting new experiments that may furnish us probative facts or

[1] Claude Bernard, *Introduction à la Médecine expérimentale* (Paris, 1865), p. 63. (Translator's note: Translated into English by H. C. Greene, *An Introduction to Experimental Medicine* [New York: Henry Schuman, 1949].)

unexpected and fruitful ones."[2] Once the experiment is done and the results clearly established, if a theory takes them over in order to generalize them, coordinate them, and draw from them new subjects for experiment, still so much the better: ". . . if one is imbued with the principles of experimental method, there is nothing to fear; for so long as the idea is a right one, it will go on being developed; when it is an erroneous idea, experiment is there to correct it."[3] But so long as the experiment lasts, the theory should remain waiting, under strict orders to stay outside the door of the laboratory; it should keep silent and leave the scientist without disturbing him while he faces the facts directly; the facts must be observed without a preconceived idea and gathered with the same scrupulous impartiality, whether they confirm or contradict the predictions of the theory. The report that the observer will give us of his experiment should be a faithful and scrupulously exact reproduction of the phenomena, and should not let us even guess what system the scientist places his confidence in or distrusts.

"Men who have an excessive faith in their theories or in their ideas are not only poorly disposed to make discoveries but they also make very poor observations. They necessarily observe with a preconceived idea and, when they have begun an experiment, they want to see in its results only a confirmation of their theory. Thus they distort observation and often neglect very important facts because they go counter to their goal. That is what made us say elsewhere that we must never do experiments in order to confirm our ideas but merely to check them. . . . But it quite naturally happens that those who believe too much in their own theories do not sufficiently believe in the theories of others. Then the dominant idea of these condemners of others is to find fault with the theories of the latter and to seek to contradict them. The setback for science remains the same. They are doing experiments only in order to destroy a theory instead of doing them in order to look for the truth. They also make poor observations because they take into the results of their experiments only what fits their purpose, by neglecting what is unrelated to it, and by very carefully avoiding whatever might go in the direction of the idea they wish to combat. Thus one is led by two parallel paths to the same result, that is to say, to falsifying science and the facts.

"The conclusion of all this is that it is necessary to obliterate one's

[2] Claude Bernard, *Introduction à la Médecine expérimentale* (Paris, 1865), p. 64.

[3] *ibid.*, p. 70.

opinion as well as that of others when faced with the decisions of the experiment; . . . we must accept the results of experiment just as they present themselves with all that is unforeseen and accidental in them."[4]

Here, for example, is a physiologist who admits that the anterior roots of the spinal nerve contain the motor nerve-fibers and the posterior roots the sensory fibers. The theory he accepts leads him to imagine an experiment: if he cuts a certain anterior root, he ought to be suppressing the mobility of a certain part of the body without destroying its sensibility; after making the section of this root, when he observes the consequences of his operation and when he makes a report of it, he must put aside all his ideas concerning the physiology of the spinal nerve; his report must be a raw description of the facts; he is not permitted to overlook or fail to mention any movement or quiver contrary to his predictions or to attribute it to some secondary cause unless some special experiment has given evidence of this cause; he must, if he does not wish to be accused of scientific bad faith, establish an absolute separation or watertight compartment between the consequences of his theoretical deductions and the establishing of the facts shown by his experiments.

Such a rule is not by any means easily followed; it requires of the scientist an absolute detachment from his own thought and a complete absence of animosity when confronted with the opinion of another person; neither vanity nor envy ought to be countenanced by him. As Bacon put it, he should never show eyes lustrous with human passions. Freedom of mind, which constitutes the sole principle of experimental method, according to Claude Bernard, does not depend merely on intellectual conditions, but also on moral conditions, making its practice rarer and more meritorious.

But if experimental method as just described is difficult to practice, the logical analysis of it is very simple. This is no longer the case when the theory to be subjected to test by the facts is not a theory of physiology but a theory of physics. In the latter case, in fact, it is impossible to leave outside the laboratory door the theory that we wish to test, for without theory it is impossible to regulate a single instrument or to interpret a single reading. We have seen that in the mind of the physicist there are constantly present two sorts of apparatus: one is the concrete apparatus in glass and metal, manipulated by him, the other is the schematic and abstract apparatus which theory substitutes for the concrete apparatus and

[4] *ibid.*, p. 67.

on which the physicist does his reasoning. For these two ideas are indissolubly connected in his intelligence, and each necessarily calls on the other; the physicist can no sooner conceive the concrete apparatus without associating with it the idea of the schematic apparatus than a Frenchman can conceive an idea without associating it with the French word expressing it. This radical impossibility, preventing one from dissociating physical theories from the experimental procedures appropriate for testing these theories, complicates this test in a singular way, and obliges us to examine the logical meaning of it carefully.

Of course, the physicist is not the only one who appeals to theories at the very time he is experimenting or reporting the results of his experiments. The chemist and the physiologist when they make use of physical instruments, e.g., the thermometer, the manometer, the calorimeter, the galvanometer, and the saccharimeter, implicitly admit the accuracy of the theories justifying the use of these pieces of apparatus as well as of the theories giving meaning to the abstract ideas of temperature, pressure, quantity of heat, intensity of current, and polarized light, by means of which the concrete indications of these instruments are translated. But the theories used, as well as the instruments employed, belong to the domain of physics; by accepting with these instruments the theories without which their readings would be devoid of meaning, the chemist and the physiologist show their confidence in the physicist, whom they suppose to be infallible. The physicist, on the other hand, is obliged to trust his own theoretical ideas or those of his fellow-physicists. From the standpoint of logic, the difference is of little importance; for the physiologist and chemist as well as for the physicist, the statement of the result of an experiment implies, in general, an act of faith in a whole group of theories.

2. *An Experiment in Physics Can Never Condemn an Isolated Hypothesis but Only a Whole Theoretical Group*

The physicist who carries out an experiment, or gives a report of one, implicitly recognizes the accuracy of a whole group of theories. Let us accept this principle and see what consequences we may deduce from it when we seek to estimate the role and logical import of a physical experiment.

In order to avoid any confusion we shall distinguish two sorts of experiments: experiments of *application*, which we shall first just mention, and experiments of *testing*, which will be our chief concern.

You are confronted with a problem in physics to be solved practically; in order to produce a certain effect you wish to make use of knowledge acquired by physicists; you wish to light an incandescent bulb; accepted theories indicate to you the means for solving the problem; but to make use of these means you have to secure certain information; you ought, I suppose, to determine the electromotive force of the battery of generators at your disposal; you measure this electromotive force: that is what I call an experiment of application. This experiment does not aim at discovering whether accepted theories are accurate or not; it merely intends to draw on these theories. In order to carry it out, you make use of instruments that these same theories legitimize; there is nothing to shock logic in this procedure.

But experiments of application are not the only ones the physicist has to perform; only with their aid can science aid practice, but it is not through them that science creates and develops itself; besides experiments of application, we have experiments of testing.

A physicist disputes a certain law; he calls into doubt a certain theoretical point. How will he justify these doubts? How will he demonstrate the inaccuracy of the law? From the proposition under indictment he will derive the prediction of an experimental fact; he will bring into existence the conditions under which this fact should be produced; if the predicted fact is not produced, the proposition which served as the basis of the prediction will be irremediably condemned.

F. E. Neumann assumed that in a ray of polarized light the vibration is parallel to the plane of polarization, and many physicists have doubted this proposition. How did O. Wiener undertake to transform this doubt into a certainty in order to condemn Neumann's proposition? He deduced from this proposition the following consequence: If we cause a light beam reflected at 45° from a plate of glass to interfere with the incident beam polarized perpendicularly to the plane of incidence, there ought to appear alternately dark and light interference bands parallel to the reflecting surface; he brought about the conditions under which these bands should have been produced and showed that the predicted phenomenon did not appear, from which he concluded that Neumann's proposition is false, viz., that in a polarized ray of light the vibration is not parallel to the plane of polarization.

Such a mode of demonstration seems as convincing and as irrefutable as the proof by reduction to absurdity customary among mathematicians; moreover, this demonstration is copied from the reduction to absurdity, experimental contradiction playing the same role in one as logical contradiction plays in the other.

Indeed, the demonstrative value of experimental method is far from being so rigorous or absolute: the conditions under which it functions are much more complicated than is supposed in what we have just said; the evaluation of results is much more delicate and subject to caution.

A physicist decides to demonstrate the inaccuracy of a proposition; in order to deduce from this proposition the prediction of a phenomenon and institute the experiment which is to show whether this phenomenon is or is not produced, in order to interpret the results of this experiment and establish that the predicted phenomenon is not produced, he does not confine himself to making use of the proposition in question; he makes use also of a whole group of theories accepted by him as beyond dispute. The prediction of the phenomenon, whose nonproduction is to cut off debate, does not derive from the proposition challenged if taken by itself, but from the proposition at issue joined to that whole group of theories; if the predicted phenomenon is not produced, not only is the proposition questioned at fault, but so is the whole theoretical scaffolding used by the physicist. The only thing the experiment teaches us is that among the propositions used to predict the phenomenon and to establish whether it would be produced, there is at least one error; but where this error lies is just what it does not tell us. The physicist may declare that this error is contained in exactly the proposition he wishes to refute, but is he sure it is not in another proposition? If he is, he accepts implicitly the accuracy of all the other propositions he has used, and the validity of his conclusion is as great as the validity of his confidence.

Let us take as an example the experiment imagined by Zenker and carried out by O. Wiener. In order to predict the formation of bands in certain circumstances and to show that these did not appear, Wiener did not make use merely of the famous proposition of F. E. Neumann, the proposition which he wished to refute; he did not merely admit that in a polarized ray vibrations are parallel to the plane of polarization; but he used, besides this, propositions, laws, and hypotheses constituting the optics commonly accepted: he admitted that light consists in simple periodic vibrations, that these vibrations are normal to the light ray, that at each point the mean kinetic energy of the vibratory motion is a measure of the intensity of light, that the more or less complete attack of the gelatine coating on a photographic plate indicates the various degrees of this intensity. By joining these propositions, and many others that would take too long to enumerate, to Neumann's proposition, Wiener was able to formulate a forecast and establish that the experiment belied it. If he attributed this solely to Neumann's proposition, if it alone bears the responsibility for the error this negative result has put in evidence, then Wiener was taking all the other propositions he invoked as beyond doubt. But this assurance is not imposed as a matter of logical necessity; nothing stops us from taking Neumann's proposition as accurate and shifting the weight of the experimental contradiction to some other proposition of the commonly accepted optics; as H. Poincaré has shown, we can very easily rescue Neumann's hypothesis from the grip of Wiener's experiment on the condition that we abandon in exchange the hypothesis which takes the mean kinetic energy as the measure of the light intensity; we may, without being contradicted by the experiment, let the vibration be parallel to the plane of polarization, provided that we measure the light intensity by the mean potential energy of the medium deforming the vibratory motion.

These principles are so important that it will be useful to apply them to another example; again we choose an experiment regarded as one of the most decisive ones in optics.

We know that Newton conceived the emission theory for optical phenomena. The emission theory supposes light to be formed of extremely thin projectiles, thrown out with very great speed by the sun and other sources of light; these projectiles penetrate all transparent bodies; on account of the various parts of the media through which they move, they undergo attractions and repulsions; when the distance separating the acting particles is very small these actions are very powerful, and they vanish when the masses between which they act are appreciably far from each other. These essential hypotheses joined to several others, which we pass over without mention, lead to the formulation of a complete theory of reflection and refraction of light; in particular, they imply the following proposition: The index of refraction of light passing from one medium into another is equal to the velocity of the light projectile within the medium it penetrates, divided by the velocity of the same projectile in the medium it leaves behind.

This is the proposition that Arago chose in order to show that the theory of emission is in contradiction with the facts. From this proposition a second follows: Light travels faster in water than in air. Now Arago had indicated an appropriate procedure for comparing the velocity of light in air with the velocity of light in water; the procedure, it is true, was inapplicable, but Foucault modified the experiment in such a way that it could be carried out; he found that the light was propagated less rapidly in water than in air. We may conclude from this, with Foucault, that the system of emission is incompatible with the facts.

I say the *system* of emission and not the *hypothesis* of emission; in fact, what the experiment declares stained with error is the whole group of propositions accepted by Newton, and after him by Laplace and Biot, that is, the whole theory from which we deduce the relation between the index of refraction and the velocity of light in various media. But in condemning this system as a whole by declaring it stained with error, the experiment does not tell us where the error lies. Is it in the fundamental hypothesis that light consists in projectiles thrown out with great speed by luminous bodies? Is it in some other assumption concerning the actions experienced by light corpuscles due to the media through which they move? We know nothing about that. It would be rash to believe, as Arago seems to have thought, that Foucault's experiment condemns once and for all the very hypothesis of emission, i.e., the assimilation of a ray of light to a swarm of projectiles. If physicists had attached some value to this task, they would undoubtedly have succeeded in founding on this assumption a system of optics that would agree with Foucault's experiment.

In sum, the physicist can never subject an isolated hypothesis to experimental test, but only a whole group of hypotheses; when the experiment is in disagreement with his predictions, what he learns is that at least one of the hypotheses constituting this group is unacceptable and ought to be modified; but the experiment does not designate which one should be changed.

We have gone a long way from the conception of the experimental method arbitrarily held by persons unfamiliar with its actual functioning. People generally think that each one of the hypotheses employed in physics can be taken in isolation, checked by experiment, and then, when many varied tests have established its validity, given a definitive place in the system of physics. In reality, this is not the case. Physics is not a machine which lets itself be taken apart; we cannot try each piece in isolation and, in order to adjust it, wait until its solidity has been carefully checked. Physical science is a system that must be taken as a whole; it is an organism in which one part cannot be made to function except when the parts that are most remote from it are called into play, some more so than others, but all to some degree. If something goes wrong, if some discomfort is felt in the functioning of the organism, the physicist will have to ferret out through its effect on the entire system which organ needs to be remedied or modified without the possibility of isolating this organ and examining it apart. The watchmaker to whom you give a watch that has stopped separates all the wheelworks and examines them one by one until he finds the part that is defective or broken. The doctor to whom a patient appears cannot dissect him in order to establish his diagnosis; he has to guess the seat and cause of the ailment solely by inspecting disorders affecting the whole body. Now, the physicist concerned with remedying a limping theory resembles the doctor and not the watchmaker.

3. *A "Crucial Experiment" Is Impossible in Physics*

Let us press this point further, for we are touching on one of the essential features of experimental method, as it is employed in physics.

Reduction to absurdity seems to be merely a means of refutation, but it may become a method of demonstration: in order to demonstrate the truth of a proposition it suffices to corner anyone who would admit the contradictory of the given proposition into admitting an absurd consequence. We know to what extent the Greek geometers drew heavily on this mode of demonstration.

Those who assimilate experimental contradiction to reduction to absurdity imagine that in physics we may use a line of argument similar to the one Euclid employed so frequently in geometry. Do you wish to obtain from a group of phenomena a theoretically certain and indisputable explanation? Enumerate all the hypotheses that can be made to account for this group of phenomena; then, by experimental contradiction eliminate all except one; the latter will no longer be a hypothesis, but will become a certainty.

Suppose, for instance, we are confronted with only two hypotheses. Seek experimental conditions such that one of the hypotheses forecasts the production of one phenomenon and the other the production of quite a different effect; bring these conditions into existence and observe what happens; depending on whether you observe the first or the second of the predicted phenomena, you will condemn the second or the first hypothesis; the hypothesis not condemned will be henceforth indisputable; debate will be cut off, and a new truth will be acquired by science. Such is the experimental test that the author of the *Novum Organum* called the *"fact of the cross*, borrowing this expression from the crosses which at an intersection indicate the various roads."

We are confronted with two hypotheses concerning the nature of light; for Newton, Laplace, or Biot light consisted of projectiles hurled with extreme speed, but for Huygens, Young, or Fresnel light consisted of vibrations whose waves are propagated within an ether. These are the only two possible hypotheses as far as one can see: either the motion is carried away by the body it excites and remains attached to it, or else it passes from one body to another. Let us pursue the first hypothesis; it declares that light travels more quickly in water than in air; but if we follow the second, it declares that light travels more quickly in air than in water. Let us set up Foucault's apparatus; we set into motion the turning mirror; we see two luminous spots formed before us, one colorless, the other greenish. If the greenish band is to the left of the colorless one, it means that light travels faster in water than in air, and that the hypothesis of vibrating waves is false. If, on the contrary, the greenish band is to the right of the colorless one, that means that light travels faster in air than in water, and that the hypothesis of emissions is condemned. We look through the magnifying glass used to examine the two luminous spots, and we notice that the greenish spot is to the right of the colorless one; the debate is over; light is not a body, but a vibratory wave motion propagated by the ether; the emission hypothesis has had its day; the wave hypothesis has been put beyond doubt, and the crucial experiment has made it a new article of the scientific credo.

What we have said in the foregoing paragraph shows how mistaken we should be to attribute to Foucault's experiment so simple a meaning and so decisive an importance; for it is not between two hypotheses, the emission and wave hypotheses, that Foucault's experiment judges trenchantly; it decides rather between two sets of theories each of which has to be taken as a whole, i.e., between two entire systems, Newton's optics and Huygens' optics.

But let us admit for a moment that in each of these systems everything is compelled to be necessary by strict logic, except a single hypothesis; consequently, let us admit that the facts, in condemning one of the two systems, condemn once and for all the single doubtful assumption it contains. Does it follow that we can find in the "crucial experiment" an irrefutable procedure for transforming one of the two hypotheses before us into a demonstrated truth? Between two contradictory theorems of geometry there is no room for a third judgment; if one is false, the other is necessarily true.

Do two hypotheses in physics ever constitute such a strict dilemma? Shall we ever dare to assert that no other hypothesis is imaginable? Light may be a swarm of projectiles, or it may be a vibratory motion whose waves are propagated in a medium; is it forbidden to be anything else at all? Arago undoubtedly thought so when he formulated this incisive alternative: Does light move more quickly in water than in air? "Light is a body. If the contrary is the case, then light is a wave." But it would be difficult for us to take such a decisive stand; Maxwell, in fact, showed that we might just as well attribute light to a periodical electrical disturbance that is propagated within a dielectric medium.

Unlike the reduction to absurdity employed by geometers, experimental contradiction does not have the power to transform a physical hypothesis into an indisputable truth; in order to confer this power on it, it would be necessary to enumerate completely the various hypotheses which may cover a determinate group of phenomena; but the physicist is never sure he has exhausted all the imaginable assumptions. The truth of a physical theory is not decided by heads or tails.

4. *Criticism of the Newtonian Method. First Example: Celestial Mechanics*

It is illusory to seek to construct by means of experimental contradiction a line of argument in imitation of the reduction to absurdity; but the geometer is acquainted with other methods for attaining certainty than the method of reducing to an absurdity; the direct demonstration in which the truth of a proposition is established by itself and not by the refutation of the contradictory proposition seems to him the most perfect of arguments. Perhaps physical theory would be more fortunate in its attempts if it sought to imitate direct demonstration. The hypotheses from which it starts and develops its conclusions would then be tested one by one; none would have to be accepted until it presented all the certainty that experimental method can confer on an abstract and general proposition; that is to say, each would necessarily be either a law drawn from observation by the sole use of those two intellectual operations called induction and generalization, or else a corollary mathematically deduced from such laws. A theory based on such hypotheses would then not present anything arbitrary or doubtful; it would deserve all the confidence merited by the faculties which serve us in formulating natural laws.

It was this sort of physical theory that Newton had in mind when, in the "General Scholium" which crowns his *Principia,* he rejected so vigorously as outside of natural philosophy any hypothesis that induction did not extract from experiment; when he asserted that in a sound physics every proposition should be drawn from phenomena and generalized by induction.

The ideal method we have just described therefore deserves to be named the Newtonian method. Besides, did not Newton follow this method when he established the system of universal attraction, thus adding to his precepts the most magnificent of examples? Is not his theory of gravitation derived entirely from the laws which were revealed to Kepler by observation, laws which problematic reasoning transforms and whose consequences induction generalizes?

This first law of Kepler's, "The radial vector from the sun to a planet sweeps out an area proportional to the time during which the planet's motion is observed," did, in fact, teach Newton that each planet is constantly subjected to a force directed toward the sun.

The second law of Kepler's, "The orbit of each planet is an ellipse having the sun at one focus," taught him that the force attracting a given planet varies with the distance of this planet from the sun, and that it is in an inverse ratio to the square of this distance.

The third law of Kepler's, "The squares of the periods of revolution of the various planets are proportional to the cubes of the major axes of their orbits," showed him that different planets would, if they were brought to the same distance from the sun, undergo in relation to it attractions proportional to their respective masses.

The experimental laws established by Kepler and transformed by geometric reasoning yield all the characteristics present in the action exerted by the sun on a planet; by induction Newton generalized the result obtained; he allowed this result to express the law according to which any portion of matter acts on any other portion whatsoever, and he formulated this great principle: "Any two bodies whatsoever attract each other with a force which is proportional to the product of their masses and in inverse ratio to the square of the distance between them." The principle of universal gravitation was found, and it was obtained, without any use having been made of any fictive hypothesis, by the inductive method the plan of which Newton outlined.

Let us again examine this application of the Newtonian method, this time more closely; let us see if a somewhat strict logical analysis will leave intact the appearance of rigor and simplicity that this very summary exposition attributes to it.

In order to assure this discussion of all the clarity it needs, let us begin by recalling the following principle, familiar to all those who deal with mechanics: We cannot speak of the force which attracts a body in given circumstances before we have designated the supposedly fixed term of reference to which we relate the motion of all bodies; when we change this point of reference or term of comparison, the force representing the effect produced on the observed body by the other bodies surrounding it changes in direction and magnitude according to the rules stated by mechanics with precision.

That posited, let us follow Newton's reasoning.

Newton first took the sun as the fixed point of reference; he considered the motions affecting the different planets by reference to the sun; he admitted Kepler's laws as governing these motions, and derived the following proposition: If the sun is the point of reference in relation to which all forces are compared, each planet is subjected to a force directed toward the sun, a force proportional to the mass of the planet and to the inverse square of its distance from the sun. Since the latter is taken as the reference point, it is not subject to any force.

In an analogous manner Newton studied the motion of the satellites and for each of these he chose as a fixed reference point the planet which the satellite accompanies, the earth in the case of the moon, Jupiter in the case of the masses moving around Jupiter. Laws just like Kepler's were taken as governing these motions, from which it follows that we can formulate the following proposition: If we take as a fixed reference point the planet accompanied by a satellite, this satellite is subject to a force directed toward the planet varying inversely with the square of the distance. If, as happens with Jupiter, the same planet possesses several satellites, these satellites, were they at the same distance from the planet, would be acted on by the latter with forces proportional to their respective masses. The planet is itself not acted on by the satellite.

Such, in very precise form, are the propositions which Kepler's laws of planetary motion and the extension of these laws to the motions of satellites authorize us to formulate. For these propositions Newton substituted another which may be stated as follows: Any two celestial bodies whatsoever exert on each other a force of attraction in the direction of the straight line joining them, a force proportional to the product of their masses and to the inverse square of the distance between them. This statement presupposes all motions and forces to be related to the same reference point; the latter is an ideal standard of reference which may well be conceived by the geometer but which does not characterize in an exact and concrete manner the position in the sky of any body.

Is this principle of universal gravitation merely a generalization of the two statements provided by Kepler's laws and their extension to the motion of satellites? Can induction derive it from these two statements? Not at all. In fact, not only is it more general than these two statements and unlike them, but it contradicts them. The student of mechanics who accepts the principle of universal attraction can calculate the magnitude and direction of the forces between the various planets and the sun when the latter is taken as the reference point, and if he does he finds that these forces are not what our first statement would require. He can determine the magnitude and direction of each of the forces between Jupiter and its satellites when we refer all the motions to the planet, assumed to be fixed, and if he does he notices that these forces are not what our second statement would require.

The principle of universal gravity, very far from being derivable by generalization and induction from the observational laws of Kepler, formally contradicts these laws. If Newton's theory is correct, Kepler's laws are necessarily false.

Kepler's laws based on the observation of celestial motions do not transfer their immediate experimental certainty to the principle of universal weight, since if, on the contrary, we admit the absolute exactness of Kepler's laws, we are compelled to reject the proposition on which Newton based his celestial mechanics. Far from adhering to Kepler's laws, the physicist who claims to justify the theory of universal gravitation finds that he has, first of all, to resolve a difficulty in these laws: he has to prove that his theory, incompatible with the exactness of Kepler's laws, subjects the motions of the planets and satellites to other laws scarcely different enough from the first laws for Tycho Brahé, Kepler, and their contemporaries to have been able to discern the deviations between the Keplerian and Newtonian orbits. This proof derives from the circumstances that the sun's mass is very large in relation to the masses of the various planets and the mass of a planet is very large in relation to the masses of its satellites.

Therefore, if the certainty of Newton's theory does not emanate from the certainty of Kepler's laws, how will this theory prove its validity? It will calculate, with all the high degree of approximation that the constantly perfected methods of algebra involve, the perturbations which at each instant remove every heavenly body from the orbit assigned to it by Kepler's laws; then it will compare the calculated perturbations with the perturbations observed by means of the most precise instruments and the most scrupulous methods. Such a comparison will not only bear on this or that part of the Newtonian principle, but will involve all its parts at the same time; with those it will also involve all the principles of dynamics; besides, it will call in the aid of all the propositions of optics, the statics of gases, and the theory of heat, which are necessary to justify the properties of telescopes in their construction, regulation, and correction, and in the elimination of the errors caused by diurnal or annual aberration and by atmospheric refraction. It is no longer a matter of taking, one by one, laws justified by observation, and raising each of them by induction and generalization to the rank of a principle; it is a matter of comparing the corollaries of a whole group of hypotheses to a whole group of facts.

Now, if we seek out the causes which have made the Newtonian method fail in this case for which it was imagined and which seemed to be the most perfect application for it, we shall find them in that double character of any law made use of by theoretical physics: This law is symbolic and approximate.

Undoubtedly, Kepler's laws bear quite directly on the very objects of astronomical observation; they are as little symbolic as possible. But in this purely experimental form they remain inappropriate for suggesting the principle of universal gravitation; in order to acquire this fecundity they must be transformed and must yield the characters of the forces by which the sun attracts the various planets.

Now this new form of Kepler's laws is a symbolic form; only dynamics gives meanings to the words "force" and "mass," which serve to state it, and only dynamics permits us to substitute the new symbolic formulas for the old realistic formulas, to substitute statements relative to "forces" and "masses" for laws relative to orbits. The legitimacy of such a substitution implies full confidence in the laws of dynamics.

And in order to justify this confidence let us not proceed to claim that the laws of dynamics were beyond doubt at the time Newton made use of them in symbolically translating Kepler's laws; that they had received enough empirical confirmation to warrant the support of reason. In fact, the laws of dynamics had been subjected up to that time to only very limited and very crude tests. Even their enunciations had remained very vague and involved; only in Newton's *Principia* had they been for the first time formulated in a precise manner. It was in the agreement of the facts with the celestial mechanics which Newton's labors gave birth to that they received their first convincing verification.

Thus the translation of Kepler's laws into symbolic laws, the only kind useful for a theory, presupposed the prior adherence of the physicist to a whole group of hypotheses. But, in addition, Kepler's laws being only approximate laws, dynamics permitted giving them an infinity of different symbolic translations. Among these various forms, infinite in number, there is one and only one which agrees with Newton's principle. The observations of Tycho Brahé, so felicitously reduced to laws by Kepler, permit the theorist to choose this form, but they do not constrain him to do so, for there is an infinity of others they permit him to choose.

The theorist cannot, therefore, be content to invoke Kepler's laws in order to justify his choice. If he wishes to prove that the principle he has adopted is truly a principle of natural classification for celestial motions, he must show that the observed perturbations are in agreement with those which had been calculated in advance; he has to show how from the course of Uranus he can deduce the existence and position of a new planet, and find Neptune in an assigned direction at the end of his telescope.

5. *Criticism of the Newtonian Method (Continued). Second Example: Electrodynamics*

Nobody after Newton except Ampère has more clearly declared that all physical theory should be derived from experience by induction only; no work has been more closely modelled after Newton's *Philosophiae naturalis Principia mathematica* than Ampère's *Théorie mathématique des phénomènes électrodynamiques uniquement déduite de l'expérience.*

"The epoch marked by the works of Newton in the history of the sciences is not only one of the most important discoveries that man has made concerning the causes of the great phenomena of nature, but it is also the epoch in which the human mind opened a new route in the sciences whose object is the study of these phenomena."

These are the lines with which Ampère began the exposition of his *Théorie mathématique*; he continued in the following terms:

"Newton was far from thinking" that the law of universal weight "could be discovered by starting from more or less plausible abstract considerations. He established the fact that it had to be deduced from observed facts, or rather from those empirical laws which, like those of Kepler, are but results generalized from a great number of facts.

"To observe the facts first, to vary their circumstances as far as possible, to make precise measurements along with this first task in order to deduce from them general laws based only on experience, and to deduce from these laws, independently of any hypothesis about the nature of the forces producing the phenomena, the mathematical value of these forces, i.e., the formula representing them—that is the course Newton followed. It has been generally adopted in France by the scientists to whom physics owes the enormous progress it has made in recent times, and it has served me as a guide in all my research on electrodynamic phenomena. I have consulted only experience in order to establish the laws of these phenomena, and I have deduced from them the formula which can only represent the forces to which they are due; I have made no investigation about the cause itself assignable to these forces, well convinced that any investigation of this kind should be preceded simply by experimental knowledge of the laws and of the determination, deduced solely from these laws, of the value of the elementary force."

Neither very close scrutiny nor great perspicacity is needed in order to recognize that the *Théorie mathématique des phénomènes électrodynamiques* does not in any way proceed according to the method prescribed by Ampère and to see that it is not "deduced only from experience" (*uniquement déduite de l'expérience*). The facts of experience taken in their primitive rawness cannot serve mathematical reasoning; in order to feed this reasoning they have to be transformed and put into a symbolic form. This transformation Ampère did make them undergo. He was not content merely with reducing the metal apparatus in which currents flow to simple geometric figures; such an assimilation imposes itself too naturally to give way to any serious doubt. Neither was he content merely to use the notion of force, borrowed from mechanics, and various theorems constituting this science; at the time he wrote, these theorems might be considered as beyond dispute. Besides all this, he appealed to a whole set of entirely new hypotheses which are entirely gratuitous and sometimes even rather surprising. Foremost among these hypotheses it is appropriate to mention the intellectual operation by which he decomposed into infinitely small elements the electric current, which, in reality, cannot be broken without ceasing to exist; then the supposition that all real electrodynamic actions are resolved into fictive actions involving the pairs that the elements of current form, one pair at a time; then the postulate that the mutual actions of two elements are reduced to two forces applied to the elements in the direction of the straight line joining them, forces equal and opposite in direction; then the postulate that the distance between two elements enters simply into the formula of their mutual action by the inverse of a certain power.

These diverse assumptions are so little self-evident and so little necessary that several of them have been criticized or rejected by Ampère's successors; other hypotheses equally capable of translating symbolically the fundamental experiments of electrodynamics have been proposed by other physicists, but none of them has succeeded in giving this translation without formulating some new postulate, and it would be absurd to claim to do so.

The necessity which leads the physicist to translate experimental facts symbolically before introducing them into his reasoning, renders the purely inductive path Ampère drew impracticable; this path is also forbidden to him because each of the observed laws is not exact but merely approximate.

Ampère's experiments have the grossest degree of approximation. He gave a symbolic translation of the facts observed in a form appropriate for the success of his theory, but how easily he might have taken advantage of the uncertainty of the observations in order to give quite a different translation! Let us listen to Wilhelm Weber:

"Ampère made a point of expressly indicating in the title of his memoir that his mathematical theory of electrodynamic phenomena is *deduced only from experiment,* and indeed in his book we find expounded in detail the simple as well as ingenious method which led him to his goal. There we find, presented with all the precision and scope desirable, the exposition of his experiments, the deductions that he draws from them for theory, and the description of the instruments he employs. But in fundamental experiments, such as we have here, it is not enough to indicate the general meaning of an experiment, to describe the instruments used in performing it, and to tell in a general way that it has yielded the result expected; it is indispensable to go into the details of the experiment itself, to say how often it has been repeated, how the conditions were modified, and what the effect of these modifications has been; in a word, to compose a sort of brief of all the circumstances permitting the reader to sit in judgment on the degree of reliability and certainty of the result. Ampère does *not* give these precise details concerning his experiments, and the demonstration of the fundamental law of electrodynamics still awaits this indispensable supplementation. The fact of the mutual attraction of two conducting wires has been verified over and over again and is beyond all dispute; but these verifications have always been made under conditions and by such means that no *quantitative* measurement was possible and these measurements are far from having reached the degree of precision required for considering the law of these phenomena demonstrated.

"More than once, Ampère has drawn from the *absence* of any electrodynamic action the same consequences as from a measurement that would have given him a result equal to *zero*, and by this artifice, with great sagacity and with even greater skill, he has succeeded in bringing together the data necessary for the establishment and demonstration of his theory; but these *negative* experiments with which we must be content in the absence of direct *positive* measurements," those experiments in which all passive resistances, all friction, all causes of error tend precisely to produce the effect we wish to observe, "cannot have all the value or demonstrative force of those positive measurements, especially when they are not obtained with the procedures and under the conditions of true measurement, which are moreover impossible to obtain with the instruments Ampère has employed."[5]

Experiments with so little precision leave the physicist with the problem of choosing between an infinity of equally possible symbolic translations, and confer no certainty on a choice they do not impose; only intuition, guessing the form of theory to be established, directs this choice. This role of intuition is particularly important in the work of Ampère; it suffices to run through the writings of this great geometer in order to recognize that his fundamental formula of electrodynamics was found quite completely by a sort of divination, that his experiments were thought up by him as afterthoughts and quite purposefully combined so that he might be able to expound according to the Newtonian method a theory that he had constructed by a series of postulates.

Besides, Ampère had too much candor to dissimulate very learnedly that what was artificial in his exposition was *entirely deduced from experiment*; at the end of his *Théorie mathématique des phénomènes électrodynamiques* he wrote the following lines: "I think I ought to remark in finishing this memoir that I have not yet had the time to construct the instruments represented in Diagram 4 of

[5] Wilhelm Weber, *Electrodynamische Maassbestimmungen* (Leipzig, 1846). Translated into French in *Collection de Mémoires relatifs à la Physique* (Société française de Physique), Vol. III: *Mémoires sur l'Electrodynamique.*

the first plate and in Diagram 20 of the second plate. The experiments for which they were intended have not yet been done." Now the first of the two sets of apparatus in question aimed to bring into existence the last of the four fundamental cases of equilibrium which are like columns in the edifice constructed by Ampère: it is with the aid of the experiment for which this apparatus was intended that we were to determine the power of the distance according to which electrodynamic actions proceed. Very far from its being the case that Ampère's electrodynamic theory was *entirely deduced from experiment*, experiment played a very feeble role in its formation: it was merely the occasion which awakened the intuition of this physicist of genius, and his intuition did the rest.

It was through the research of Wilhelm Weber that the very intuitive theory of Ampère was first subjected to a detailed comparison with the facts; but this comparison was not guided by the Newtonian method. Weber deduced from Ampère's theory, taken as a whole, certain effects capable of being calculated; the theorems of statics and of dynamics, and also even certain propositions of optics, permitted him to conceive an apparatus, the electrodynamometer, by means of which these same effects may be subjected to precise measurements; the agreement of the calculated predictions with the results of the measurements no longer, then, confirms this or that isolated proposition of Ampère's theory, but the whole set of electrodynamical, mechanical, and optical hypotheses that must be invoked in order to interpret each of Weber's experiments.

Hence, where Newton had failed, Ampère in his turn just stumbled. That is because two inevitable rocky reefs make the purely inductive course impracticable for the physicist. In the first place, no experimental law can serve the theorist before it has undergone an interpretation transforming it into a symbolic law; and this interpretation implies adherence to a whole set of theories. In the second place, no experimental law is exact but only approximate, and is therefore susceptible to an infinity of distinct symbolic translations; and among all these translations the physicist has to choose one which will provide him with a fruitful hypothesis without his choice being guided by experiment at all.

This criticism of the Newtonian method brings us back to the conclusions to which we have already been led by the criticism of experimental contradiction and of the crucial experiment. These conclusions merit our formulating them with the utmost clarity. Here they are:

To seek to separate each of the hypotheses of theoretical physics from the other assumptions on which this science rests in order to subject it in isolation to observational test is to pursue a chimera; for the realization and interpretation of no matter what experiment in physics imply adherence to a whole set of theoretical propositions.

The only experimental check on a physical theory which is not illogical consists in comparing the *entire system of the physical theory with the whole group of experimental laws*, and in judging whether the latter is represented by the former in a satisfactory manner.

6. *Consequences Relative to the Teaching of Physics*

Contrary to what we have made every effort to establish, it is generally accepted that each hypothesis of physics may be separated from the group and subjected in isolation to experimental test. Of course, from this erroneous principle false consequences are deduced concerning the method by which physics should be taught. People would like the professor to arrange all the hypotheses of physics in a certain order, to take the first one, enounce it, expound its experimental verifications, and then when the latter have been recognized as sufficient, declare the hypothesis accepted. Better still, people would like him to formulate this first hypothesis by inductive generalization of a purely experimental law; he would begin this operation again on the second hypothesis, on the third, and so on until all of physics was constituted. Physics would be taught as geometry is: hypotheses would follow one another as theorems follow one another; the experimental test of each assumption would replace the demonstration of each proposition; nothing which is not drawn from facts or immediately justified by facts would be promulgated.

Such is the ideal which has been proposed by many teachers, and which several perhaps think they have attained. There is no lack of authoritative voices inviting them to the pursuit of this ideal. M. Poincaré says: "It is important not to multiply hypotheses excessively, but to make them only one after the other. If we construct a theory based on multiple hypotheses, and experiment condemns the theory, which one among our premises is it necessary to change? It will be impossible to know. And if, on the other hand, the experiment succeeds, shall we think we have verified all these hypotheses at the same time? Shall we think we have determined several unknowns with a single equation?"[6]

[6] H. Poincaré, *Science et Hypothèse*, p. 179.

In particular, the purely inductive method whose laws Newton formulated is given by many physicists as the only method permitting one to expound rationally the science of nature. Gustave Robin says: "The science we shall make will be only a combination of simple inductions suggested by experience. As to these inductions, we shall formulate them always in propositions easy to retain and *susceptible of direct verification*, never losing sight of the fact that *a hypothesis cannot be verified by its consequences*."[7] This is the Newtonian method recommended if not prescribed for those who plan to teach physics in the secondary schools. They are told: "The procedures of mathematical physics are not adequate for secondary-school instruction, for they consist in starting from hypotheses or from definitions posited a priori in order to deduce from them conclusions which will be subjected to experimental check. This method may be suitable for specialized classes in mathematics, but it is wrong to apply it at present in our elementary courses in mechanics, hydrostatics, and optics. Let us replace it by the inductive method."[8]

The arguments we have developed have established more than sufficiently the following truth: It is as impracticable for the physicist to follow the inductive method whose practice is recommended to him as it is for the mathematician to follow that perfect deductive method which would consist in defining and demonstrating everything, a method of inquiry to which certain geometers seem passionately attached, although Pascal properly and rigorously disposed of it a long time ago. Therefore, it is clear that those who claim to unfold the series of physical principles by means of this method are naturally giving an exposition of it that is faulty at some point.

Among the vulnerable points noticeable in such an exposition, the most frequent and, at the same time, the most serious, because of the false ideas it deposits in the minds of students, is the "fictitious experiment." Obliged to invoke a principle which has not really been drawn from facts or obtained by induction, and averse, moreover, to offering this principle for what it is, namely, a postulate, the physicist invents an imaginary experiment which, were it carried out with success, would possibly lead to the principle whose justification is desired.

[7] G. Robin, *Oeuvres scientifiques. Thermodynamique générale* (Paris, 1901), Introduction, p. xii.

[8] Note on a lecture of M. Joubert, inspector-general of secondary-school instruction, *L'Enseignement secondaire*, April 15, 1903.

To invoke such a fictitious experiment is to offer an experiment to be done for an experiment done; this is justifying a principle not by means of facts observed but by means of facts whose existence is predicted, and this prediction has no other foundation than the belief in the principle supported by the alleged experiment. Such a method of demonstration implicates him who trusts it in a vicious circle; and he who teaches it without making it exactly clear that the experiment cited has not been done commits an act of bad faith.

At times the fictitious experiment described by the physicist could not, if we attempted to bring it about, yield a result of any precision; the very indecisive and rough results it would produce could undoubtedly be put into agreement with the proposition claimed to be warranted; but they would agree just as well with certain very different propositions; the demonstrative value of such an experiment would therefore be very weak and subject to caution. The experiment that Ampère imagined in order to prove that electrodynamic actions proceed according to the inverse square of the distance, but which he did not perform, gives us a striking example of such a fictitious experiment.

But there are worse things. Very often the fictitious experiment invoked is not only not realized but incapable of being realized; it presupposes the existence of bodies not encountered in nature and of physical properties which have never been observed. Thus Gustave Robin, in order to give the principles of chemical mechanics the purely inductive exposition that he wishes, creates at will what he calls witnessing bodies (*corps témoins*), bodies which by their presence alone are capable of agitating or stopping a chemical reaction.[9] Observation has never revealed such bodies to chemists.

The unperformed experiment, the experiment which would not be performed with precision, and the absolutely unperformable experiment do not exhaust the diverse forms assumed by the fictitious experiment in the writings of physicists who claim to be following the experimental method; there remains to be pointed out a form more illogical than all the others, namely, the absurd experiment. The latter claims to prove a proposition which is contradictory if regarded as the statement of an experimental fact.

The most subtle physicists have not always known how to guard against the intervention of the absurd experiment in their expositions. Let us quote, for instance, some lines taken from J. Bertrand: "If we accept it as an experimental fact that electricity is carried

[9] G. Robin, *op.cit.*, p. ii.

to the surface of bodies, and as a necessary principle that the action of free electricity on the points of conductors should be null, we can deduce from these two conditions, supposing they are strictly satisfied, that electrical attractions and repulsions are inversely proportional to the square of the distance."[10]

Let us take the proposition "There is no electricity in the interior of a conducting body when electrical equilibrium is established in it," and let us inquire whether it is possible to regard it as the statement of an experimental fact. Let us weigh the exact sense of the words figuring in the statement, and particularly, of the word *interior*. In the sense we must give this word in this proposition, a point interior to a piece of electrified copper is a point taken within the mass of copper. Consequently, how can we go about establishing whether there is or is not any electricity at this point? It would be necessary to place a testing body there, and to do that it would be necessary to take away beforehand the copper that is there, but then this point would no longer be within the mass of copper; it would be outside that mass. We cannot without falling into a logical contradiction take our proposition as a result of observation.

What, therefore, is the meaning of the experiments by which we claim to prove this proposition? Certainly, something quite different from what we make them say. We hollow out a cavity in a conducting mass and note that the walls of this cavity are not charged. This observation proves nothing concerning the presence or absence of electricity at points deep within the conducting mass. In order to pass from the experimental law noted to the law stated we play on the word *interior*. Afraid to base electrostatics on a postulate, we base it on a pun.

If we simply turn the pages of the treatises and manuals of physics we can collect any number of fictitious experiments; we should find there abundant illustrations of the various forms that such an experiment can assume, from the merely unperformed experiment to the absurd experiment. Let us not waste time on such a fastidious task. What we have said suffices to warrant the following conclusion: The teaching of physics by the purely inductive method such as Newton defined it is a chimera. Whoever claims to grasp this mirage is deluding himself and deluding his pupils. He is giving them, as facts seen, facts merely foreseen; as precise observations, rough reports; as performable procedures, merely ideal experiments; as

[10] J. Bertrand, *Leçons sur la Théorie mathématique de l'Electricité* (Paris, 1890), p. 71.

experimental laws, propositions whose terms cannot be taken as real without contradiction. The physics he expounds is false and falsified.

Let the teacher of physics give up this ideal inductive method which proceeds from a false idea, and reject this way of conceiving the teaching of experimental science, a way which dissimulates and twists its essential character. If the interpretation of the slightest experiment in physics presupposes the use of a whole set of theories, and if the very description of this experiment requires a great many abstract symbolic expressions whose meaning and correspondence with the facts are indicated only by theories, it will indeed be necessary for the physicist to decide to develop a long chain of hypotheses and deductions before trying the slightest comparison between the theoretical structure and the concrete reality; also, in describing experiments verifying theories already developed, he will very often have to anticipate theories to come. For example, he will not be able to attempt the slightest experimental verification of the principles of dynamics before he has not only developed the chain of propositions of general mechanics but also laid the foundations of celestial mechanics; and he will also have to suppose as known, in reporting the observations verifying this set of theories, the laws of optics which alone warrant the use of astronomical instruments.

Let the teacher therefore develop, in the first place, the essential theories of the science; without doubt, by presenting the hypotheses on which these theories rest, it is necessary for him to prepare their acceptance; it is good for him to point out the data of common sense, the facts gathered by ordinary observation or simple experiments or those scarcely analyzed which have led to formulating these hypotheses. To this point, moreover, we shall insist on returning in the next chapter; but we must proclaim loudly that these facts sufficient for suggesting hypotheses are not sufficient to verify them; it is only after he has constituted an extensive body of doctrine and constructed a complete theory that he will be able to compare the consequences of this theory with experiment.

Instruction ought to get the student to grasp this primary truth: Experimental verifications are not the base of theory but its crown. Physics does not make progress in the way geometry does: the latter grows by the continual contribution of a new theorem demonstrated once and for all and added to theorems already demonstrated; the former is a symbolic painting in which continual retouching gives greater comprehensiveness and unity, and the *whole* of which gives a picture resembling more and more the *whole* of the experimental facts, whereas each detail of this picture cut off and isolated from the whole loses all meaning and no longer represents anything.

To the student who will not have perceived this truth, physics will appear as a monstrous confusion of fallacies of reasoning in circles and begging the question; if he is endowed with a mind of high accuracy, he will repel with disgust these perpetual defiances of logic; if he has a less accurate mind, he will learn by heart here words with inexact meaning, these descriptions of unperformed and unperformable experiments, and lines of reasoning which are sleight-of-hand passes, thus losing in such unreasoned memory work the little correct sense and critical mind he used to possess.

The student who, on the other hand, will have seen clearly the ideas we have just formulated will have done more than learned a certain number of propositions of physics; he will have understood the nature and true method of experimental science.[11]

7. *Consequences Relative to the Mathematical Development of Physical Theory*

Through the preceding discussions the exact nature of physical theory and of its relations with experiment emerge more and more clearly and precisely.

The materials with which this theory is constructed are, on the one hand, the mathematical symbols serving to represent the various quantities and qualities of the physical world, and, on the other hand, the general postulates serving as principles. With these materials theory builds a logical structure; in drawing the plan of this structure it is hence bound to respect scrupulously the laws that logic imposes on all deductive reasoning and the rules that algebra prescribes for any mathematical operation.

The mathematical symbols used in theory have meaning only under very definite conditions; to define these symbols is to enumerate these conditions. Theory is forbidden to make use of these signs outside these conditions. Thus, an absolute temperature by definition can be positive only, and by definition the mass of a body is invariable; never will theory in its formulas give a zero or negative

[11] It will be objected undoubtedly that such teaching of physics would be hardly accessible to young minds; the answer is simple: Do not teach physics to minds not yet ready to assimilate it. Mme. de Sévigné used to say, speaking of young children: "Before you give them the food of a truckdriver, find out if they have the stomach of a truckdriver."

value to absolute temperature, and never in its calculations will it make the mass of a given body vary.

Theory is in principle grounded on postulates, that is to say, on propositions that it is at leisure to state as it pleases, provided that no contradiction exists among the terms of the same postulate or between two distinct postulates. But once these postulates are set down it is bound to guard them with jealous rigor. For instance, if it has placed at the base of its system the principle of the conservation of energy, it must forbid any assertion in disagreement with this principle.

These rules bring all their weight to bear on a physical theory that is being constructed; a single default would make the system illogical and would oblige us to upset it in order to reconstruct another; but they are the only limitations imposed. IN THE COURSE OF ITS DEVELOPMENT, *a physical theory is free to choose any path it pleases provided that it avoids any logical contradiction; in particular, it is free not to take account of experimental facts.*

This is no longer the case WHEN THE THEORY HAS REACHED ITS COMPLETE DEVELOPMENT. When the logical structure has reached its highest point it becomes necessary to compare the set of mathematical propositions obtained as conclusions from these long deductions with the set of experimental facts; by employing the adopted procedures of measurement we must be sure that the second set finds in the first a sufficiently similar image, a sufficiently precise and complete symbol. If this agreement between the conclusions of theory and the facts of experiment were not to manifest a satisfactory approximation, the theory might well be logically constructed, but it should nonetheless be rejected because it would be contradicted by observation, because it would be *physically* false.

This comparison between the conclusions of theory and the truths of experiment is therefore indispensable, since only the test of facts can give physical validity to a theory. But this test by facts should bear exclusively on the conclusions of a theory, for only the latter are offered as an image of reality; the postulates serving as points of departure for the theory and the intermediary steps by which we go from the postulates to the conclusions do not have to be subject to this test.

We have in the foregoing pages very thoroughly analyzed the error of those who claim to subject one of the fundamental postulates of physics directly to the test of facts through a procedure such as a crucial experiment; and especially the error of those who accept as principles only "inductions consisting exclusively in erecting into general laws not the interpretation but *the very result of a very large number of experiments.*"[12]

There is another error lying very close to this one; it consists in requiring that all the operations performed by the mathematician connecting postulates with conclusions should have *a physical meaning,* in wishing "to reason only about *performable operations,*" and in "introducing only magnitudes accessible to experiment."[13]

According to this requirement any magnitude introduced by the physicist in his formulas should be connected through a process of measurement to a property of a body; any algebraic operation performed on these magnitudes should be translated into concrete language by the employment of these processes of measurement; thus translated, it should express a real or possible fact.

Such a requirement, legitimate when it comes to the final formulas at the end of a theory, has no justification if applied to the intermediary formulas and operations establishing the transition from postulates to conclusions.

Let us take an example.

J. Willard Gibbs studied the theory of the dissociation of a perfect composite gas into its elements, also regarded as perfect gases. A formula was obtained expressing the law of chemical equilibrium internal to such a system. I propose to discuss this formula. For this purpose, keeping constant the pressure supporting the gaseous mixture, I consider the absolute temperature appearing in the formula and I make it vary from 0 to $+\infty$.

If we wish to attribute a physical meaning to this mathematical operation, we shall be confronted with a host of objections and difficulties. No thermometer can reveal temperatures below a certain limit, and none can determine temperatures high enough; this symbol which we call "absolute temperature" cannot be translated through the means of measurement at our disposal into something having a concrete meaning unless its numerical value remains between a certain minimum and a certain maximum. Moreover, at temperatures sufficiently low this other symbol which thermodynamics calls "a perfect gas" is no longer even an approximate image of any real gas.

These difficulties and many others, which it would take too long to enumerate, disappear if we heed the remarks we have formulated. In the construction of the theory, the discussion we have just given is only an intermediary step, and there is no justification for seeking

[12] G. Robin, *op.cit.*, p. xiv.

[13] *loc.cit.*

a physical meaning in it. Only when this discussion shall have led us to a series of propositions, shall we have to submit these propositions to the test of facts; then we shall inquire whether, within the limits in which the absolute temperature may be translated into concrete readings of a thermometer and the idea of a perfect gas is approximately embodied in the fluids we observe, the conclusions of our discussion agree with the results of experiment.

By requiring that mathematical operations by which postulates produce their consequences shall always have a physical meaning, we set unjustifiable obstacles before the mathematician and cripple his progress. G. Robin goes so far as to question the use of the differential calculus; if Professor Robin is intent on constantly and scrupulously satisfying this requirement, he would practically be unable to develop any calculation; theoretical deduction would be stopped in its tracks from the start. A more accurate idea of the method of physics and a more exact line of demarcation between the propositions which have to submit to factual test and those which are free to dispense with it would give back to the mathematician all his freedom and permit him to use all the resources of algebra for the greatest development of physical theories.

8. *Are Certain Postulates of Physical Theory Incapable of Being Refuted by Experiment?*

We recognize a correct principle by the facility with which it straightens out the complicated difficulties into which the use of erroneous principles brought us.

If, therefore, the idea we have put forth is correct, namely, that comparison is established necessarily between the *whole* of theory and the *whole* of experimental facts, we ought in the light of this principle to see the disappearance of the obscurities in which we should be lost by thinking that we are subjecting each isolated theoretical hypothesis to the test of facts.

Foremost among the assertions in which we shall aim at eliminating the appearance of paradox, we shall place one that has recently been often formulated and discussed. Stated first by G. Milhaud in connection with the "*pure bodies*" of chemistry,[14] it has been developed at length and forcefully by H. Poincaré with regard to principles of mechanics;[15] Edouard Le Roy has also formulated it with great clarity.[16]

[14] G. Milhaud, "La Science rationnelle," *Revue de Métaphysique et de Morale*, IV (1896), 280. Reprinted in *Le Rationnel* (Paris, 1898), p. 45.

[15] H. Poincaré, "Sur les Principes de la Mécanique," *Bibliothèque du Congrès*

That assertion is as follows: Certain fundamental hypotheses of physical theory cannot be contradicted by any experiment, because they constitute in reality *definitions*, and because certain expressions in the physicist's usage take their meaning only through them.

Let us take one of the examples cited by Le Roy:

When a heavy body falls freely, the acceleration of its fall is constant. Can such a law be contradicted by experiment? No, for it constitutes the very definition of what is meant by "falling freely." If while studying the fall of a heavy body we found that this body does not fall with uniform acceleration, we should conclude not that the stated law is false, but that the body does not fall freely, that some cause obstructs its motion, and that the deviations of the observed facts from the law as stated would serve to discover this cause and to analyze its effects.

Thus, M. Le Roy concludes, "laws are verifiable, taking things strictly . . . , because they constitute the very criterion by which we judge appearances as well as the methods that it would be necessary to utilize in order to submit them to an inquiry whose precision is capable of exceeding any assignable limit."

Let us study again in greater detail, in the light of the principles previously set down, what this comparison is between the law of falling bodies and experiment.

Our daily observations have made us acquainted with a whole category of motions which we have brought together under the name of motions of heavy bodies; among these motions is the falling of a heavy body when it is not hindered by any obstacle. The result of this is that the words "free fall of a heavy body" have a meaning for the man who appeals only to the knowledge of common sense and who has no notion of physical theories.

On the other hand, in order to classify the laws of motion in question the physicist has created a theory, the theory of weight, an important application of rational mechanics. In that theory, intended to furnish a symbolic representation of reality, there is also the question of "free fall of a heavy body," and as a consequence of the hypotheses supporting this whole scheme free fall must necessarily be a uniformly accelerated motion.

The words "free fall of a heavy body" now have two distinct mean-

International de Philosophie, III: *Logique et Histoire des Sciences* (Paris, 1901), p. 457; "Sur la valeur objective des théories physiques," *Revue de Métaphysique et de Morale*, x (1902), 263; *La Science et l'Hypothèse*, p. 110.

[16] E. Le Roy, "Un positivisme nouveau," *Revue de Métaphysique et de Morale*, IX (1901), 143-144.

ings. For the man ignorant of physical theories, they have their *real* meaning, and they mean what common sense means in pronouncing them; for the physicist they have a *symbolic* meaning, and mean "uniformly accelerated motion." Theory would not have realized its aim if the second meaning were not the sign of the first, if a fall regarded as free by common sense were not also regarded as uniformly accelerated, or *nearly* uniformly accelerated, since common-sense observations are essentially devoid of precision, according to what we have already said.

This agreement, without which the theory would have been rejected without further examination, is finally arrived at: a fall declared by common sense to be nearly free is also a fall whose acceleration is nearly constant. But noticing this crudely approximate agreement does not satisfy us; we wish to push on and surpass the degree of precision which common sense can claim. With the aid of the theory that we have imagined, we put together apparatus enabling us to recognize with sensitive accuracy whether the fall of a body is or is not uniformly accelerated; this apparatus shows us that a certain fall regarded by common sense as a free fall has a slightly variable acceleration. The proposition which in our theory gives its symbolic meaning to the words "free fall" does not represent with sufficient accuracy the properties of the real and concrete fall that we have observed.

Two alternatives are then open to us.

In the first place, we can declare that we were right in regarding the fall studied as a free fall and in requiring that the theoretical definition of these words agree with our observations. In this case, since our theoretical definition does not satisfy this requirement, it must be rejected; we must construct another mechanics on new hypotheses, a mechanics in which the words "free fall" no longer signify "uniformly accelerated motion," but "fall whose acceleration varies according to a certain law."

In the second alternative, we may declare that we were wrong in establishing a connection between the concrete fall we have observed and the symbolic free fall defined by our theory, that the latter was too simplified a scheme of the former, that in order to represent suitably the fall as our experiments have reported it the theorist should give up imagining a weight falling freely and think in terms of a weight hindered by certain obstacles like the resistance of the air, that in picturing the action of these obstacles by means of appropriate hypotheses he will compose a more complicated scheme than a free weight but one more apt to reproduce the details of the experiment; in short, in accord with the language we have previously established (Ch. IV, Sec. 3), we may seek to eliminate by means of suitable "corrections" the "causes of error," such as air resistance, which influenced our experiment.

M. Le Roy asserts that we shall prefer the second to the first alternative, and he is surely right in this. The reasons dictating this choice are easy to perceive. By taking the first alternative we should be obliged to destroy from top to bottom a very vast theoretical system which represents in a most satisfactory manner a very extensive and complex set of experimental laws. The second alternative, on the other hand, does not make us lose anything of the terrain already conquered by physical theory; in addition, it has succeeded in so large a number of cases that we can bank with interest on a new success. But in this confidence accorded the law of fall of weights, we see nothing analogous to the certainty that a mathematical definition draws from its very essence, that is, to the kind of certainty we have when it would be foolish to doubt that the various points on a circumference are all equidistant from the center.

We have here nothing more than a particular application of the principle set down in Section 2 of this chapter. A disagreement between the concrete facts constituting an experiment and the symbolic representation which theory substitutes for this experiment proves that some part of this symbol is to be rejected. But which part? This the experiment does not tell us; it leaves to our sagacity the burden of guessing. Now among the theoretical elements entering into the composition of this symbol there is always a certain number which the physicists of a certain epoch agree in accepting without test and which they regard as beyond dispute. Hence, the physicist who wishes to modify this symbol will surely bring his modification to bear on elements other than those just mentioned.

But what impels the physicist to act thus is *not* logical necessity. It would be awkward and ill inspired for him to do otherwise, but it would not be doing something logically absurd; he would not for all that be walking in the footsteps of the mathematician mad enough to contradict his own definitions. More than this, perhaps some day by acting differently, by refusing to invoke causes of error and take recourse to corrections in order to reestablish agreement between the theoretical scheme and the fact, and by resolutely carrying out a reform among the propositions declared untouchable by common consent, he will accomplish the work of a genius who opens a new career for a theory.

Indeed, we must really guard ourselves against believing forever warranted those hypotheses which have become universally adopted conventions, and whose certainty seems to break through experimental contradiction by throwing the latter back on more doubtful assumptions. The history of physics shows us that very often the human mind has been led to overthrow such principles completely, though they have been regarded by common consent for centuries as inviolable axioms, and to rebuild its physical theories on new hypotheses.

Was there, for instance, a clearer or more certain principle for thousands of years than this one: In a homogeneous medium, light is propagated in a straight line? Not only did this hypothesis carry all former optics, catoptrics, and dioptrics, whose elegant geometric deductions represented at will an enormous number of facts, but it had become, so to speak, the physical definition of a straight line. It is to this hypothesis that any man wishing to make a straight line appeals, the carpenter who verifies the straightness of a piece of wood, the surveyor who lines up his sights, the geodetic surveyor who obtains a direction with the help of the pinholes of his alidade, the astronomer who defines the position of stars by the optical axis of his telescope. However, the day came when physicists tired of attributing to some cause of error the diffraction effects observed by Grimaldi, when they resolved to reject the law of the rectilinear propagation of light and to give optics entirely new foundations; and this bold resolution was the signal of remarkable progress for physical theory.

9. *On Hypotheses Whose Statement Has No Experimental Meaning*

This example, as well as others we could add from the history of science, should show that it would be very imprudent for us to say concerning a hypothesis commonly accepted today: "We are certain that we shall never be led to abandon it because of a new experiment, no matter how precise it is." Yet M. Poincaré does not hesitate to enunciate it concerning the principles of mechanics.[17]

To the reasons already given to prove that these principles cannot be reached by experimental refutation, M. Poincaré adds one which seems even more convincing: Not only can these principles not be refuted by experiment because they are the universally ac-

[17] H. Poincaré, "Sur les Principes de la Mécanique," *Bibliothèque du Congrès international de Philosophie,* Sec. III: "Logique et Histoire des Sciences" (Paris, (1901), pp. 475, 491.

cepted rules serving to discover in our theories the weak spots indicated by these refutations, but also, they cannot be refuted by experiment because *the operation which would claim to compare them with the facts would have no meaning.*

Let us explain that by an illustration.

The principle of inertia teaches us that a material point removed from the action of any other body moves in a straight line with uniform motion. Now, we can observe only relative motions; we cannot, therefore, give an experimental meaning to this principle unless we assume a certain point chosen or a certain geometric solid taken as a fixed reference point to which the motion of the material point is related. The fixation of this reference frame constitutes an integral part of the statement of the law, for if we omitted it, this statement would be devoid of meaning. There are as many different laws as there are distinct frames of reference. We shall be stating one law of inertia when we say that the motion of an isolated point assumed to be seen from the earth is rectilinear and uniform, and another when we repeat the same sentence in referring the motion to the sun, and still another if the frame of reference chosen is the totality of fixed stars. But then, one thing is indeed certain, namely, that whatever the motion of a material point is, when seen from a first frame of reference, we can always and in infinite ways choose a second frame of reference such that seen from the latter our material point appears to move in a straight line with uniform motion. We cannot, therefore, attempt an experimental verification of the principle of inertia; false when we refer the motions to one frame of reference, it will become true when selection is made of another term of comparison, and we shall always be free to choose the latter. If the law of inertia stated by taking the earth as a frame of reference is contradicted by an observation, we shall substitute for it the law of inertia whose statement refers the motion to the sun; if the latter in its turn is contraverted, we shall replace the sun in the statement of the law by the system of fixed stars, and so forth. It is impossible to stop this loophole.

The principle of the equality of action and reaction, analyzed at length by M. Poincaré,[18] provides room for analogous remarks. This principle may be stated thus: "The center of gravity of an isolated system can have only a uniform rectilinear motion."

This is the principle that we propose to verify by experiment. "Can we make this verification? For that it would be necessary for

[18] *ibid.*, pp. 472ff.

isolated systems to exist. Now, these systems do not exist; the only isolated system is the whole universe.

"But we can observe only relative motions; the absolute motion of the center of the universe will therefore be forever unknown. We shall never be able to know if it is rectilinear and uniform or, better still, the question has no meaning. Whatever facts we may observe, we shall hence always be free to assume our principle is true."

Thus many a principle of mechanics has a form such that it is absurd to ask one's self: "Is this principle in agreement with experiment or not?" This strange character is not peculiar to the principles of mechanics; it also marks certain fundamental hypotheses of our physical or chemical theories.[19]

For example, chemical theory rests entirely on the "law of multiple proportions"; here is the exact statement of this law:

Simple bodies *A*, *B*, and *C* may by uniting in various proportions form various compounds M, M', ... The masses of the bodies *A*, *B*, and *C* combining to form the compound *M* are to one another as the three numbers *a*, *b*, and *c*. Then the masses of the elements *A*, *B*, and *C* combining to form the compound M' will be to one another as the numbers xa, yb, and zc (x, y, and z being three whole numbers).

Is this law perhaps subject to experimental test? Chemical analysis will make us acquainted with the chemical composition of the body M' not exactly but with a certain approximation. The uncertainty of the results obtained can be extremely small; it will never be strictly zero. Now, in whatever relations the elements *A*, *B*, and *C* are combined within the compound M', we can always represent these relations, with as close an approximation as you please, by the mutual relations of three products xa, yb, and zc, where x, y, and z are whole numbers; in other words, whatever the results given by the chemical analysis of the compound M', we are always sure to find three integers x, y, and z thanks to which the law of multiple proportions will be verified with a precision greater than that of the experiment. Therefore, no chemical analysis, no matter how refined, will ever be able to show the law of multiple proportions to be wrong.

In like manner, all crystallography rests entirely on the "law of rational indices" which is formulated in the following way:

A trihedral being formed by three faces of a crystal, a fourth face

[19] P. Duhem, *Le Mixte et la Combinaison chimique: Essai sur l'évolution d'une idée* (Paris, 1902), pp. 159-161.

cuts the three edges of this trihedral at distances from the summit which are proportional to one another as three given numbers, the parameters of the crystal. Any other face whatsoever should cut these same edges at distances from the summit which are to one another as xa, yb, and zc, where x, y, and z are three integers, the indices of the new face of the crystal.

The most perfect protractor determines the direction of a crystal's face only with a certain degree of approximation; the relations among the three segments that such a face makes on the edges of the fundamental trihedral are always able to get by with a certain error; now, however small this error is, we can always choose three numbers x, y, and z such that the mutual relations of these segments are represented with the least amount of error by the mutual relations of the three numbers xa, yb, and zc; the crystallographer who would claim that the law of rational indices is made justifiable by his protractor would surely not have understood the very meaning of the words he is employing.

The law of multiple proportions and the law of rational indices are mathematical statements deprived of all physical meaning. A mathematical statement has physical meaning only if it retains a meaning when we introduce the word "nearly" or "approximately." This is not the case with the statements we have just alluded to. Their object really is to assert that certain relations are *commensurable* numbers. They would degenerate into mere truisms if they were made to declare that these relations are approximately commensurable, for any incommensurable relation whatever is always approximately commensurable; it is even as near as you please to being commensurable.

Therefore, it would be absurd to wish to subject certain principles of mechanics to *direct* experimental test; it would be absurd to subject the law of multiple proportions or the law of rational indices to this *direct* test.

Does it follow that these hypotheses placed beyond the reach of direct experimental refutation have nothing more to fear from experiment? That they are guaranteed to remain immutable no matter what discoveries observation has in store for us? To pretend so would be a serious error.

Taken in isolation these different hypotheses have no experimental meaning; there can be no question of either confirming or contradicting them by experiment. But these hypotheses enter as essential foundations into the construction of certain theories of rational mechanics, of chemical theory, of crystallography. The object of these theories is to represent experimental laws; they are schematisms intended essentially to be compared with facts.

Now this comparison might some day very well show us that one of our representations is ill adjusted to the realities it should picture, that the corrections which come and complicate our schematism do not produce sufficient concordance between this schematism and the facts, that the theory accepted for a long time without dispute should be rejected, and that an entirely different theory should be constructed on entirely different or new hypotheses. On that day some one of our hypotheses, which taken in isolation defied direct experimental refutation, will crumble with the system it supported under the weight of the contradictions inflicted by reality on the consequences of this system taken as a whole.[20]

In truth, hypotheses which by themselves have no physical meaning undergo experimental testing in exactly the same manner as other hypotheses. Whatever the nature of the hypothesis is, we have seen at the beginning of this chapter that it is never in isolation contradicted by experiment; experimental contradiction always bears as a whole on the entire group constituting a theory without any possibility of designating which proposition in this group should be rejected.

There thus disappears what might have seemed paradoxical in the following assertion: Certain physical theories rest on hypotheses which do not by themselves have any physical meaning.

10. *Good Sense Is the Judge of Hypotheses Which Ought to Be Abandoned*

When certain consequences of a theory are struck by experimental contradiction, we learn that this theory should be modified but we are not told by the experiment what must be changed. It leaves to the physicist the task of finding out the weak spot that impairs the whole system. No absolute principle directs this inquiry, which different physicists may conduct in very different ways without having the right to accuse one another of illogicality. For instance, one may be obliged to safeguard certain fundamental

[20] At the International Congress of Philosophy held in Paris in 1900, M. Poincaré developed this conclusion: "Thus is explained how experiment may have been able to edify (or suggest) the principles of mechanics, but will never be able to overthrow them." Against this conclusion, M. Hadamard offered various remarks, among them the following: "Moreover, in conformity with a remark of M. Duhem, it is not *an* isolated hypothesis but the whole group of the hypotheses of mechanics that we can try to verify experimentally." *Revue de Métaphysique et de Morale*, VIII (1900), 559.

hypotheses while he tries to reestablish harmony between the consequences of the theory and the facts by complicating the schematism in which these hypotheses are applied, by invoking various causes of error, and by multiplying corrections. The next physicist, disdainful of these complicated artificial procedures, may decide to change some one of the essential assumptions supporting the entire system. The first physicist does not have the right to condemn in advance the boldness of the second one, nor does the latter have the right to treat the timidity of the first physicist as absurd. The methods they follow are justifiable only by experiment, and if they both succeed in satisfying the requirements of experiment each is logically permitted to declare himself content with the work that he has accomplished.

That does not mean that we cannot very properly prefer the work of one of the two to that of the other. Pure logic is not the only rule for our judgments; certain opinions which do not fall under the hammer of the principle of contradiction are in any case perfectly unreasonable. These motives which do not proceed from logic and yet direct our choices, these "reasons which reason does not know" and which speak to the ample "mind of finesse" but not to the "geometric mind," constitute what is appropriately called good sense.

Now, it may be good sense that permits us to decide between two physicists. It may be that we do not approve of the haste with which the second one upsets the principles of a vast and harmoniously constructed theory whereas a modification of detail, a slight correction, would have sufficed to put these theories in accord with the facts. On the other hand, it may be that we may find it childish and unreasonable for the first physicist to maintain obstinately at any cost, at the price of continual repairs and many tangled-up stays, the worm-eaten columns of a building tottering in every part, when by razing these columns it would be possible to construct a simple, elegant, and solid system.

But these reasons of good sense do not impose themselves with the same implacable rigor that the prescriptions of logic do. There is something vague and uncertain about them; they do not reveal themselves at the same time with the same degree of clarity to all minds. Hence, the possibility of lengthy quarrels between the adherents of an old system and the partisans of a new doctrine, each camp claiming to have good sense on its side, each party finding the reasons of the adversary inadequate. The history of physics would furnish us with innumerable illustrations of these quarrels at all times and in all domains. Let us confine ourselves to the tenacity and ingenuity with which Biot by a continual bestowal of corrections and accessory hypotheses maintained the emissionist doctrine in optics, while Fresnel opposed this doctrine constantly with new experiments favoring the wave theory.

In any event this state of indecision does not last forever. The day arrives when good sense comes out so clearly in favor of one of the two sides that the other side gives up the struggle even though pure logic would not forbid its continuation. After Foucault's experiment had shown that light traveled faster in air than in water, Biot gave up supporting the emission hypothesis; strictly, pure logic would not have compelled him to give it up, for Foucault's experiment was *not* the crucial experiment that Arago thought he saw in it, but by resisting wave optics for a longer time Biot would have been lacking in good sense.

Since logic does not determine with strict precision the time when an inadequate hypothesis should give way to a more fruitful assumption, and since recognizing this moment belongs to good sense, physicists may hasten this judgment and increase the rapidity of scientific progress by trying consciously to make good sense within themselves more lucid and more vigilant. Now nothing contributes more to entangle good sense and to disturb its insight than passions and interests. Therefore, nothing will delay the decision which should determine a fortunate reform in a physical theory more than the vanity which makes a physicist too indulgent towards his own system and too severe towards the system of another. We are thus led to the conclusion so clearly expressed by Claude Bernard: The sound experimental criticism of a hypothesis is subordinated to certain moral conditions; in order to estimate correctly the agreement of a physical theory with the facts, it is not enough to be a good mathematician and skillful experimenter; one must also be an impartial and faithful judge.

CHAPTER VII

THE CHOICE OF HYPOTHESES

1. *What the Conditions Imposed by Logic on the Choice of Hypotheses Reduce To*

We have carefully analyzed the various operations through which a physical theory is constructed; we have particularly subjected to severe criticism the rules which permit us to compare the conclusions of theory with experimental laws; we are now free to go back to the very foundations of theory and say what they should be, knowing what they have to bear. Hence, we are going to ask the question: What are the conditions imposed by logic on the choice of hypotheses on which a physical theory is to be based?

Moreover, the different problems we have investigated above and the solutions which we have offered to them dictate, so to speak, the answers to us.

Does logic demand that our hypotheses be the consequences of some cosmological system, or at least, that they agree with the consequences of such a system? By no means. Our physical theories do not pride themselves on being explanations; our hypotheses are *not* assumptions about the very nature of material things. Our theories have as their sole aim the economical condensation and classification of experimental laws; they are autonomous and independent of any metaphysical system. The hypotheses on which we build them do not, therefore, need to borrow their materials from this or that philosophical doctrine; they do *not* claim the authority of a metaphysical school and have nothing to fear from its critics.

Does logic require our hypotheses to be simply experimental laws generalized by induction? Logic cannot lay down requirements which it is impossible to satisfy. Now, we have recognized that it is impossible to construct a theory by a purely inductive method. Newton and Ampère failed in this, and yet these two mathematicians had boasted of allowing nothing in their systems which was not drawn entirely from experiment. Therefore, we shall not be averse to admitting among the fundamental bases of our physics postulates not furnished by experiment.

Does logic insist on our not introducing hypotheses except one by one, subjecting each one of them, before declaring it admissible, to a thorough test of its solidity? That again would be an absurd demand. Any experimental test puts into play the most diverse parts of physics and appeals to innumerable hypotheses; it never tests a given hypothesis by isolating it from the others. Logic cannot summon each hypothesis in turn to try out a role we expect it to play, for such a tryout is impossible.

What then are the conditions logically imposed on the choice of hypotheses to serve as the base of our physical theory? These conditions are three in number.

In the first place, a hypothesis shall not be a self-contradictory proposition, for the physicist does not intend to utter nonsense.

In the second place, the different hypotheses which are to support physics shall not contradict one another. Physical theory, indeed, is not to be resolved into a mass of disparate and incompatible models; it aims to preserve with jealous care a logical unity, for an intuition we are powerless to justify, but which it is impossible for us to be blind to, shows us that only on this condition will theory tend towards its ideal form, namely, that of a natural classification.

In the third place, hypotheses shall be chosen in such a manner that from them *taken as a whole* mathematical deduction may draw consequences representing with a sufficient degree of approximation the *totality* of experimental laws. In fact, the proper aim of physical theory is the schematic representation by means of mathematical symbols of the laws established by the experimenter; any theory one of whose consequences is in plain contradiction with an observed law should be mercilessly rejected. But it is not possible to compare an isolated consequence of theory with an isolated experimental law. The two systems must be taken in their integrity: the entire system of theoretical representations on the one hand, and the entire system of observed data on the other. As such they are to be compared to each other and their resemblance judged.

2. *Hypotheses Are Not the Product of Sudden Creation, but the Result of Progressive Evolution. An Example Drawn from Universal Attraction*

The requirements imposed by logic on the hypotheses which are to support a physical theory reduce to these three conditions. So long as he respects them, the theorist enjoys complete freedom, and he may lay the foundations of the system he is going to construct in any way he pleases.

Will not such freedom be the most embarrassing of all vexations?

Well now! Confronting the physicist there extends, farther than one can see, the innumerable multitude and unordered pack of experimental laws with nothing yet available to summarize, classify, and coordinate them. He has to formulate principles whose consequences will yield a simple, clear, and orderly representation of this frightening total of observational data; but before he can judge whether the consequences of his hypotheses attain their object, before he can recognize whether they yield a methodic classification of, and a picture resembling, the experimental laws, he must constitute the entire system from his presuppositions; and when he asks logic to guide him in this difficult task, to designate which hypotheses he should choose, which he should reject, he receives merely this prescription to avoid contradiction, a prescription that is exasperating in the extreme latitude it allows to his hesitations. Can such unlimited freedom be useful to a man? Is his mind powerful enough to create a physical theory all out of one piece?

Surely no. Thus history shows us that no physical theory has ever been created out of whole cloth. The formation of any physical theory has always proceeded by a series of retouchings which from almost formless first sketches have gradually led the system to more finished states; and in each of these retouchings, the free initiative of the physicist has been counselled, maintained, guided, and sometimes absolutely dictated by the most diverse circumstances, by the opinions of men as well as by what the facts teach. A physical theory is not the sudden product of a creation; it is the slow and progressive result of an evolution.

When several taps of the beak break the shell of an egg from which the chick escapes, a child may imagine that this rigid and immobile mass, similar to the white shells he picks up on the edge of a stream, has suddenly taken life and produced the bird who runs away with a chirp; but just where his childish imagination sees a sudden creation, the naturalist recognizes the last stage of a long development; he thinks back to the first fusion of two microscopic nuclei in order to review next the series of divisions, differentiations, and reabsorptions which, cell by cell, have built up the body of the chick.

The ordinary layman judges the birth of physical theories as the child the appearance of the chick. He believes that this fairy whom he calls by the name of science has touched with his magic wand the forehead of a man of genius and that the theory immediately appeared alive and complete, like Pallas Athena emerging fully armed from the forehead of Zeus. He thinks it was enough for Newton to see an apple fall in an orchard in order that the effects of falling bodies, the motions of the earth, the moon, and the planets and their satellites, the trips of comets, the ebb and flow of the ocean, should all suddenly come to be summarized and classified in that one proposition: Any two bodies attract each other proportionally to the product of their masses and inversely to the square of their mutual distance.

Those who have a deeper insight into the history of physical theories know that in order to find the germ of this doctrine of universal gravitation, we must look among the systems of Greek science; they know the slow metamorphoses of this germ in the course of its millenary evolution; they enumerate the contributions of each century to the work which will receive its viable form from Newton; they do not forget the doubts and gropings through which Newton himself passed before producing a finished system; and at no moment in the history of universal attraction do they perceive any phenomenon resembling a sudden creation; nor one instance in which the human mind, free from the impetus of any motive alien to the appeal of past doctrines and to the contradictions of present experiments, would have used all the freedom which logic grants it in forming hypotheses.

We cannot expound here in great detail the history of the efforts by which mankind has prepared the memorable discovery of universal attraction; one volume would hardly suffice for that. However, we should at least like to outline it in a rough sketch in order to show through what vicissitudes this fundamental hypothesis passed before being clearly formulated.

As soon as man thought of studying the physical world, a class of phenomena was bound because of its universality and importance to engage his attention; *weight* was bound to be the object of the first thoughts of physicists.

Let us not stop to recall what the philosophers of ancient Hellas were able to say about the heavy and the light, but let us take as the starting point of the history we wish to run through the physics taught by Aristotle. Besides, let us retain, of the evolution sketched a long time ago, but which we shall follow starting from that point, only what prepared the way for the Newtonian theory, by neglecting systematically everything not tending to that goal.

To Aristotle, all bodies were mixtures composed in various proportions of the four elements, earth, water, air, and fire; of these four elements, the first three were heavy; the earth was heavier than water, which was heavier than air; only fire was light. Mixtures were more or less heavy or light according to the proportion of the elements forming them.

What does this amount to? A heavy body is a body endowed with such a "substantial form" that it moves by itself towards a mathematical point, the center of the universe, every time it is not prevented from doing so; and in order to prevent this motion, there must be underneath it a solid support or a fluid heavier than it. A lighter fluid would not prevent its motion, *for the heavier tends to be placed below the lighter.* A light body is, accordingly, a body whose substantial form is such that it moves by itself away from the center of the world.

If bodies are endowed with such substantial forms, it is because each tends to occupy its "natural place," this place being closer to the center of the world as the body is richer in heavy elements, and is as much removed from this point as the mixture is penetrated by lighter elements. The location of each element in its natural place would bring about an order in the world in which each element would have reached the perfection of its form; if, therefore, the substantial form of any element or of any mixture is endowed with one of these qualities called heaviness or lightness, the explanation is that the order of the world returns by a "natural motion" to its perfection every time a "violent motion" momentarily disturbs it. In particular, it is this tendency of every heavy body toward its natural place, toward the center of the universe, which explains the rotundity of the earth and the perfect sphericity of the surface of the seas. Aristotle had already sketched a mathematical demonstration of this scheme that Adrastus, Pliny the Elder, Theon of Smyrna, Simplicius, and Saint Thomas Aquinas reproduced and developed. Thus, conforming to the great principle of Aristotelian metaphysics, the efficient cause of the motion of heavy bodies is at the same time its final cause; it is identified not with a violent attraction exerted by the center of the universe, but with a natural tendency experienced by each body toward the place most favorable for its own preservation and for the harmonious disposition of the world.

Such are the hypotheses on which Aristotle's theory of weight is based, and which the commentators of the school of Alexandria, the Arabs, and the occidental philosophers of the Middle Ages developed and made precise. Julius Caesar Scaliger expounded it at length,[1] and John B. Benedetti gave a particularly clear formulation of it,[2] taken up by Galileo himself in his early writings.[3]

Moreover, this doctrine was made precise in the course of the meditations of the Scholastic philosophers. Weight is not a tendency in a body to place all of itself at the center of the universe, which would be absurdly impossible, or to place no-matter-which of its points there; in every heavy body there is a very definite point which wishes to be united with the center of the universe, and this point is the center of gravity of the body. It is not any point whatsoever of the earth but the center of gravity of the terrestrial mass which must be at the center of the world in order for the earth to remain stationary. The gravitation is exerted between two points, thus resembling the action between two poles which has for so long a time represented the properties of magnets.

Contained in germ in a passage of Simplicius, commenting on the *De Caelo* of Aristotle, this doctrine was formulated at length in the middle of the fourteenth century by one of the doctors who illustrate the nominalistic school of the Sorbonne, namely, Albert of Saxony. After Albert of Saxony, and according to his teaching, the doctrine was adopted and expounded by the most powerful minds of the school, by Timon the Jew, by Marsilius of Inghen, by Peter of Ailly, and by Nifo.[4]

After suggesting to Leonardo da Vinci several of his most original thoughts,[5] the doctrine of Albert of Saxony extended its powerful influence well beyond the Middle Ages. Guido Ubaldo del Monte formulated it clearly: "When we say that a heavy body desires by a natural propensity to place itself in the center of the universe, we wish to express the fact that this heavy body's own center of gravity desires to unite with the center of the universe."[6] This doctrine of Albert of Saxony dominated the mind of many a physicist even in the middle of the seventeenth century. It inspired the arguments,

[1] *Julii Caesaris Scaligeri "Exotericarum exercitationum liber* XV: *De subtilitate adversus Cardanum"* (Paris, 1557), Problem IV.

[2] *J. Baptistae Benedicti "Diversarum speculationum liber. Disputationes de quibusdam placitis Aristotelis"* (Turin, 1585), Ch. XXXV, p. 191.

[3] *Le Opere di Galileo Galilei*, Vol. I: *De Motu* (reprinted faithfully from the national edition; Florence, 1890), p. 252. This work, composed by Galileo about 1590, was published only in our time by Antonio Favaro.

[4] The detailed history of this doctrine will be found in our work, *Les Origines de la Statique*, Ch. XV: "Les propriétés mécaniques du centre de gravité.—D'Albert de Saxe à Torricelli."

[5] See P. Duhem, "Albert de Saxe et Léonard de Vinci," *Bulletin italien*, V (1905), pp. 1, 113.

[6] *Guidi Ubaldi e Marchionibus Montis "In duos Archimedis aequiponderantium libros paraphrasis scholiis illustrata"* (Pisa, 1588), p. 10.

appearing strange to those unfamiliar with this doctrine of Albert, with which Fermat supported his geostatic proposition.[7] In 1636 Fermat wrote to Roberval, who disputed the legitimacy of his arguments: "The first objection consists in the fact that you do not wish to grant that the midpoint of a line joining two equal weights falling freely proceeds to unite with the center of the world. In this it certainly seems to me that you do violence to the natural light and to first principles."[8] The propositions formulated by Albert of Saxony had ended by taking their place among the number of self-evident truths.

The Copernican revolution, by destroying the geocentric system, overthrew the very foundations on which this theory of weight rested.

The earth, heavy body par excellence, no longer tended to place itself at the center of the universe. Physicists were to base the theory of gravity on new hypotheses; what considerations were going to suggest these hypotheses to them? Considerations based on analogy. They were going to compare the falling of weights to the earth with the movement of iron toward the magnet.

Moreover, iron and its ores are related to the magnet; thus, when they are placed in the neighborhood of a magnet, the perfection of the universe requires them to go and join this body; that is why their substantial form is altered in the neighborhood of the magnet, why they acquire the "magnetic virtue" through which they rush towards the magnet.

Such is the unanimous teaching of the Aristotelian schoolmen, particularly of Averroes and Saint Thomas, on the subject of magnetic action.

This action was studied more closely in the thirteenth century; it was noted that every magnet possesses two poles, that opposite poles, so called, attract one another but like poles repel each other. In 1269 Peter of Maricourt, better known as Petrus Peregrinus, gave a description of magnetic action which is a marvel of clarity and experimental sagacity.[9]

But these new discoveries only confirm the Aristotelian doctrine

[7] See P. Duhem, *Les Origines de la Statique*, Ch. xvi: "La Doctrine d'Albert de Saxe et les Géostaticiens."

[8] Pierre de Fermat, *Oeuvres*, published because of the editorial labors of P. Tannery and C. Henry, Vol. ii: *Correspondance*, p. 31.

[9] *Epistola Petri Peregrini Maricurtensis ad Sygerum de Foucaucourt militem, de magnete* (Paris, August 8, 1269); printed by G. Gasser in Augsburg, 1558. Reprinted in *Neudrucke von Schriften und Karten über Meteorologie und Erdmagnetismus*, ed. G. Hellman, No. 10: *Rara Magnetica* (Berlin: Asher, 1896).

by making it precise. If we break a magnetic stone, the faces of the broken stone have poles of unlike designation; the substantial forms of the two fragments are such that these fragments go towards each other and tend to fuse again. Magnetic virtue then is what tends to preserve the integrity of the magnet or else, when this magnet has been broken, to restore the single magnet having its poles arranged like the original magnet.[10]

Gravitation has an analogous explanation. Terrestrial elements are endowed with a substantial form such that they remain united to the earth, of which they are a part, and preserve its spherical shape. Leonardo da Vinci, precursor of Copernicus, had already announced[11] that "the Earth is not at the middle of the circle of the sun, or at the middle of the world, but is really at the middle of its elements, which accompany it and are united to it." All parts of the earth tend towards the center of gravity of the earth and in that way the surfaces of the seas are assured a spherical form, a form whose image is in the dewdrop.

Copernicus, at the beginning of the first book of his treatise on the revolutions of the heavens, expressed himself nearly in the same terms as Leonardo da Vinci and even used the same comparisons.[12] "The earth is spherical, for all its parts tend towards its center of gravity." Water and earth both tend to it, and this gives the form of a portion of a sphere to the surface of water; the sphere would be perfect if there were a sufficient quantity of bodies of water. Moreover, the sun, the moon, and the planets also have a spherical shape which is to be explained in the case of each of these celestial bodies as it is explained in the case of the earth:

"I think that gravity is nothing else than a certain natural appetition given to the parts of the earth by divine providence of the Architect of the Universe in order that they may be restored to their unity and to their integrity by reuniting in the shape of a sphere. It is credible that the same affection is in the sun, moon, and other errant bodies in order that, through the agency of this affection, they may persist in the rotundity with which they appear to us."[13]

Is this weight a universal weight? Is a mass belonging to a heavenly body attracted at the same time by the center of gravity

[10] *ibid.*, First Part, Ch. IX.

[11] *Les Manuscrits de Léonard de Vinci*, ed. C. Ravaisson-Mollien, MS. F of the Bibliothèque de l'Institut, Fol. 41, verso. This notebook bears the notice: "Begun at Milan, Sept. 12, 1508."

[12] *Nicolai Copernici "De revolutionibus orbium coelestium" libri sex* (Nuremberg, 1543), Book I, Chs. I, II, III.

[13] *ibid.*, Book I, Ch. IX.

of this body and by the centers of gravity of the other heavenly bodies? Nothing in the writings of Copernicus indicates that he admitted such a tendency; everything in the writings of his disciples shows that the tendency toward the center of a heavenly body is, in their opinion, an appropriate property of the parts of this body. In 1626 Mersenne summarized their teaching when, after giving the definition, "The center of the universe is that point toward which all heavy bodies tend in a straight line and is the common center of the heavy bodies," he added: "We assume it but cannot demonstrate it, for there probably exists a particular center of gravity in each of the particular systems forming the universe or, in other terms, in each of the great celestial bodies."[14]

On the subject of this teaching Mersenne, however, expressed a suspicion in favor of the hypothesis of a universal gravitation: "We assume that all heavy bodies desire the center of the world and bear towards it in a straight line with natural motion. This proposition is one that nearly everybody grants although it is not demonstrated at all; who knows whether the parts of a heavenly body wrested from it may not gravitate toward this body and return to it as stones detached from the earth and carried by it would come back toward the earth? Who knows whether terrestrial stones nearer to the moon than to the earth would not descend toward the moon rather than toward the earth?"[15] In this last sentence Mersenne was showing himself tempted, as we shall see, to follow Kepler's doctrine rather than that of Copernicus.

Galileo held more faithfully and more closely to the Copernican theory of the gravity particular to each heavenly body. On the "First Day" of the famous *Dialogue on the Two Main Systems of the World* he professed, through the voice of the interlocutor Salviati, that "the parts of the earth are moved not in order to go to the center of the world but in order to be reunited to their whole; that is why they have a natural inclination toward the center of the terrestrial globe, an inclination by which they conspire to form and preserve it. . . .

"As the parts of the earth all conspire in a common accord to form the whole to which they belong, the result is that they converge on all sides with equal inclination, and in order to be unified as much as possible with one another they take the shape of the sphere. Consequently, ought we not believe that if the moon, the

[14] Marin Mersenne, *Synopsis mathematica* (Paris: Rob. Stephani, 1626), *Mechanicorum libri*, p. 7.

[15] *ibid.*, p. 8.

sun, and the other large bodies making up the world are all of the same round shape, it is for no other reason than the concordant instinct and natural convergence of all their parts? So that when one of these parts is by some violence separated from its whole, is it not reasonable to believe that it would return to it spontaneously and by natural instinct?"

Surely the divergence of such a doctrine from that of Aristotle is profound. Aristotle energetically rejected the doctrine of the ancient philosophers of nature who like Empedocles saw in weight a sympathy of like for like; in the fourth book of his *De Caelo* he declared that heavy objects fall not in order to be one with the earth, but in order to be one with the center of the universe, and that if the earth torn from its place should be retained in the orbit of the moon, stones would not fall to the earth but to the center of the world.

And yet the Copernicans preserved all that they could of Aristotle's doctrine; for them, as for the Stagyrite, gravity was a tendency inherent in the heavy body, and not a violent attraction exerted by an alien body; for them, as for the Stagyrite, this tendency longed for a mathematical point, the center of the earth or of the heavenly body to which the body studied belongs; for them, as for the Stagyrite, this tendency of all parts toward a point was the reason for the spherical shape of each of the heavenly bodies.

Galileo went much further even and carried the teaching of Albert of Saxony over to the Copernican system. Defining the center of gravity of a body, he said in his famous work *Della Scienza Meccanica*: "Thus it is this point which tends to be one with the universal center of heavy things, that is to say, the earth's center." And this thought guided him when he formulated the principle: A group of heavy bodies is in equilibrium when the center of gravity of this group is as near as possible to the center of the earth.

Copernican physics consisted then essentially in denying the tendency of each element to go toward its natural place and in substituting for this propensity the natural sympathy of the parts of the same whole seeking to reconstitute that whole. About the time when Copernicus was employing this sympathy in order to explain the gravitation peculiar to each heavenly body, Fracastoro formulated the general theory of sympathy: When two parts of the same whole are separated from each other, each sends toward the other an emanation of its substantial form, a species propagated into the intervening space; by the contact of this species each of the parts tends toward the other so that they may be united in a single whole;

thus the mutual attractions of similars are explained, the sympathy of iron for the magnet being the type of such explanation.[16]

In conformity with the example of Fracastoro most physicians and astrologers (rarely was one not both at the same time) readily invoked such sympathies. Moreover, we shall see that the role of physicians and astrologers was not of negligible importance in the development of the doctrine of universal attraction.

Nobody gave this doctrine of sympathies more extensive development than William Gilbert. In the work, so capital for the theory of magnetism, with which he brought to a close the scientific work of the sixteenth century, Gilbert expressed concerning gravitation ideas similar to those which Copernicus had voiced: "The simple and straight motion downward considered by the Aristotelians, the motion of a heavy body, is a movement of reunion (*coacervatio*) of disjunct parts being directed, on account of the matter of which they are formed, in straight lines toward the body of the earth, these lines leading to the center by the shortest path. The motions of the isolated magnetic parts of the earth are, in addition to the motion which reunites them to the whole, the movements which unite them among themselves, and which make them turn and direct them toward the whole in view of the symphony and harmony of form."[17] "This rectilinear motion which is only an inclination toward its principle, does not belong solely to the parts of the earth, but also to the parts of the sun, to those of the moon, and to those of the other celestial globes."[18] However, it is *not* that this virtue of attraction is a universal gravitation; it is a virtue proper to each heavenly body, as magnetism is to the earth or to the magnet: "let us give the reason, now, for this coition and for this movement which bestirs all nature. . . . It is a special and particular substantial form belonging to primary and principal globes; it is a proper entity and essence of their homogeneous and uncorrupted parts which we may call primary, radical, and astral form; it is not Aristotle's first form, but that special form by which the globe preserves and arranges what belongs to its nature. There is such a form in each of the globes, in the sun, in the moon, in the stars; there is also one in the earth constituting that very magnetic power which we call primary vigor. There is then a magnetic nature be-

[16] *Hieronymi Fracastorii "De sympathia et antipathia rerum," liber unus.* Reprinted in *Hieronymi Fracastorii "Opera omnia"* (Venice, 1555).

[17] *Gulielmi Gilberti Colcestrensis, medici Londinensis, "De magnete, magneticis corporibus, et de magno magnete Tellure, physiologia nova"* (London, 1600), p. 225.

[18] *ibid.*, p. 227.

longing to the earth, and which, for a fundamental reason indeed worthy of exciting our wonder, resides in each of its true parts. . . . There is in the earth a magnetic vigor belonging to it, as there is a substantial form in the sun and one in the moon; the moon disposes in a lunar manner the fragments which might be detached from it, in accord with its form and the limits imposed on it; a fragment of the sun is carried toward the sun, as the magnet toward the earth or to another magnet, by its natural inclination and as though it were excited by lust."[19]

These thoughts are scattered through the book of Gilbert on the magnet; amply developed they assume a dominant importance in his work on the system of the world, a work which his brother published after his death.[20] The leading idea of this work is condensed in the following passage: "Everything terrestrial is reunited to the earth; likewise, everything homogeneous with the sun tends toward the sun, all lunar things toward the moon, and the same for the other bodies forming the universe. Each of the parts of such a body adheres to its whole and does not spontaneously detach itself from it; if it were snatched from it, not only would it make an effort to return to it but it would be called and enticed by the globe's virtues. If it were not so, if the parts could separate themselves spontaneously, and if they did not return to their origin, the whole world would soon be dissipated in confusion. It is not a question of an appetite which brings the parts toward a certain place, a certain space, a certain term, but of a propensity toward the body, toward a common source, toward the mother where they were begotten, toward their origin, in which all these parts will be united and preserved, and in which they will remain at rest, safe from every peril."[21]

The magnetic philosophy of Gilbert made numerous adepts among physicists; let us be content with a mere reference to Francis Bacon,[22] whose opinions are a confused reflection of the doctrines of his contemporary scientist, and let us turn at once to the true creator of universal gravitation, namely, Kepler.

Even while proclaiming on more than one occasion his admiration for Gilbert and declaring himself in favor of the magnetic philosophy, Kepler went ahead and changed all its principles; he

[19] *ibid.*, p. 65.

[20] *Gulielmi Gilberti Colcestrensis, medici Regii, "De mundo nostro sublunari philosophia nova"* (Amsterdam, 1651). Gilbert died in 1603.

[21] *ibid.*, p. 115.

[22] Bacon, *Novum Organum*, Book II, Ch. XLVIII, Arts. 7, 8, 9.

replaced the tendencies of the parts of a heavenly body toward its center by their mutual attractions; he declared that this attraction proceeds from a single and universal virtue whether among the parts of the moon or of the earth; he left to one side any consideration relative to the final causes attaching this virtue to the preservation of the form of each heavenly body; in short, he went ahead and opened up all the roads to be followed by the doctrine of universal gravitation.

First of all, Kepler denied any attractive or repulsive power to any mathematical point, whether it be the center of the earth, as Copernicus thought, or the center of the universe, as Aristotle thought: "The action of fire does not consist in gaining the surface bounding the world but in fleeing from the center, not the center of the universe but the center of the earth; and not this center insofar as it is a point but insofar as it is in the middle of a body, a body which is opposed to the nature of fire desiring to expand. I shall say furthermore that the flame does not flee but is driven out by the heavier air as an inflated bladder would be by water. . . . If we were to place the earth at rest in some place and bring near it a larger earth, the first one would become a weight in relation to the second one and would be attracted by the latter as a stone is attracted by the earth. Gravity is not an action but a passivity of the stone which is attracted."[23]

"A mathematical point, whether it be the center of the world or some other point, cannot in fact move weights; nor can it be the object toward which they tend. Let physicists prove then that such a force can belong to a point which is not a body, and which is conceived only in an entirely relative way!

"It is impossible for the substantial form of a stone, putting the stone's body into motion, to seek a mathematical point like the center of the world without caring about the body in which this point lies. Let physicists demonstrate then that natural things have some sympathy for what does not exist!

". . . . Here is the *true doctrine of gravity*: Gravity is a mutual affection among related bodies which tends to unite and conjoin them; the magnetic faculty is a property of the same order; the earth attracts the stone, rather than the stone tending toward the earth. Even if we placed the center of the earth at the center of the world, it would not be toward this center of the world that weights would be carried, but toward the center of the round body to which

[23] *Joannis Kepleris "Littera ad Herwartum,"* March 28, 1605. Reprinted in *Joannis Kepleri astronomi "Opera omnia,"* ed. C. Frisch, II, 87.

they are related, that is, toward the center of the earth. Thus, no matter where the earth is transported it is always toward it that heavy bodies are borne, thanks to the faculty animating it. If the earth were not round, heavy bodies on all sides would not be borne straight to the center of the earth, but depending on whether they came from one place or another they would be borne toward different points. If in some place in the universe we were to put two stones close to each other and beyond the sphere of influence of any body related to them, these stones in the manner of two magnets would come and meet in a place in between, and the paths they would follow in order to meet would be in inverse ratio to their masses."[24]

This "true doctrine of gravity" soon spread in Europe and found favor with many a mathematician. In 1626 Mersenne made allusion to it in his *Synopsis mathematica.* On August 16, 1636, Etienne Pascal and Roberval wrote a letter to Fermat for the primary purpose of disputing the old principle of Albert of Saxony, jealously maintained by the mathematician of Toulouse, "that if two equal weights are joined by a straight line, firm and weightless, and if in that arrangement they may descend again freely, they will never come to rest until the middle of the line (which is the center of gravity of the ancients) is united to the common center of heavy things." They objected to this principle, as follows: "It may also be the case and it is very probable that gravity is a mutual attraction or a natural desire of bodies to come together, as is clear in the case of the iron and the magnet where we find that if the magnet is arrested, the iron being free will go and seek it; if the iron is arrested, the magnet will go toward it; and if both are free, they will draw near each other reciprocally, so that in any case the stronger of the two will take the shorter path."[25]

Do terrestrial bodies have no other *magnetic faculty* than the power which brings them back to the ground from which they have been taken and which constitutes their gravity?

The movement which swells the waters of the sea and produces the tide follows so exactly the moon's transit of the meridian that the moon had to be regarded as the cause of this phenomenon as soon as its laws had been recognized at all correctly; the observations of Eratosthenes, Seleucus, Hipparchus, and especially Posi-

[24] *Joannis Kepleri "De motibus stellae Martis commentarii"* (Prague, 1609). Reprinted in *J. Kepleri "Opera omnia,"* III, 151.

[25] Pierre de Fermat, *Oeuvres*, ed. P. Tannery and C. Henry, II, 35.

donius[26] assured the ancient philosophers of a complete enough knowledge of these laws for Cicero, Pliny the Elder, Strabo, and Ptolemy not to be afraid to state that the phenomenon of the tides depended on the course of the moon. But this dependence was soon established by the detailed description of the diverse vicissitudes of the tide given by the Arab astronomer Albumasar in the ninth century in his *Introductorium magnum ad Astronomiam.*

The moon, then, determines the rising of the waters of the ocean. But in what manner does it determine it?

Ptolemy and Albumasar did not hesitate to invoke a particular virtue, a special influence of the moon on the waters of the sea. Such an explanation was not intended to please the true disciples of Aristotle; whatever has been said in this regard, the fact is that faithful Aristotelians, whether Arabs or masters of occidental Scholasticism, strongly repudiated explanations which invoked occult powers inaccessible to the senses: the action of the magnet on iron was about the only one of these mysterious virtues they were willing to accept; they would not at all admit that heavenly bodies could exercise any influence which does not proceed from their motion or from their light. Therefore, it is from the light of the moon, from the heat that this light may create, from the currents that this heat may cause in the atmosphere, and from the ebullition that this may produce within the waters of the sea that the explanation of ebb and flow was sought by Avicenna, Averroes, Robert Grosseteste, Albertus Magnus, and Roger Bacon.

A very shaky explanation it was, and one which too many obvious objections would ruin in advance. Already Albumasar had observed that the moon's light was negligible in ocean tide, since this tide is produced as well under a new moon as under a full moon, and since it takes place the same way whether the moon is at its zenith or at its nadir. The somewhat childish explanation which Robert Grosseteste had proposed in order to remove this last objection, despite Roger Bacon's enthusiastic vote for it, could not hurt Albumasar's argument. From the thirteenth century on, the best of the Scholastics, including Saint Thomas, admitted the possibility of astral influences other than light; just at that time William of Auvergne in his work *De Universo* compared the action of the moon on the waters of the sea to the action of the magnet on iron.

The magnetic theory of tides was known by the great physicists

[26] See Roberto Almagia, "Sulla dottrina della marea nell' antichita classica e nel medio evo," *Atti del Congresso internazionale di Scienze historiche,* Rome, April 1-9, 1903, XII, 151.

who in the middle of the fourteenth century distinguished the nominalist school of the Sorbonne. Albert of Saxony and Timon the Jew expounded it in their *Questions* on Aristotle's *De Caelo* and *Meteors*, but they hesitated to grant it their wholehearted support; they knew too well the validity of Albumasar's objections to acquiesce unqualifiedly to the explanations of Albertus Magnus and Roger Bacon; and yet this occult magnetic attraction exerted by the moon on the seas is contrary to their Aristotelian rationalism.

The virtue that the tides manifest was, on the other hand, made to order for the astrologers who found in it the undeniable proof of the influences that the heavenly bodies exert on sublunar things. This hypothesis was in no less favor among the physicians who compared the role played by heavenly bodies in the tidal phenomenon with the role they attributed to them in crises of disease; did not Galen attach the "critical days of pituitary diseases" to the phases of the moon?

At the end of the fifteenth century Giovanni Pico della Mirandola took up again without compromise the Aristotelian thesis of Avicenna and Averroes: he denied the power of heavenly bodies to act here below except by their light; he rejected all judiciary astrology as illusory; he repudiated the medical doctrine of critical days; and at the same time he declared the magnetic theory of tides erroneous.[27]

The challenge hurled at the astrologers and physicians by Pico della Mirandola was immediately met by a physician from Siena, Lucius Bellantius, in a book which had a steady succession of editions.[28] In Book III of this work the author, examining what Pico della Mirandola had said about tides, wrote these lines: "The rays by which the moon chiefly acts when it attracts and swells the waters of the sea are not the rays of moonlight, for at the time of conjunctions there would be no ebb and flow whereas we can and do notice them then; it is by means of virtual rays of influence that the moon attracts the sea as the magnet attracts iron. With the aid of these rays we can easily resolve all the objections concerning this matter."

The book of Lucius Bellantius was undoubtedly the signal for a renewal of support for the magnetic theory of tides: in the middle of the sixteenth century this theory was generally accepted.

[27] *Joannis Pici Mirandulae "Adversus astrologos"* (Bologna, 1495).

[28] *Lucii Bellantii Senensis "Liber de astrologia veritate et in disputationes Joannis Pici adversus astrologos responsiones"* (Bologna, 1495; Florence, 1498; Venice, 1502; Basel, 1504).

Cardan included in his classification of seven simple motions ". . . a new, different nature which is made up of some obedience of things like that of water on account of the moon, like that of iron on account of the magnet, the so-called Hercules stone."[29]

Julius Caesar Scaliger adopted the same opinion: "Iron is moved by the magnet without being in contact with it; why should not the sea likewise follow a very eminent heavenly body?"[30]

Duret mentioned the opinion of Lucius Bellantius, without adopting it, however: "This author assures us that the moon attracts the waters of the sea not by the rays of its light, but by the virtue and power of certain of its occult properties, just as the magnet does to iron."[31]

Finally, Gilbert professed that "the moon does not act on the sea through its rays or through its light. How then does it act? Through the joint action or conspiracy of the two bodies and, to explain my thought with the aid of an analogy, through magnetic attraction."[32]

Moreover, this action of the moon on the sea's waters belongs to those sympathetic propensities of like for like in which the Copernicans sought the explanation of gravity. Every body has a substantial form such that it tends to unite itself to another body of the same nature; therefore, it is natural for the sea water to try to rejoin the moon, which for astrologers as well as for physicians is preeminently the humid celestial body.

Ptolemy in his *Opus quadripartitum* and Albumasar in his *Introductorium magnum* attribute to Saturn the property of creating cold; to Jupiter, temperate weather; to Mars, burning heat; to the moon, humidity. Hence, the moon's action on the waters of the sea is a sympathy between two bodies of the same family, a "cognate virtue," as the Arab author said.

These doctrines were preserved by the physicians and astrologers of the Middle Ages and of the Renaissance: "We cannot doubt," said Cardan, "the influence exerted by the celestial bodies; it is an occult action governing all perishable things. And yet certain disrespectful and ambitious minds, much more impious than Eratosthenes, dare to deny it. . . . Do we not see that among terrestrial

[29] *Les livres d'Hiérome Cardanus, médecin milanois, intitulés de la subtilité et subtiles inventions*, translated from Latin into French by Richard Le Blanc (Paris, 1556), p. 35.

[30] *Julii Caesaris Scaligeri "Exercitationes . . . ,"* Problem LII.

[31] Claude Duret, *Discours de la vérité des causes et effets de divers cours, mouvemens, flux, et reflux de la mer océane, mer meditérannée et autres mers de la Terre* (Paris, 1600), p. 204.

[32] *Gulielmi Gilberti . . . "De mundo nostro . . . ,"* p. 307.

substances there are some like the magnet whose qualities exert manifest actions? . . . Why should we refuse such actions to the eternal and very eminent body of heaven? . . . On account of its size and the quantity of light it diffuses, the sun is the principal commander of all things. The moon comes next, for the same reasons, for it appears to us the biggest heavenly body after the sun although it really is not so. Above all the moon commands humid things, fish, waters, the marrow and brain of animals, and among roots, garlic and onion which especially contain moisture."[33]

Even Kepler, who rose so energetically against the unwarranted claims of judicial astrology, was not afraid to write: "Experience proves that everything containing humidity swells when the moon rises and shrinks when the moon sets."[34]

Kepler boasted of having been the first to upset this opinion according to which the tide would be an effort of sea water to unite with the moon's humors. "As certain as the ebb and flow of the sea is it certain that the humidity of the moon is foreign to the cause of this phenomenon. I am the first, so far as I know, to have revealed, in my prolegomenon to *De motibus stellae Martis*, the process by which the moon causes the ebb and flow of the sea. It consists in this: The moon does not act like a humid or humidifying celestial body, but like a mass related to the mass of the earth; it attracts the waters of the sea by a magnetic action, not because they are humors but because they are endowed with terrestrial substance, a substance to which they also owe their gravity."[35]

The tide is indeed a propensity of like to unite with like not in that they both participate in the nature of water but in that they both participate in the nature of the masses making up our globe. Thus the moon's attraction does not exert itself solely on the waters covering the earth but also on the solid parts and on the earth as a whole; and conversely, the earth exerts a magnetic attraction on the moon's heavy bodies. "If the moon and the earth were not retained in their respective orbits by an animal force or by some equivalent force, the earth would climb toward the moon and the moon would descend toward the earth until these two heavenly bodies were joined. If the earth ceased attracting the waters cover-

[33] *Hieronymi Cardani "De rerum varietate" libri* XVII (Basel, 1557), Book II, Ch. XIII.

[34] *Joannis Kepleri "De fundamentis Astrologiae"* (Prague, 1602), Thesis XV. Reprinted in *J. Kepleri "Opera omnia,"* I, 422.

[35] *J. Kepleri "Notae in librum Plutarchi de facie in orbe Lunae"* (Frankfurt, 1634). Reprinted in *J. Kepleri "Opera omnia,"* VIII, 118.

ing it, the sea waves would all rise and flow toward the body of the moon."[36]

These opinions have enticed more than one physicist: on September 1, 1631, Mersenne wrote to Jean Rey: "I do not at all doubt that the stones thrown up by a man on the moon would fall back on the moon even though he should have his head turned in our direction; for stones fall back on the earth because they are nearer to it than to other systems."[37] But Jean Rey did not welcome with favor this Keplerian manner of looking at the matter; on the first day of the year 1632, he replied to Mersenne: "You do not at all doubt, you say, that stones thrown upward by a man on the moon would fall back on said moon even though he were facing us. I see nothing surprising in that; if I must speak frankly, I have a contrary opinion, for I presuppose that you mean to be speaking of stones taken from here (thus, perhaps, there might not be any on the moon). Now, such stones have no other inclination than to be borne to their center, namely, that of the earth; they would come toward us with the man who would be throwing them if he were one of our earth's creatures, justifying in that way the truth of the saying: *Nescio qua natale solum dulcedine cunctos allicit* (*our native soil has a certain charm and attraction for all of us*). And if they happened to be attracted by the moon as by a magnet (which you ought to suspect as well as the earth), you have in that case the earth and the moon endowed with the same magnetic faculty attracting the same body, and converging on the latter conjointly because they attract one another mutually, or better still, because they concur in uniting with one another, as I see two magnetic spheres, made to swim in a basin of water, drawing near each other. For there is no ground of objection in the distance being too great; the influences that the moon casts on the earth and those the earth must cast on the moon, since the earth serves the latter as a moon according to your opinion—these influences make us see clearly that each is in the sphere of activity of the other."[38]

Still it is this objection that Descartes voiced; questioned by Mersenne on the point of "knowing whether a body weighs more or less when it is near the center of the earth than far from it," Des-

[36] *Joannis Kepleri "De motibus stellae Martis"* (1609). Reprinted in *J. Kepleri "Opera omnia,"* III, 151.

[37] Jean Rey, *Essays de . . . , Docteur en médecine, sur la recherche de la cause pour laquelle l'estain et le plomb augmentent de poids quand on les calcine* (new edition increased by the correspondence of Mersenne and Jean Rey; Paris, 1777), p. 109.

[38] *ibid.*, p. 122.

cartes employed the following argument, really appropriate to prove that bodies far from the earth weigh less then those near it: "The planets which are not self-luminous, for example, the moon, Venus, Mercury, etc., being, as is likely, bodies of the same matter as the earth . . . , it seems that these planets should therefore be heavy and fall toward the earth were it not the case that their great distance removes their inclination to do so."[39]

Despite the difficulties that physicists encountered during the first part of the seventeenth century in explaining why the mutual gravitation of the earth and moon does not cause them to fall toward each other, belief in such gravitation went on spreading and becoming stronger. Descartes, we have seen, thought that a similar gravitation could exist between the earth and the other planets like Venus and Mercury. Francis Bacon had pushed farther; he had imagined that the sun could exert an action of the same nature on the different planets. In the *Novum Organum* the distinguished chancellor put in a special category "the *magnetic* motion which, belonging to the class of motions of *minor aggregation* but operating sometimes at great distances and on considerable masses, merits special investigation under this heading, especially when it does not begin by contact as most other motions of aggregation do, and is limited to raising bodies or swelling them without producing anything else. If it is true that the moon attracts waters and that under its influence nature sees humid masses swell . . . if the sun enchains Venus and Mercury and does not allow them to go farther away than a certain distance, it seems indeed that these motions belong neither to the species of *major aggregation* nor to the species of *minor aggregation*, but that tending to an average and imperfect aggregation, they should constitute a species apart."[40]

The hypothesis that the sun might be exerting on the planets an action analogous to the one that the earth and the planets respectively exert on their own parts, and similar even to the action between the earth and the planets, was bound to appear a very daring supposition; it implied, in fact, that there existed a natural analogy between the sun and the planets, and many a physicist was bound to refuse this postulate; we find in the writings of Gassendi evidence of the repugnance felt by more than one mind toward admitting

[39] R. Descartes, *Correspondence*, ed. P. Tannery and C. Adam, Letter CXXIX (July 13, 1638), Vol. II, p. 225.

[40] *F. Baconis "Novum Organum"* (London, 1620), Book II, Ch. XXVIII, Art. 9.

the postulate. Notice under what circumstances this repugnance of Gassendi manifested itself:

The Copernicans, who had so readily attributed gravitation to a mutual sympathy of terrestrial bodies and who had employed a similar sympathy among the different parts of a celestial body in order to explain the spherical shape of that body, generally refused to recognize the magnetic attraction exerted by the moon on the waters of the sea. They clung to quite a different theory of tides; the source of this theory was at the origin of their system and it seemed to them to be a particularly convincing proof of it.

In 1544 the works of Caelio Calcagnini appeared at Basel;[41] the author had died three years before, just when Joachim Rheticus in his *Narratio Prima* informed the world of Copernicus' system before the great Polish astronomer had published his *De revolutionibus orbium caelestium libri sex*. The works of Calcagnini contained a dissertation, already old, entitled *Quod Caelum stet, Terra vero moveatur, vel de perenni motu Terrae*.[42] Without admitting yet the annual motion of the earth around the sun, this precursor of Copernicus already was attributing the daily motion of the heavenly bodies to the earth's rotation. In this dissertation the following passage was to be read: "Necessarily, the farther a thing is from the center, the more rapidly it moves. In that way is resolved an enormous difficulty, which was the object of numerous lengthy investigations and which, it is said, was the despair of Aristotle to the point of causing his death. It was the question of the cause producing at perfectly fixed intervals of time that remarkable oscillation of the sea. . . . The difficulty is resolved without trouble if we take into account the opposing impulsions animating the earth, first causing one part to descend, then raising it, the former producing a depression of waters, the latter projecting them upwards."[43]

Galileo was to take up this theory, making it precise and detailed, a theory which tries to explain the ebb and flow of the ocean through actions brought about by the earth's rotation.

The explanation was untenable, for it demanded that the interval between two high tides should be equal to half a sidereal day, whereas the most obvious observations show that it is equal to half a lunar day. Galileo, however, persisted in giving this explanation

[41] *Caelii Calcagnini Ferrarensis "Opera aliquot"* (Basel, 1544).

[42] This dissertation, addressed to Bonaventura Pistophilius, is not dated; it is followed in the *Opera* of Calcagnini by another dissertation addressed to the same person and dated January 1525. It is probable that the first dissertation was written prior to that date.

[43] *Caelii Calcagnini* . . . , p. 392.

as one of the best proofs of the earth's motion, and those who with him accepted the reality of this motion gladly repeated this argument, for example, Gassendi in the work *De motu impresso a motore translato,* which he published in Paris in 1641.

Naturally, the opponents of Copernicus held out for the explanation of tides through lunar attraction, an explanation which did not imply terrestrial rotation.

Among the most ardent adversaries of the system of Copernicus, Morin must be mentioned; with equal ardor he tried to restore judicial astrology and to forecast horoscopes. Thinking he saw in Gassendi's work a personal attack, Morin replied with a libellous tract entitled: *Alae telluris tractae*; in this work he opposed Galileo's theory with the magnetic theory of tides.

The difference of level between high and low tide is very large at the time of a full moon or a new moon; it is much less when the moon is in its first or last quarter. This alternation of "live waters" and "dead waters" had been very embarrassing up to then for the magnetic philosophers.

Morin gave an explanation for it which he drew, he said, from the principles of astrology. This alternation is explained by the concourse of the sun and the moon: in their conjunctions as in their oppositions their forces are directed in the same straight line passing through the earth, and it is "a vulgar axiom that united virtues are stronger than dispersed virtues."

Morin fell back on principles of judiciary astrology in order to affirm the role played by the sun in the variations of tide, and it is indeed to the indisputable credit of the astrologers that they prepared all the materials for the Newtonian theory of tides, whereas the defenders of rational scientific methods, Aristotelians, Copernicans, atomists, and Cartesians, have in emulation fought its advent.

The principles invoked by Morin were, besides, very old ones; already Ptolemy in his *Opus quadripartitum* had admitted that the position of the sun in relation to the moon could either strengthen or weaken the influence of the latter; and this opinion had been transmitted from generation to generation down to Gaspard Contarini, who taught that "the sun exerts some action apt to raise or appease the waters of the sea";[44] down to Duret, according to whom "it is quite apparent that the sun and the moon labor powerfully

[44] *Gasparis Contarini "De elementis eorumque mixtionibus" libri* II (Paris, 1548).

in that emotion and agitation of the waves of the sea";[45] down to Gilbert, who called to the aid of the moon "the auxiliary troops of the sun," and who declared the sun capable of "increasing the lunar powers at the time of the new moon and the full moon."[46]

Faithful to their rationalism, the Scholastic Aristotelians tried to explain the alternation of live and dead waters without attributing any occult virtue to the sun. Albertus Magnus claimed he was invoking only the variation of the light received by the moon from the sun according to the relative position of these two heavenly bodies.[47] In an attempt at a rational explanation of the same kind, Timon the Jew glimpsed, at least, a great truth, for he admitted the coexistence of two tides, a lunar and a solar tide; he attributed the first to the generation of water caused by the cold of the moon, and the second to the boiling caused by the heat of the sun.[48]

But it is to the physicians and the astrologers of the sixteenth century that we must attribute the precise and fruitful idea of decomposing the total tide into two tides of the same nature though of unequal intensity, one produced by the moon and the other by the sun, and to explain the diverse vicissitudes of ebb and flow by the agreement or disagreement of these two tides.

This idea was formally enounced in 1528 by a Dalmatian nobleman, Frederick Grisogon of Zara, whom Hannibal Raymond introduces to us as a "great physician, philosopher, and astrologer."

In a work devoted to the critical days of diseases,[49] he laid down this principle: "The sun and the moon draw toward them the rising of the sea so that the maximum rise is perpendicularly below each of them; therefore there are for each of them two maxima of rise, one below the heavenly body and the other in the opposite part that we call the nadir of this heavenly body." And Frederick Grisogon circumscribed the terrestrial sphere by two ellipsoids of revolution, one whose major axis is directed toward the sun and the other whose major axis is in the direction of the moon. Each of these two ellipsoids represents the shape that the sea would take if it

[45] Claude Duret, *op.cit.*, p. 236.

[46] *Gulielmi Gilberti . . . "De mundo nostro . . . ,"* pp. 309 and 313.

[47] *Alberti Magni "De causis proprietatum elementorum" liber unus*, Tract. II, Ch. VI. Reprinted in *B. Alberti Magni "Opera omnia"* (London, 1651), v, 306.

[48] *"Quaestiones super quatuor libros meteorum" compilatae per doctissimum philosophum professorem Thimonem* (Paris, 1516 and 1518), Book II, question ii.

[49] *Federici Chrisogoni nobilis Jadertini "De artificioso modo collegiandi, pronosticandi et curandi febres et de prognosticis aegritudinum per dies criticos necnon de humana felicitate, ac denique de fluxu et refluxu maris"* (Venice: printed by Joan. A. de Sabio, 1528).

were subjected to the action of only one heavenly body; by compounding them the diverse peculiarities of the tide are explained.

The theory of Frederick Grisogon of Zara did not take long to spread. In 1557 the distinguished mathematician, physician, and astrologer Jerome Cardan expounded a summary of it.[50] About the same time Federico Delfini was teaching at Padua a theory of the tides derived from the same principle.[51] Thirty years later Paolo Gallucci reproduced the theory of Frederick Grisogon[52] while Annibale Raimondo expounded and commented on the two doctrines of Grisogon and Delfino.[53] Finally, just at the end of the sixteenth century Claude Duret impudently reproduced Delfino's doctrine under his own name.[54]

The hypothesis of the sun's action on the waters of the sea, an action entirely similar to that exerted by the moon, had already passed its test and had already provided a very satisfactory theory of the ebb and flow when Morin helped himself to its use in his libel against Gassendi.

Gassendi rose energetically against the idea of a magnetic virtue through which the moon would attract the earth's waters; but he rebutted still more violently the new hypothesis formulated by Morin: "Usually moisture is held to be the phenomenon proper to the moon, and it belongs to the sun not to promote this phenomenon but to prevent it. But Morin likes to have the sun second the actions of the moon; he declares the actions of the sun and moon corroborate each other. He therefore supposes that the actions of the sun as well as those of the moon are conditioned by the same specific nature, as they say; with regard to the phenomenon we are studying, if the action of the moon attracts waters, it ought to be the same for the action of the sun."[55]

That year 1643, when Gassendi declared the invalidity of the hypothesis that the moon and sun could exert analogous attractions, was the one in which this hypothesis was formulated anew, but generalized and broadened into the assumption of a universal gravitation. This grandiose assumption was due to Roberval who,

[50] *Hieronymi Cardani "De rerum varietate" libri* XVII (Basel, 1557), Book II, Cap. XIII.

[51] *Federici Delphini "De fluxu et refluxu aquae maris"* (Venice, 1559; 2nd ed., Basel, 1577).

[52] *Pauli Gallucii "Theatrum mundi et temporis"* (1588), p. 70.

[53] Annibale Raimondo, *Trattato del flusso e reflusso del mare* (Venice, 1589).

[54] Claude Duret, *op.cit.*

[55] *Gassendi "Epistolae tres de motu impresso a motore translato"* (Paris, 1643), Letter III, Art. XVI. Reprinted in *Opuscula philosophica* (London, 1658), III, 534.

not daring to present it too openly under his name, gave himself as only the editor and annotator of a work which he said was composed by Aristarchus of Samos.[56]

Roberval asserted: "A certain property or accident inheres in all the fluid matter that fills the space included between the heavenly bodies, and inheres in each of their parts; through the force of this property this matter is united into a single, continuous body whose parts by an incessant effort are borne toward one another and mutually attracted to one another, to the point of being closely cohesive and unable to be separated except by a greater force. That being posited, if this matter were alone and not joined to the sun or other bodies, it would be concentrated into a perfect globe; it would take on exactly the shape of a sphere, and could never remain in equilibrium except by taking that shape. In this shape the center of action would coincide with the center of form. Toward this center all the parts of matter would tend through their own effort or appetite and through the mutual attraction of the whole; it would not be, as the ignorant imagine, through the virtue of the same center but through the virtue of the whole system whose parts are equally disposed around this center. . . .

"Inherent in the entire system of the earth and its elements, and in each of the parts of this system is a certain accident or property similar to the property we have attributed to the system of the world taken in its entirety; through the force of this property all the parts of this system are united into a single mass, are borne toward each other, and attract each other mutually; they cohere closely and can be separated only by a greater force. But the diverse parts of earthly elements participate unequally in this property or

[56] *Aristarchi Samii "De Mundi systemate, partibus et motibus cujusdem" liber singularis,* ed. P. de Roberval (Paris, 1644). This work was reprinted by Mersenne in 1647 in Volume III of his *Cogitata physico-mathematica.* I think that if we were to interpret Roberval's thought exactly, we should not see in his system a theory of universal gravitation: parts of the interplanetary fluid would attract only parts of the same fluid; terrestrial parts would attract only terrestrial parts; parts of the system of Venus, only parts of the same system; etc. However, there would be a mutual attraction between the system of the earth and the system of the moon, between the system of Jupiter and the satellites of that heavenly body. The application of Archimedes' principle made by Roberval to the equilibrium of a planetary system within the interplanetary fluid would then be erroneous; but a similar error frequently occurs in the mathematical works of the sixteenth century and is present even in the early writings of Galileo. In any case, Descartes in his criticism of Roberval's system understood him to be assuming universal gravitation. (See letter of Descartes to Mersenne dated April 20, 1646, in R. Descartes, *Correspondance,* ed. P. Tannery and C. Adam, IV, 399.)

accident; for the denser the part, the more it participates in this property. . . . In the three bodies called earth, water, and air, this property is what we usually term gravity or levity, since for us levity is only a smaller gravity compared to a larger gravity."

Roberval repeated similar considerations concerning the sun and the other celestial bodies so that a hundred years after the publication of the six books of Copernicus on the celestial revolutions, the hypothesis of universal gravitation was formulated.

However, a lacuna made this hypothesis incomplete: According to what law does the mutual attraction of two material parts become attenuated when the distance between these two bodies is increased? No answer was given by Roberval to this question. But this answer could not take long to be formulated; or, it would be better to say, it was not formulated yet because it was not held in doubt by anyone.

The analogy between the influences emanating from astral bodies and the light emitted by them was really a commonplace for the physicians and astrologers of the Middle Ages and Renaissance; most of the Scholastic Aristotelians pushed this analogy to the point of making it into an identity or indissoluble connection. Scaliger was already under compunction to protest against this extreme: "Heavenly bodies can act without the aid of light. The magnet does well without light; how much more splendidly will heavenly bodies act!"[57]

Whether identical with light or not, all the virtues and all the *species* of its substantial form that a body emits in the space around it have to be propagated or, as was said in the Middle Ages, "multiplied" according to the same laws. In the thirteenth century Roger Bacon undertook to give a general theory of this propagation;[58] in any homogeneous medium it is effected by following rectilinear rays[59] and, to use the modern expression, by "spherical waves." If he had been as good a mathematician as he expected physicists to be, Bacon might easily have drawn the following conclusion from his reasoning:[60] The force of such a species is always in inverse ratio to the square of the distance from the source from which it emanates. Such a law was the natural corollary of the admitted analogy between the propagation of these virtues and that of light.

[57] *Julii Caesaris Scaligeri "Exercitationes . . . ,"* Problem LXXXV.

[58] *Rogerii Bacconnis Angli "Specula mathematica in qua de specierum multiplicatione, earumdemque in inferioribus virtute agitur"* (Frankfurt, 1614).

[59] *ibid.*, Dist. II, Chs. I, II, III.

[60] *ibid.*, Dist. III, Ch. II.

Perhaps no astronomer has insisted more on this analogy than Kepler did. The rotation of the sun was for him the cause of the revolution of the planets: the sun sends out to its planets a certain quality, a certain resemblance of its motion, a certain species of motion (*species motus*) which is to lead them towards their whole. This *species motus* or this power of moving (*virtus movens*) is not identical with solar light, but it has a certain kinship with it; it makes use, perhaps, of solar light as an instrument or vehicle.[61]

Now, the intensity of the light emitted by a heavenly body varies in inverse ratio to the square of the distance from this body; knowledge of this proposition appears to go back to antiquity; it is found in a work on optics attributed to Euclid, and Kepler gave a demonstration of it.[62] The analogy would have it that the power of moving (*virtus movens*) emanating from the sun should vary in inverse ratio to the square of the distance from that heavenly body. But the dynamics which Kepler used is still the ancient dynamics of Aristotle; the force moving a movable body is proportional to the speed of the latter; hence the law of areas which Kepler discovered taught him the following proposition: The moving power to which a planet is subjected varies inversely simply with its distance from the sun.

This mode of variation, hardly in conformity with the analogy of the species of motion coming from the sun or of the light emitted by it, does go contrary to Kepler; he tried to adjust it to this analogy by means of this observation in particular: Light spreads out in space in all directions, whereas the *virtus motrix* is propagated solely in the plane of the sun's equator. The intensity of the former is inverse to the square of the distance from the source, the intensity of the latter is inverse simply to the distance traversed; these two distinct laws express the same truth in one case as in the other: The total quantity of light or of "species of motion" which is propagated does not suffer any loss in the course of propagation.[63]

The very explanations of Kepler show us with what force, to his mind, the law of the inverse square of the distances is imposed,

[61] *Joannis Kepleri "De motibus . . . ,"* Ch. XXXIV (reprinted in *J. Kepleri "Opera omnia,"* III, 302); *"Epitome Astronomiae Copernicanae,"* Book IV, Part II, Art. 3 (reprinted in *J. Kepleri "Opera omnia,"* VI, 374).

[62] *Joannis Kepleri "Ad vitellium paralipomena quibus Astronomiae pars optica traditur"* (Frankfurt, 1604), Ch. I, Prop. IX. Reprinted in *J. Kepleri "Opera omnia,"* II, 133.

[63] *Joannis Kepleri "De motibus . . . ,"* Ch. XXXVI (reprinted in *J. Kepleri "Opera omnia,"* III, 302, 309); *"Epitome Astronomiae Copernicanae,"* Book IV, Part II, Art. 3 (reprinted in *J. Kepleri "Opera omnia,"* VI, 349).

first of all, on the intensity of a quality when a body emits this quality in every direction around it. This law appeared to his contemporaries to be endowed with the same self-evidence. Ismael Bullialdus established it first of all for light;[64] he did not hesitate to extend it to the power of motion (*virtus motrix*) that the sun, according to Kepler, exerts on the planets: "This virtue, by which the sun seizes or hooks the planets and which is like the hands of the body for the sun, is emitted in a straight line into the whole space the world occupies; it is like a *species* of the sun turning with the body of that heavenly orb; being corporeal it diminishes and becomes weaker as the distance increases, and the ratio of this diminution is for light inverse to the square of the distance."[65]

The power of motion mentioned by Bullialdus, and by Kepler too, is not directed along the radial line from the sun to the planet, but is normal to that line. It is not a force of attraction similar to the one admitted by Roberval, and later by Newton; but we see clearly that the physicists of the seventeenth century dealing with the attraction of two bodies were from the very start led to suppose it to be inverse to the square of the distance between the two bodies.

The works of Father Athanasius Kircher on the magnet offer us a second example of the law.[66] The analogy between the light emitted by a source and the virtue emanating from each of the poles of a magnet urged him to adopt a law of decrease in the inverse ratio of the square of the distance for the intensity of either quality; if he did not arm himself with this hypothesis in the case of magnetism or of light, it is because the hypothesis assures a diffusion to infinity of both these virtues, whereas he accepted for any virtue a sphere of action beyond which it is completely annulled.

Thus, from the first half of the seventeenth century all the materials which were to be used in constructing the hypothesis of universal attraction were assembled, cut, and ready to be put into operation; but it was not yet suspected what an extension this work would have. The "magnetic virtue" through which the diverse parts of matter are borne toward one another was employed to explain the falling of heavy bodies and the ebb of the sea. Nobody yet

[64] *Ismaelis Bullialdi "De natura lucis"* (Paris, 1638), Prop. XXXVII, p. 41.

[65] *Ismaelis Bullialdi "Astronomia Philolaïca"* (Paris, 1645), p. 23.

[66] *Athanasii Kircherii "Magnes, sive de arte magnetica"* (Rome, 1641), Book I, Props. XVII, XIX, XX. In Proposition XX Kircher spoke of a decrease in the inverse ratio of the *distance*; that is simply a lapse proceeding from the fact that Kircher, reasoning with spherical areas, represented them by arcs of a circle. The author's thought is nonetheless very clear.

thought of drawing from it the representation of the motions of the heavenly bodies; quite on the contrary, when the physicists approached the problem of celestial mechanics, this attractive force embarrassed them considerably.

The reason is that the science whose principles should have helped them, dynamics, was still in its infancy. Subject still to the teachings of Aristotle in his *De Caelo,* physicists pictured the action causing a planet to turn around the sun on the model of a horse in harness: directed at each moment by the speed of the moving body, the action is proportional to that speed. It was by means of this principle that Cardan compared the power of the "vital principle" moving Saturn to the power of the "vital principle" moving the moon.[67] It was still a very naïve calculation but it was the first model of the reasoning which was to help compose celestial mechanics.

Imbued with the principles which guided Cardan in the course of his calculations, the mathematicians of the sixteenth century and of the first half of the seventeenth century were ignorant of the fact that a heavenly body once thrown into uniform motion in a circle no longer needs to be drawn in the direction of its motion; on the contrary, it requires a pull towards the center of the circle to retain it on its trajectory and to prevent it from flying off on a tangent.

These two problems, then, dominated celestial mechanics: to apply to each planet a force perpendicular to the radius vector from the sun, a force harnessed, so to speak, to this radius vector as a work horse is to the arm of the lever it causes to turn; and to avoid an attraction of the sun on a planet which would, it seemed, precipitate these two bodies toward each other.

Kepler found the power of motion (*virtus motrix*) in a quality or species of motion (*species motus*) emanating from the sun; of the magnetic attraction, so clearly invoked by him to explain gravity and tides, he said nothing when he was dealing with the heavenly bodies. Descartes replaced the *species motus* with the dragging effect due to the vortex of ether. "But Kepler had prepared this matter so well that the adjustment Descartes made between the corpuscular philosophy and the Copernican astronomy was not very difficult."[68]

In order to prevent attraction from throwing the planets into the

[67] *Hieronymi Cardani "Opus novum de proportionibus"* (Basel, 1570), Prop. CLXIII, p. 165.

[68] G. W. Leibniz, letters to Molanus (?), in Leibniz, *Philos. Schriften,* ed. Gerhardt, IV, 301.

sun, Roberval plunged the whole system of the world into an ethereal medium subject to the same attractions and dilated more or less by the heat of the sun. Each planet surrounded by its elements occupies within this medium a position of equilibrium assigned to it by Archimedes' principle; in addition, the sun's motion engenders, by friction within this ether, a vortex which drags the planets exactly like the *species motus* employed by Kepler.

The system of Borelli smacks of both Roberval's and Kepler's influence.[69] Borelli like Kepler sought the force which drags each planet around its trajectory in a power or virtue emanating from the sun, transported by its light and having an intensity inverse to the distance between the two bodies. Like Roberval he assumed that there is "in each planet a *natural instinct* through which it seeks to draw near the sun in a straight line. In the same way we see that every heavy body has a natural instinct to draw near our earth, impelled as it is by the weight which makes it akin to the earth; so also we notice that iron is borne in a straight line toward the magnet."[70]

Borelli compared this force carrying a planet toward the sun to weight. He does not seem to have identified it with the latter; in that respect his system is inferior to that of Roberval. It is also inferior to it in that he assumed the attraction experienced by the planet to be independent of the distance of that heavenly body from the sun. But it surpasses Roberval's system in one point: in order to balance that force and prevent the planet from rushing into the sun, he no longer appealed to the pressure of a fluid within which the planet would float by virtue of Archimedes' principle; he employed the illustration of the sling whose stone moving in a circle strongly tends to stretch the string; he balanced the instinct by which the planet is borne toward the sun by setting up in opposition to it the centrifugal tendency, that of every revolving body to get away from its center of revolution;[71] he calls it repelling force (*vis repellens*) and assumes it to be inverse to the radius of the orbit.

Borelli's idea differs profoundly from the opinions at which his immediate predecessors had halted. Was its generation, however, original with him? Might he not have found some germ of the idea in his reading? Aristotle reported to us that Empedocles explained the stationary position of the earth by means of the rapid rotation

[69] *Alphonsi Borelli "Theoriae Medicorum planetarum ex causis physicis deductae"* (Florence, 1665). Cf. Ernest Goldbeck, *Die Gravitations-hypothese bei Galilei und Borelli* (Berlin, 1897).

[70] *Alphonsi Borelli . . .*, p. 76.

[71] *ibid.*, p. 74.

of the heavens; "thus does it happen with water contained in a bucket which is being swung around; even when the bottom of the bucket is above the water, the water does not fall; the rotation prevents it from doing so."[72] And Plutarch, in a work widely read by the ancient astronomers which Kepler translated and commented on, expressed himself as follows: "Its very motion and the violence of its revolution help keep the moon from falling on the earth, just as the objects placed in a sling are prevented from falling by their being turned round in a circle. Motion according to nature (weight) drags all things, with the exception of those things in which another motion suppresses this; hence, weight does not move the moon because its circular motion makes weight lose its power."[73] Plutarch could not state more clearly the hypothesis which Borelli was to adopt.

This recourse to a centrifugal force was none the less a stroke of genius. Borelli unfortunately could not profit from the idea which had presented itself to him; he did not know the exact laws of this centrifugal force, even in the case in which the moving body describes a circle with uniform motion. All the more reason was there for his inability to calculate it in the case in which this moving body moves in an ellipse in conformity with Kepler's laws. Thus, he could not by a conclusive deduction derive these laws from the hypotheses he formulated.

In 1674 the physicist Hooke was secretary of the Royal Society in London; he in turn approached the problem which engaged the efforts of Kepler, Roberval, and Borelli.[74] He knew that "any body once set into motion persists in moving indefinitely in a straight line with uniform motion until other forces come and deflect its path into a circle, an ellipse, or some other more complex curve." He knew also what forces will determine the trajectories of the various celestial bodies: "All celestial bodies without exception exert a power of attraction or of weight directed towards their center by virtue of which not only do they retain their own parts and prevent them from escaping into space, as we see the Earth does, but they also attract all the other celestial bodies in the sphere of their activity. Whence it follows, for instance, that not only do the Sun and Moon act on the course and motion of the Earth as the Earth acts on them, but that Mercury, Venus, Mars, Jupiter, and

[72] Aristotle, Περὶ οὐρανοῦ, B, αγ. (Book II, 13.)

[73] Plutarch, Περὶ τοῦ ἐμφαινομένου προσώπου τῷ κυκλῳ τῆς σελήνης, Z.

[74] Robert Hooke, *An Attempt to Prove the Annual Motion of the Earth.* (London, 1674).

Saturn also have a considerable influence on the Earth's motion by virtue of their attractive power, in the same way as the Earth has a powerful influence on the motion of these bodies." Finally, Hooke knew that "powers of attraction are exerted with all the more energy as the bodies on which they act draw closer to the center from which those powers emanate." He confessed that "he has not yet determined by experiment what the successive degrees of this increase are for different distances." But he assumed at that time that the intensity of this attractive power followed the inverse ratio of the square of the distance, although he did not state this law before 1678. His affirmation of this is all the more probable since at the same time his colleague Wren, of the Royal Society, was already in possession of this law, according to the testimony of Newton and Halley. Hooke and Wren had no doubt each obtained it from the comparison between gravity and light, a comparison which caused Halley, about the same time, to suspect it.

Hooke, therefore, was in possession by 1672 of all the postulates which would serve in constructing the system of universal attraction, but he could not take advantage of these postulates. The difficulty which halted Borelli halted him in turn: he did not know how to deal with curvilinear motion producing a force variable in magnitude and direction. He was compelled to publish his hypotheses, though they were sterile, hoping that a more skillful mathematician would make them fruitful: "This idea, if followed up as it deserves to be, cannot fail to be very useful to astronomers for reducing all celestial motions to a rule with certainty, something which, I believe, will never be established in any other way. Those who know the theory of the oscillations of the pendulum and of circular motion will easily understand the foundation for the general principle I state, and they will know how to find in nature the way to establish its true physical character."

The indispensable instrument for the accomplishment of such a task is the knowledge of the general laws relating a curvilinear motion to the forces producing it. Now at the time when Hooke's essay appeared, these laws had just been formulated, and it was, in fact, the study of the oscillations of the pendulum which led to their discovery. In 1673 Huygens published his treatise on the pendulum clock;[75] the theorems at the end of this treatise provide the means for solving, at least for circular trajectories, the problems which could not have been broached by Borelli or Hooke.

[75] *Christiani Hugenii "De horologio oscillatorio"* (Paris, 1673).

Huygens' work gave a new and fruitful impetus to investigations concerning the mechanical explanation of the motion of heavenly bodies. In 1689 Leibniz took up again a theory analogous to that of Borelli: Each celestial body is subject to an attractive force directed toward the sun, to a centrifugal force in the opposite direction whose magntitude is to be obtained from the theorems of Huygens, and finally, to an impetus from the ethereal medium bathing it, an impetus which Leibniz assumed to be normal to the radial vector in inverse ratio to the length of this line; this impetus plays exactly the same role as the power of motion (*virtus motrix*) invoked by Kepler and Borelli; it is simply its translation in the system of Descartes and Roberval. With the aid of the rules formulated by Huygens, Leibniz calculated the force with which a planet should gravitate toward the sun if its motion is governed by Kepler's laws, and he found this force to be inversely proportional to the square of the radial vector.[76]

In 1684 Halley for his part applied Huygens' theorems to Hooke's hypotheses. By assuming the orbits of the different planets to be circular, he noted that the proportionality, discovered by Kepler, between the squares of the periods of revolution and the cubes of the diameters presupposes the different planets to be subjected to forces proportional to their masses and to the inverse squares of their distances from the sun.

But at the time when Halley was making these attempts, which he would not publish, and before Leibniz formulated his theory, Newton was communicating to the Royal Society in London the first results of his reflections on celestial mechanics; in 1686 he presented to it his *Philosophiae naturalis principia mathematica* in which is developed in all its richness the theory of which only remnants were glimpsed by Hooke, Wren, and Halley.

Prepared by the repeated efforts of physicists, this theory was not suddenly revealed to Newton. By 1665 or 1666, seven or eight years before Huygens offered his work on the pendulum clock, *De horologio oscillatorio*, Newton through his own efforts discovered the laws of uniform circular motion; he compared these laws, as Halley was to do in 1684, with Kepler's third law and recognized as a result of this comparison that the sun attracted equal masses of different planets with a force inversely proportional to the square of the distances. But he wanted a more precise check on his theory; he wished to be sure that by diminishing in a certain proportion

[76] *Leibnitii "Tentamen de motuum caelestium causis" Acta Eruditorum* (Leipzig, 1689).

the weight which we note on the earth's surface, we obtain exactly the force capable of balancing the centrifugal force which tends to drag the moon. Now, the dimensions of the earth were not well known and gave Newton a value for gravity at the place occupied by the moon which was a value higher by one sixth than the result expected. A strict observer of experimental method, Newton did not publish a theory which went contrary to observation; he disclosed nothing to anyone of the results of his reflections until 1682. At that time Newton learned the results of the new geodetic measurements made by Picard; he was able to take up his calculations again, and this time the result was thoroughly satisfactory; the doubts of the great mathematician vanished, and he dared to produce his admirable system. It had taken him twenty years of constant reflection to achieve the work to which so many physicists since Leonardo da Vinci and Copernicus had brought their contribution.

The most diverse considerations and the most disparate doctrines arose in turn to make their bid for the construction of celestial mechanics: common experience revealing gravity, as well as the scientific measurements of Tycho Brahé and of Picard; the observational laws formulated by Kepler, the vortices of the Cartesians and atomists, as well as the rational dynamics of Huygens; the metaphysical doctrines of the Aristotelians, as well as the systems of the physicians and dreams of astrologers; comparisons of weight with magnetic action, as well as the affinities between the light and the mutual actions of heavenly bodies. In the course of this long and laborious birth, we can follow the slow and gradual transformations through which the theoretical system evolved; but at no time can we see a sudden and arbitrary creation of new hypotheses.

3. *The Physicist Does Not Choose the Hypotheses on Which He Will Base a Theory; They Germinate in Him without Him*

The evolution which produced the system of universal gravity slowly unfolded itself in the course of centuries; thus we have been able to follow step by step the process through which the idea gradually rose to the degree of perfection given to it by Newton. At times the evolution which is to terminate in the construction of a theoretical system is extremely condensed, and a few years suffice to lead the hypotheses which are to carry this theory from the state in which they are barely outlined to that in which they are completed.

Thus, in 1819, Oersted discovered the action of an electric current on the magnetized needle; in 1820 Arago informed the Académie des Sciences of this experiment; on September 18, 1820 the Académie heard the reading of a memoir in which Ampère presented the mutual actions of currents that he had just demonstrated; and on December 23, 1823 it welcomed another memoir in which Ampère gave electrodynamics and electromagnetism their definitive form. A hundred and forty years separate *De revolutionibus orbium caelestium libri sex* (Corpernicus' main work) from *Philosophiae naturalis principia mathematica* (Newton's *Principia*); less than four years separate the publication of Oersted's experiment from the memorable reading of Ampère's paper. But if space permitted us in this book to relate in detail the history of electrodynamic doctrines in the course of those four years,[77] we would there find again all the characteristics we have met in the evolution of celestial mechanics. We should *not* find the genius of Ampère embracing suddenly a vast experimental domain already constituted and by a free and creative decision choosing the system of hypotheses which would represent these data of observation. We should notice the hesitations, the gropings and the gradual progress obtained by a series of partial retouchings which we have seen during the three half-centuries separating Copernicus from Newton. The history of electrodynamics strongly resembles the history of universal attraction. The multiple efforts and the repeated attempts constituting the warp of these two histories succeed one another more rapidly in the first than in the second; this was due to the fortunate circumstance of Ampère's productivity which for four years gave the Académie des Sciences a paper to hear nearly every month; it was also due to the galaxy of mathematical scientists, able physicists, and men of genius who tried with him to construct a new doctrine, for to the name of Ampère the history of electrodynamics should attach not only the name of Oersted but also the names of Arago, Humphry Davy, Biot, Savart, La Rive, Becquerel, Faraday, Fresnel, and Laplace.

At times the history of the gradual evolution producing a system of physical hypotheses remains and will ever remain unknown. It is condensed in a small number of years and concentrated in a single mind; the discoverer did not inform us, as Ampère did, about the ideas which germinated in him, as they made their appearance;

[77] The reader desiring to reconstruct this history will find all the necessary documents in *Collection de Mémoires relatifs à la Physique*, published by the Société française de physique, Vols. II and III: *Mémoires sur l'Electrodynamique* (1885 and 1887).

imitating the long patience of Newton, he waited for his theory to assume a more perfect form before publishing it. We may be certain that it was not in this final form that his discovery initially presented itself to his mind, that this form is the result of innumerable improvements and retouchings, and that in each of the latter the free choice of the discoverer was guided or conditioned, in a manner more or less conscious to him, by an infinity of external and internal circumstances.

Furthermore, however rapid and condensed the evolution of a theory may be, it is always possible to note that a long period of preparation preceded its appearance; between the first sketch and the perfect form the intermediate stages may escape us to the point that we imagine we are viewing a free and sudden creation; but a preliminary labor has made favorable the ground in which the seed fell; it has made possible this accelerated development, and this labor was followed up in the course of centuries.

Oersted's experiment sufficed to provoke an intense and almost feverish industry which in four years brought electrodynamics to maturity, but that was because at the time this seed was deposited within the science of the nineteenth century, the latter was remarkably well prepared to receive it, nourish it, and develop it. Newton had already announced that electrical and magnetic attractions ought to follow laws analogous to those of universal gravity; this supposition had been transformed into an experimental truth by Cavendish and Coulomb for electrical attractions, by Tobias Mayer and Coulomb for magnetic phenomena; physicists were thus accustomed to resolving all forces acting at a distance into elementary actions inversely proportional to the square of the distances separating the elements between which they are exerted. Furthermore, the analysis of various problems posed by astronomy had introduced mathematicians to the difficulties which the composition of such forces offers. The gigantic mathematical effort of the eighteenth century had just been summarized by the celestial mechanics of Laplace; the methods invented in order to deal with the motions of heavenly bodies looked in every direction in terrestrial mechanics for an opportunity to prove their fecundity, and mathematical physics made progress with astonishing rapidity. In particular, Poisson developed, with the aid of the analytical procedures conceived by Laplace, the mathematical theory of static electricity and of magnetism, while Fourier found in the study of the propagation of heat wonderful opportunities to make use of the same procedures. Electrodynamic and electromagnetic phenomena could be made clear to physicists and to mathematicians as well, the latter being armed to take possession of them and reduce them in theory.

Contemplation of a set of experimental laws does not, therefore, suffice to suggest to the physicist what hypotheses he should choose in order to give a theoretical representation of these laws; it is also necessary that the thoughts habitual with those among whom he lives and the tendencies impressed on his own mind by his previous studies come and guide him, and restrict the excessively great latitude left to his choice by the rules of logic. How many parts of physics retain to this day a merely empirical form until circumstances prepare the genius of a physicist to conceive the hypotheses which will organize them into a theory!

On the other hand, when the processes of universal science have prepared minds sufficiently to receive a theory, it arises in a nearly inevitable manner and, very often, physicists not knowing each other and pursuing their reflections at a great distance from each other generate the theory at the same time. One would say that the idea is in the air, carried from one country to another by a gust of wind, and is ready to fertilize any genius who is disposed to welcome it and develop it, as with pollen giving birth to a fruit wherever it meets a ripe calyx.

In the course of his studies, the historian of the sciences constantly has opportunities to observe this simultaneous emergence of the same doctrine in countries far from one another, but no matter how frequently this phenomenon occurs, he can never contemplate it without astonishment.[78] We have already had the opportunity to see the system of universal gravity germinate in the minds of Hooke, Wren, and Halley at the same time that it was being organized in the mind of Newton. Similarly, in the middle of the nineteenth century we see the principle of the equivalence of heat and work formulated almost simultaneously by Robert Mayer in Germany, by Joule in England, and by Colding in Denmark; each of them, however, did not know the reflections of his rivals, and none of them suspected that the same idea had a few years before attained a precocious maturity in France in the genius of Sadi Carnot.

We could multiply illustrations of this surprising simultaneity of discoveries, but we shall limit ourselves to one more example which seems to us particularly striking.

The phenomenon of total reflection which light can experience

[78] Cf. F. Mentré, "La simultanéité des découvertes scientifiques," *Revue scientifique*, 5th series, II (1904), p. 555.

on the surface separating two media is not easily understood in the theoretic structure constituting the system of waves. Fresnel had in 1823 given the proper formulas for representing this phenomenon, but he had obtained them by means of one of the strangest and most illogical divinations mentioned in the history of physics.[79] The ingenious experimental verifications that he gave did not leave any doubt about the accuracy of his formulas, but they only made it more desirable to hope for a logically admissible hypothesis which would attach them to the general theory of optics. For thirteen years physicists could not discover such a hypothesis; at last, the very simple but very unforeseen and original consideration of the "evanescent wave" came and supplied it to them. Now, the remarkable thing is that the idea of an evanescent wave presented itself simultaneously to the minds of four different mathematicians, too far from one another to communicate to each other the thoughts haunting them. Cauchy first formulated the hypothesis of the evanescent wave in a letter addressed to Ampère in 1836;[80] in 1837 Green communicated the idea to the Philosophical Society of Cambridge,[81] and in Germany F.-E. Neumann published it in the *Annalen* of Poggendorff;[82] finally, from 1841 to 1845 MacCullagh made it the subject of three notes presented to the Academy at Dublin.[83]

This illustration appears to us to be a very apt one for throwing full light on the conclusion with which we shall stop: Logic leaves the physicist who would like to make a choice of a hypothesis with a freedom that is almost absolute; but this absence of any guide or rule cannot embarrass him, for, in fact, the physicist does not choose the hypothesis on which he will base a theory; he does not choose it any more than a flower chooses the grain of pollen which will fertilize it; the flower contents itself with keeping its corolla wide open to the breeze or to the insect carrying the generative dust of the fruit; in like manner, the physicist is limited to opening his thought through attention and reflection to the idea which is to take seed in him without him. When Newton was asked how he went about making a discovery, he replied: "I keep the subject

[79] Augustin Fresnel, *Oeuvres complètes*, I, 782.

[80] Augustin Cauchy, *Comptes rendus*, II (1836), 364. Reprinted in *Poggendorff's Annalen*, IX (1836), 39.

[81] George Green, *Transactions of the Cambridge Mathematical Society*, VI (1838), 403. Reprinted in *Mathematical Papers*, p. 321.

[82] F.-E. Neumann, in *Poggendorff's Annalen*, X (1837), 510.

[83] J. MacCullagh, *Proceedings of the Irish Royal Academy*, Vols. II, III. Reprinted in MacCullagh, *Collected Papers*, pp. 187, 218, 250.

constantly before me, and I wait until the first glimmer of light begins to dawn slowly and gradually, and changes into full daylight and clarity."[84]

It is only when the physicist begins to see clearly a new hypothesis received but not chosen by him that his free and laborious activity comes into play; for now it is a matter of combining this hypothesis with those already admitted, of obtaining numerous and varied consequences, and of comparing them carefully with experimental laws. It is up to him to accomplish these tasks quickly and accurately; it is not up to him to conceive a brand new idea, but it is very much up to him to develop this idea and to make it bear fruit.

4. *On the Presentation of Hypotheses in the Teaching of Physics*

Logic does not give the teacher who wishes to expound the hypotheses on which physical theories are based any more clues than it gives the discoverer of them. It teaches him only that the group of physical hypotheses constitutes a system of principles whose consequences ought to represent the collection of laws established by experimenters. Accordingly, a truly logical exposition of physics would begin with a statement of *all* the hypotheses which will be used in various theories; this would be followed by deductions of a good many consequences of these hypotheses; and the conclusion would confront this multitude of consequences with the multitude of experimental laws they should represent.

Clearly such a mode of exposition of physics, which would be the only perfectly logical one, is absolutely impracticable, and therefore it is certain that no instruction in physics can be offered in a form that is perfectly satisfactory from the logical point of view. *Any exposition of physical theories will be obliged to compromise between the requirements of logic and the intellectual needs of the student.*

We have already indicated that the teacher will have to be content with formulating, first of all, a certain more or less extensive group of hypotheses, and deducing from them a certain number of consequences which he will subject without delay to the test of facts. This test, evidently, will not be fully convincing; it will imply confidence in certain propositions proceeding from consequences not yet formulated. The student would undoubtedly be shocked

[84] Reply quoted by Jean Baptiste Biot in the article "Newton," which he wrote for the *Biographie universelle* of Michaud.

by the vicious circles he will notice, if he were not duly warned in advance, and if he did not know that the verification of formulas thus attempted is precocious, anticipating the delays imposed by strict logic on any application of a theory.

For instance, a teacher who has laid down the group of hypotheses on which general and celestial mechanics rest, and who has deduced a certain number of chapters of these two sciences, will not wait until he has dealt with thermodynamics, optics, and the theories of electricity and magnetism in order to compare his theories with various experimental laws. Yet in making this comparison he may happen to use an astronomical telescope, take account of expansion, and correct causes of error from electricity or magnetism, thus starting to use theories he has not yet expounded. The student who is not forewarned will complain of the paradox; however, he will stop being astonished when he has understood that these verifications are introduced to him in advance in order to make clear as soon as possible, through examples, the theoretical propositions expounded to him, but that they should logically come much later when he possesses the entire system of theoretical physics.

This practical impossibility of expounding the system of physics in the very form that strict logic would require and this necessity of keeping a kind of balance between what logic claims and what the understanding of the student can assimilate make the teaching of this science particularly delicate. In fact, the teacher is really allowed to teach a lesson to which the punctilious logician would object, but this toleration is subject to certain conditions: the student should know that the lesson he receives is not exempt from lacunas and assertions not yet justified, and he ought to see clearly where these lacunas are and what these assertions are; in short, the instruction with which he is to be satisfied, though necessarily halting and incomplete, should not cause false ideas to germinate in his mind.

Therefore, the constant concern of the teacher will be to combat the false idea, so ready to slip into such instruction.

No isolated hypothesis and no group of hypotheses separated from the rest of physics is capable of an absolutely autonomous experimental verification; no crucial experiment can decide between two and only two hypotheses. The teacher, however, cannot wait until all hypotheses have been stated before subjecting certain of them to the test of observation: he cannot possibly avoid presenting certain experiments, Foucault's experiment or Otto Wiener's experiment, for instance, as implying adherence to a certain hypothesis to the prejudice of a contrary one; but he will have to indicate carefully to what point the test he is describing anticipates theories not yet expounded, and how the so-called crucial experiment implies the prior acceptance of a good many propositions which we have agreed not to argue about any longer.

No system of hypotheses can be obtained by experimental induction alone; however, induction may indicate to some extent the path leading to certain hypotheses, and it is not forbidden to say so in the form of a remark. For instance, it is not forbidden in beginning an exposition of celestial mechanics to take the laws of Kepler and show how the mechanical translation of these laws leads to statements which seem to be appealing to the later hypothesis of universal attraction, but once these statements are obtained, it will be necessary to observe closely at what point they differ from the hypothesis later substituted for them.

In particular, every time we ask experimental induction to suggest a hypothesis, we shall have to be on our guard against offering an experiment not carried out for an experiment done, a purely imaginary experiment for a feasible experiment; needless to say, we shall above all have to strictly proscribe appeal to an impossible experiment.

5. *Hypotheses Cannot Be Deduced from Axioms Provided by Common-Sense Knowledge*

Among the considerations often surrounding the introduction of a physical hypothesis, there are some deserving close attention; though very much in favor among a great number of physicists, these considerations are, if we do not watch out, particularly dangerous and fertile in yielding false ideas. They consist in justifying the introduction of certain hypotheses with the aid of so-called self-evident propositions obtained from common sense.

A hypothesis may happen to find in the teachings of common sense some analogies or illustrations; the hypothesis may happen to be a proposition of common sense made clearer and more precise by analysis. In these various cases, the teacher will be able, needless to say, to mention these relations of resemblance between the hypotheses on which theory rests and the laws that everyday experience reveals; the choice of these hypotheses will appear all the more natural and all the more satisfying to the mind.

But mention of such relations of resemblance requires the most careful precautions, for it is very easy to be deceived about the real resemblance between a proposition of common sense and a statement of theoretical physics. Very often the analogy is entirely superficial, between words rather than ideas; it would disappear if we were to make a translation of the symbolic statement in which the theory is formulated, that is, if we were to transform each one of the terms employed in this statement by substituting, according to Pascal's advice, the definition for the defined; we should then see at what point the resemblance between the two propositions we have imprudently brought together is artificial and purely verbal.

In those unsound popularizations in which the minds of our generation look for the adulterated science with which they intoxicate themselves, we very frequently read arguments in which the consideration of "energy" provides so-called intuitive premises. Most of the time these premises are really puns, playing on the ambiguity of the word energy; people take judgments true in the common sense of the word energy, in the sense in which they say that the crossing of Africa by a company of explorers under Marchand took a great expenditure of energy, and these judgments are carried over as a whole to energy understood in the sense given to the term by thermodynamics, namely, to the function of the state of a system whose total differential is for each elementary change equal to the excess of external work over the heat released.

Also, not very long ago, those who take delight in such verbal tricks deplored the fact that the principle of the increase of entropy was much more abstruse and difficult to understand than the principle of the conservation of energy; yet, the two principles require very similar mathematical calculations. But the term entropy has a meaning only in the language of the physicist; it is unknown in the common language; thus, it does not lend itself to equivocations. Of late, we no longer hear these plaints with regard to the obscurity in which the second law of thermodynamics would remain immersed; it is regarded today as clear and capable of being popularized. Why? Because its name has been changed. People now call it the law of the "dissipation" or "degradation of energy"; Now, those who are not physicists, but wish to appear so, understand these words also. They lend them, it is true, a meaning which is not the one which physicists attribute to them; but what do they care? The door has been opened now to many a specious discussion which they take for reasoning but which is only a play on words. That is exactly what they hoped for.

The employment of Pascal's valuable rule causes these deceptive analogies to disappear as a gust of wind dissipates a mirage.

Those who claim to obtain from the fund of common sense the

hypotheses which will support their theories may also be victims of another illusion.

The fund of common sense is not a treasure buried in the soil to which no coin can ever come to be added; it is the capital of an enormous and prodigiously active association formed by the union of human minds. From century to century this capital is transformed and increased. Theoretical science contributes its very great share to these transformations and to this increase of wealth: this science is constantly diffused by instruction, by conversation, by books and periodicals; it penetrates to the bottom of common-sense knowledge; it awakens its attention to phenomena hitherto neglected; it teaches it to analyze notions which had remained confused. It thus enriches the patrimony of truths common for all men or, at least, for all those who have reached a certain degree of intellectual culture. Should a teacher then come desiring to expound a physical theory, he will find among the truths of common sense some propositions admirably suited to justify his hypothesis. He will believe that he has obtained the latter from the primary and necessary demands of our reason, that is to say, that he has *deduced* them from genuine *axioms*; in fact, he will simply have withdrawn from the fund of common-sense knowledge the money that theoretical science had itself deposited in that treasury, in order to return it to theoretical science.

We find a striking example of this serious error and vicious circle in the exposition of the principles of mechanics given by many an author. We shall borrow the following exposition from Euler, but what we shall quote from the arguments set forth by this great mathematician could be found repeated in a great many more recent writings:

"In the first chapter," said Euler, "I demonstrate the universal laws of nature observed by a body when it is free to move and is not acted upon by any force. If such a body is at rest at a given instant, it will persevere forever in its state of rest; if it is in motion, it will move forever in a straight line with constant speed: these two laws may be conveniently united under the name of the law of the conservation of state. From this it follows that conservation of state is an essential property of all bodies, and that all bodies taken as such have a force or faculty of persevering perpetually in their state, a force which is no other than the force of inertia. . . . Since every body by its very nature perseveres constantly in the same state whether of rest or of motion, it is clear that it will be necessary to attribute to external forces any circumstance in which a body will not follow this law and will move with non-uniform motion or else in a curved line. . . . Thus are constituted the true principles of mechanics by means of which we are to explain everything concerning the alteration of motion. As these principles have hitherto been merely confirmed in a slight manner, I have demonstrated them in such a way that they may be understood not only as being certain but also as necessarily true."[85]

If we pursue the reading of Euler's treatise, we find at the beginning of Chapter II the following passages:

"*Definition*: Power is the force which takes a body at rest and sets it into motion, or which alters its motion. Gravity is a force or power of this kind; in fact, if a body is free from any restraint, gravity takes it out of rest in order to make it fall and communicates to it a motion of descent, accelerating it constantly.

"*Corollary*: Every body left to itself remains at rest or moves with a rectilinear and uniform motion. Every time, therefore, that a free body which was at rest happens to be set into motion or else to move with non-uniform motion or with non-rectilinear motion, the cause of this should be attributed to a certain power; for anything which can disturb a body's motion we call power."

Euler introduced the following sentence to us as a *definition*: "Power is the force setting a body into motion or altering its motion." What must we understand by this? Did Euler simply wish to give a nominal definition which is absolutely arbitrary, depriving the word power of any previously acquired meaning? In that case, the deduction that he puts before us will be logically impeccable, but it will simply be a syllogistic construction without any contact with reality. That is *not* what Euler intended to accomplish in his work; it is clear that in stating the sentence we have just quoted, he took the word power or force in the sense it has in current and non-scientific language; the example of weight immediately cited by him is surely evidence of this. However, because he attributed to the word power not a new and arbitrarily defined meaning but the meaning that everybody attaches to it, Euler might borrow from his predecessors, especially from Varignon, the theorems of statics which he used.

This definition is therefore not a definition of the name but of the nature of power; taking this word in the sense in which everyone understands it, Euler proposed to indicate the essential charac-

[85] *Leonhardi Euleri "Mechanica sive motus scientia, analytic exposita"* (Petropolus [now Leningrad], 1736), preface to Vol. I.

ter of power so that all the other properties of force would be obtained from this character. The sentence we have quoted is really not so much a definition as a proposition whose self-evidence is postulated by Euler, as an *axiom*. This axiom and other analogous axioms would permit him simply to prove that the laws of mechanics are not only true but necessary.

Now, is it clear merely in the light of common sense that a body in the absence of any force acting on it moves perpetually in a straight line with constant speed? Or that a body subject to a constant weight constantly accelerates the velocity of its fall? On the contrary, such opinions are remarkably far from common-sense knowledge; in order to give birth to them, it has taken the accumulated efforts of all the geniuses who for two thousand years have dealt with dynamics.[86]

The sort of thing everyday experience teaches us is that a horse cab which is not harnessed remains stationary, that a horse working with a constant effort leads a vehicle with a constant speed, that in order to make the vehicle run more rapidly the horse must work up a greater effort or else be hitched to another horse. How then should we translate what such observations teach us concerning power or force? We should formulate the following propositions:

A body which is not subjected to any power remains stationary.

A body subjected to a constant power moves with constant speed.

When we increase the power moving a body, we increase the speed of the body.

Such are the characteristics which common sense attributes to force or power; such are the hypotheses we should have to take as the bases of dynamics if we wished to found this science on the evidence of common sense.

Now, these characters are those attributed by Aristotle to power (δύναμις) or force (ισχύς);[87] this dynamics is the dynamics of the Stagyrite. In such a dynamics, when we ascertain that the falling of weights is an accelerated motion, we do not conclude from that fact that the weights are subject to a constant force, but that their weight increases proportionately as they descend.

The principles of Aristotelian dynamics seemed, besides, so cer-

[86] Cf. E. Wohlwill, "Die Entdeckung der Beharrungs gesetzes" *Zeitschrift für Völkerpsychologie und Sprachwissenschaft*, Vol. XIV (1883) and Vol. XV (1884); and P. Duhem, *De l'accélération produite par une force constante* (Congrès d'Histoire des sciences, Geneva, 1904).

[87] Aristotle, Φυσικῆς ἀκροάσεως Η, ε; Περὶ Ουρανοῦ Γ, β.

tain and their roots seemed immersed so deeply in the hard soil of common-sense knowledge that in order to extirpate them and grow in their place those hypotheses to which Euler attributed immediate self-evidence, it took one of the longest and most persistent efforts that the history of the human mind divulges to us: it was necessary for Alexander of Aphrodisias, Themistius, Simplicius, John of Philopon, Albert of Saxony, Nicholas of Cusa, Leonardo da Vinci, Cardan, Tartaglia, Julius Caesar Scaliger, and Giovanni Batista Benedetti to break the path for Galileo, Descartes, Beeckman, and Gassendi.

Thus, the propositions which Euler regarded as axioms whose self-evidence is overwhelming and on the basis of which he wished to establish a dynamics not only true but necessary are in reality propositions which dynamics alone has taught us and very slowly and painfully substituted for the false evidence of common sense.

The vicious circle in which Euler's deduction turns cannot be avoided by those who imagine they are justifying the hypotheses on which a physical theory rests by means of axioms having universal assent; the so-called axioms they invoke have been drawn from the very laws they wish to deduce from them.[88]

It is therefore altogether illusory to wish to take the teachings of common sense as the foundation of the hypotheses supporting theoretical physics. By going that way, you do not reach the dynamics of Descartes and Newton, but the dynamics of Aristotle.

We do not say that the teachings of common sense are not very true and very certain; it is very true and certain that an unharnessed coach does not go ahead and that it goes faster when harnessed with two horses than with one alone. We have repeatedly said: These certainties and truths of common sense are in the last analysis the source of all truth and all scientific certainty. But we have also said that the observations of common sense are certain to the extent and degree to which they are deficient in detail and precision; the laws of common sense are very true but on the express condition that the general terms which such laws link together should belong to those abstractions which emerge from concrete phenomena spontaneously and naturally, that is to say, unanalyzed abstractions

[88] The reader will perhaps compare what we have just said with the criticisms addressed by Ernst Mach to the demonstration, proposed by Daniel Bernouilli, for justifying the law of the parallelogram of forces. E. Mach, *La Mécanique, exposé historique et critique de son développement* (Paris, 1904), p. 45. (Translator's note: Mach, *The Science of Mechanics, a Critical and Historical Account of its Development*, translated from the German by T. J. McCormack [2d ed.; La Salle, Ill.: Open Court, 1902], p. 42.)

taken as wholes, like the general idea of a coach or the general idea of a horse.

It is a serious error to take laws which link such complex ideas, so rich in content and so little analyzed, and to wish to translate them immediately by means of symbolic formulas, products of an extreme simplification and analysis, constituting the language of mathematics; it is an odd illusion to take the idea of constant motive power as equivalent to the idea of a horse, and the idea of absolutely free motion as a representation of the idea of a coach. The laws of common sense are judgments concerning extremely complex general ideas that we conceive pertinent to our daily observations; the hypotheses of physics are relations between mathematical symbols brought to the highest degree of simplification. It is absurd not to be aware of the extremely different natures of these two kinds of propositions; it is absurd to imagine the second related to the first as a corollary to a theorem.

It is in the reverse order that we should make the transition from the hypotheses of physics to the laws of common sense. From the set of simple hypotheses serving as bases of physical theories we shall obtain more or less remote consequences, and the latter will provide a schematic representation of the laws revealed by common experience. The more perfect the theories, the more complicated will this representation be; and yet the common observations that are to be represented will always infinitely surpass the representation in complexity. Far from our being able to obtain dynamics from the laws that common sense is aware of by watching a horse and coach roll by, all the resources of dynamics scarcely suffice to give us anything but a very simplified picture of the motion of this coach.

The plan to obtain from common-sense knowledge the demonstration of hypotheses on which physical theories rest is motivated by the desire to construct physics in imitation of geometry; in fact, the axioms from which geometry is derived with such perfect rigor, the "demands" that Euclid formulated at the beginning of his *Elements* are propositions whose self-evident truth is affirmed by common sense. But we have seen on several occasions how dangerous it is to establish an alliance between mathematical method and the method that physical theories follow; how, underneath their entirely external resemblance, which is due to the borrowing of mathematical language by physics, these two methods reveal themselves to be profoundly different. We must return again to the distinction between these two methods.

Most of the abstract and general ideas which arise spontaneously in us on the occasion of our perceptions are complex and unanalyzed conceptions; there are some, however, which almost without any effort reveal themselves to be clear and simple: they are the various ideas grouped around the notions of number and shape. Common experience leads us to link these ideas by laws which, for one thing, have the immediate certainty of common-sense judgments and, for another thing, have an extremely great definiteness and precision. It has, therefore, been possible to take a certain number of these judgments as premises for deductions in which the indisputable truth of common-sense knowledge is inseparably united with the perfect clarity of the chains of syllogisms. That is how arithmetic and geometry were constituted.

But the mathematical sciences are very exceptional sciences: they are fortunate enough to deal with ideas which emerge from our daily perceptions through the spontaneous work of abstraction and generalization, and which still appear afterwards as clear, pure, and simple.

This good fortune is refused physics. The notions provided by the perceptions with which it has to deal are infinitely confused and complex notions, the study of which requires a long and painful work of analysis. The men of genius who have created theoretical physics have realized that in order to put order and clarity into this work it is necessary to look for these qualities in the only sciences which are by their nature orderly and clear, that is, the mathematical sciences. But nevertheless, they have not been able to make clarity and order come into physics and become fused immediately with self-evident certainty, as they have in arithmetic and geometry. All they have been able to do is to confront the multitude of laws obtained directly from observation, laws that are confused, complex, and disorderly but endowed with a certainty directly ascertainable, and to draw a symbolic representation of these laws, an admirably clear and orderly representation, but one which we can no longer even properly say is true.

Common sense rules in the domain of laws of observation; it alone, through our natural means of perceiving and judging our perceptions, decides what is true and what is false. In the domain of schematic representation, mathematical deduction is sovereign mistress, and everything has to be ordered by the rules she imposes. But between the two domains there is established a continual circulation and exchange of propositions and ideas. Theory asks observation to test one of its consequences by submitting it to the facts; observation suggests to theory modification of an old hypothesis or statement of a new hypothesis. In the intermediary zone across which these exchanges are effected and through which communication between observation and theory is assured, common sense and mathematical logic make their influences felt concurrently and the procedures belonging to each are mingled together in an inextricable manner.

This double movement, which alone permits physics to unite the certainties of common-sense findings with the clarity of mathematical deduction, has been depicted as follows by Edouard Le Roy:

"In short, necessity and truth are the two extreme poles of science. But these two poles do not coincide; they are like the red and violet of the spectrum. In the continuum between them, the only reality actually lived through, truth and necessity vary inversely with respect to one another, toward whichever of the two poles we are facing and directing ourselves. . . . If we choose to go toward the necessary, we turn our back on the true, we labor to eliminate everything empirical or intuitive, we tend to schematism, mere discourse, and formal games with meaningless symbols. On the other hand, in order to conquer truth, we must reverse the direction of the procedure that must be adopted; qualitative and concrete representations recover their preeminent rights, and we then see discursive necessity dissolve gradually into living contingency. Finally, it is not in the same parts or respects that science is necessary and also true, or that it is rigorous and also objective."[89]

The vigor in which this is expressed perhaps exceeds somewhat the thought itself of the author; in any case, for it to express our thought faithfully, it is sufficient to substitute the words "order" and "clarity" for the words "rigor" and "necessity" employed by M. Le Roy.

It is quite correct, then, to declare that physical science flows from two sources: one the certainty of common sense, and the other the clarity of mathematical deduction; and physical science is both certain and clear because the streams which spring from these two sources run together and mingle their waters intimately.

In geometry the clear knowledge produced by deductive logic and the certainty stemming from common sense are so exactly juxtaposed that we cannot discern that mixed zone in which all our

[89] Edouard Le Roy, "Sur quelques objections adressées à la nouvelle philosophie," *Revue de Métaphysique et de Morale* (1901), p. 319.

means of knowing operate simultaneously and in rivalry; that is why the mathematician when he deals with the physical sciences is in danger of being unaware of the existence of this zone, and why he wishes to construct physics in imitation of his preferred science, on axioms immediately obtained from common-sense knowledge. In the pursuit of this ideal, which Ernst Mach so correctly calls a "false rigor,"[90] he runs the great risk of reaching only demonstrations bristling with paradoxes and intertwined with the fallacy of begging the question.

6. *The Importance in Physics of the Historical Method*

How will the teacher responsible for expounding physics forewarn his students against the dangers of such a method? How will he be able to get them to survey in a glance the enormous extent of the territory separating the domain of ordinary experience, in which common-sense laws govern, from the theoretical domain ordered by clear principles? How will he be able at the same time to make them follow the double movement through which the mind establishes continual and mutual communication between these two domains, between the empirical knowledge which, deprived of theory, would reduce physics to formless matter, and mathematical theory which, separated from observation and detached from the testimony of the senses, would offer science only a form devoid of matter?

But why must we seek to represent this method all in one dose? Do we not confront a student who in childhood knew nothing of physical theories and who at an adult age has attained a full knowledge of all the hypotheses on which these theories rest? This student whose education has been pursued for thousands of years is mankind. Why in the intellectual development of each man should we not imitate the progress through which man's knowledge of science has been formed? Why should we not in teaching prepare the introduction of each hypothesis by means of a summary but faithful exposition of the vicissitudes which preceded its adoption by science?

The legitimate, sure, and fruitful method of preparing a student to receive a physical hypothesis is the historical method. To retrace the transformations through which the empirical matter accrued while the theoretical form was first sketched; to describe the long collaboration by means of which common sense and deductive logic

[90] E. Mach, *La Mécanique* . . . , p. 80. (Translator's note: *The Science of Mechanics* . . . , p. 82.)

analyzed this matter and modelled that form until one was exactly adapted to the other: that is the best way, surely even the only way, to give to those studying physics a correct and clear view of the very complex and living organization of this science.

No doubt it is impossible to take up again step by step the slow, hesitant, groping march by which the human mind attained a clear view of each physical principle; that would require too much time. To enter instruction the evolution of each hypothesis must be foreshortened and condensed; it must be reduced in the ratio of the duration of a man's education to the duration of the development of science. With the help of such abbreviation, the metamorphoses through which a creature passes from the embryonic to the adult state reproduce, naturalists say, the real or ideal line through which this creature is attached to the primary trunk of living creatures.

This abbreviation is, moreover, nearly always easy, provided that we really decide to neglect all merely accidental facts, e.g., the name of an author, date of discovery, and episode or anecdote, in order to dwell only on historical facts appearing essential to the physicist's eyes, and only on circumstances in which the theory was enriched by a new principle or saw an obscurity or erroneous idea disappear.

This importance which the history of the methods by which discoveries are made acquires in the study of physics is an additional mark of the great difference between physics and geometry.

In geometry, where the clarity of deductive method is fused directly with the self-evidence of common sense, instruction can be offered in a completely logical manner. It is enough for a postulate to be stated for a student to grasp immediately the data of common-sense knowledge that such a judgment condenses; he does not need to know the road by which this postulate has penetrated into science. The history of mathematics is, of course, a legitimate object of curiosity, but it is not essential to the understanding of mathematics.

It is not the same with physics. There, we have seen, it is forbidden to be purely and completely logical in teaching. Consequently, the only way to relate the formal judgments of theory to the factual matter which these judgments are to represent, and still avoid the surreptitious entry of false ideas, is to justify each essential hypothesis through its history.

To give the history of a physical principle is at the same time to make a logical analysis of it. The criticism of the intellectual processes that physics puts into play is related indissolubly to the exposition of the gradual evolution by which deduction perfects a theory and makes of it a more precise and better-ordered representation of laws revealed by observation.

Besides, the history of science alone can keep the physicist from the mad ambitions of dogmatism as well as the despair of Pyrrhonian skepticism.

By retracing for him the long series of errors and hesitations preceding the discovery of each principle, it puts him on guard against false evidence; by recalling to him the vicissitudes of the cosmological schools and by exhuming doctrines once triumphant from the oblivion in which they lie, it reminds him that the most attractive systems are only provisional representations, and not definitive explanations.

And, on the other hand, by unrolling before him the continuous tradition through which the science of each epoch is nourished by the systems of past centuries, through which it is pregnant with the physics of the future; by mentioning to him the predictions that theory has formulated and experiment realized: by these it creates and fortifies in him that conviction that physical theory is not merely an artificial system, suitable today and useless tomorrow, but that it is an increasingly more natural classification and an increasingly clearer reflection of realities which experimental method cannot contemplate directly.

Every time the mind of the physicist is on the point of going to some extreme, the study of history rectifies him by means of an appropriate correction. In order to define the role that history plays with respect to the physicist, we may borrow from history the following words of Pascal: "When he praises himself, I lower him; when he lowers himself, I praise him."[91] History thus maintains him in that state of perfect equilibrium in which he can soundly judge the aim and structure of physical theory.

[91] B. Pascal, *Pensées*, ed. Havet, Art. 8.

Appendix

PHYSICS OF A BELIEVER[1]

1. Introduction

The *Revue de Métaphysique et de Morale* a little more than a year ago published an article in which the opinions I emitted on different occasions concerning physical theories were expounded and discussed.[2] The author of this article, Abel Rey, has taken the trouble to study assiduously even the smallest writings in which I had expounded my thought, and he has followed the course of this thought with a great concern for accuracy; thus, he has drawn for his readers a picture whose fidelity has keenly impressed me; and surely, I shall not be bargaining with M. Rey by offering him acknowledgement of my appreciation in exchange for the sympathy with which his understanding has assimilated what I had published.

And yet (is there anyone who does not find something about which to complain in his own portrait, however accurate the painter may have been?), it seemed to me that M. Rey had solicited somewhat more than exactly the premises which I had set down and that he has drawn conclusions from them which were not altogether contained in them. I should like to apply some restrictions to these conclusions.

M. Rey terminates his article as follows:

"Our intention here has been to examine only the scientific philosophy of M. Duhem and not his scientific work itself. In order to find and formulate precisely the expression of this philosophy . . . , it seems that we may propose the following formula: In its tendencies toward a qualitative conception of the material universe, in its challenging distrust with regard to a complete explanation of this universe by itself, of the sort mechanism imagines it has, and in its animadversions, more pronounced than genuine, with respect to an integral scientific skepticism, Duhem's scientific philosophy is that of a believer."

Of course, I believe with all my soul in the truths that God has revealed to us and that He has taught us through His Church; I have never concealed my faith, and that He in whom I hold it will

[1] An article published in the *Annales de Philosophie chrétienne*, 77th Year, 4th Series, Vol. I (Oct. and Nov. 1905), p. 44 and p. 133.

[2] Abel Rey, "La philosophie scientifique de M. Duhem," *Revue de Métaphysique et de Morale*, XII (July 1904), 699.

keep me from ever being ashamed of it, I hope from the bottom of my heart: in this sense, it is permissible to say that the physics I profess is the physics of a believer. But surely it is not in this sense that M. Rey meant the formula by which he characterized this physics; rather did he mean that the beliefs of the Christian had more or less consciously guided the criticism of the physicist, that they had inclined his reason to certain conclusions, and that these conclusions were hence to appear suspect to minds concerned with scientific rigor but alien to the spiritualist philosophy or Catholic dogma; in short, that one must be a believer, not to mention being a perspicacious one, in order to adopt altogether the principles as well as the consequences of the doctrine that I have tried to formulate concerning physical theories.

If that were the case, I should have been singularly pursuing the wrong course and failed of my aim. In fact, I have constantly aimed to prove that physics proceeds by an autonomous method absolutely independent of any metaphysical opinion; I have carefully analyzed this method in order to exhibit through this analysis the proper character and exact scope of the theories which summarize and classify its discoveries; I have denied that these theories have any ability to penetrate beyond the teachings of experiment or any capacity to surmise realities hidden under data observable by the senses; I have thereby denied these theories the power to draw the plan of any metaphysical system, as I have denied metaphysical doctrines the right to testify for or against any physical theory. If all these efforts have terminated only in a conception of physics in which religious faith is implicitly and almost clandestinely postulated, then I must confess I have been strangely mistaken about the result to which my work was tending.

Before admitting such a mistake, I should like to be allowed to glance again at this work as a whole, to fix my gaze particularly on the parts in which the seal of the Christian faith was believed noticeable, and to recognize whether, against my intention, this seal is really impressed therein or else, on the contrary, whether an illusion, easy to dissipate, has not led to the taking of certain characteristics not belonging to the work as the mark of a believer. I hope that this inquiry, by clearing up confusions and ambiguities, will put the following conclusion beyond doubt: Whatever I have said of the method by which physics proceeds, or of the nature and scope that we must attribute to the theories it constructs, does not in any way prejudice either the metaphysical doctrines or the religious beliefs of anyone who accepts my words. The believer and the nonbeliever may both work in common accord for the progress of physical science such as I have tried to define it.

2. Our Physical System Is Positivist in Its Origins

We should like to prove that the system of physics which we propose is subjected in all its parts to the most rigorous requirements of positive method, and that it is positivist in its conclusions as well as in its origins.

First, of what preoccupations is the constitution of our system the result? Is our conception of physical theory the work of a believer who is uneasy about the disparity between the teachings of his church and the lessons of reason? Does it arise from an effort that faith in divine things would have attempted in order to attach itself to the doctrines of human science (*fides quaerens intellectum*)?* If so, the nonbeliever may conceive legitimate suspicions regarding such a system; he may fear that some proposition oriented toward Catholic beliefs has, unawares even to the author, slipped through the close meshes of rigorous criticism, so ready is the human mind to think true what it wishes! On the other hand, these suspicions would cease to have any ground if the scientific system occupying us were born within the very matrix of experiment and were forced on the author outside of any metaphysical or theological concern, and almost despite himself, through the daily practice and teaching of the science.

Here then we are going to relate how we were led to teach concerning the aim and structure of physical theory an opinion that is said to be brand new; we shall do so in all sincerity, not because we have the vanity to believe the career of our thought interesting in itself, but in order that the knowledge of the origins of the doctrine may make for a more exact judgment of its logical validity, for it is this validity that is in question.

Let us take ourselves back twenty-five years to the time when we received our first initiation, as physicist-to-be, in the mathematics classes of the Collège Stanislas. The man who gave us this initiation, Jules Moutier, was an ingenious theorist; his critical sense, ever alert and extremely perspicacious, distinguished with sure accuracy the weak point of many a system which others accepted without dispute; proof of his inquiring mind is not lacking, and physical chemistry owes one of its most important laws to him. It was this teacher who planted in us the seed of our admiration for physical theory and the desire to contribute to its progress. Naturally he

* Translator's note: "Faith inquiring into the intellect."

oriented our first tendencies in the same direction to which his own preferences brought him. Now, although he appealed in his investigations to the most diverse methods, each in turn, it was to the mechanical attempts at explanation that Moutier returned most often with a sort of predilection. Like most of the theorists of his time he saw the ideal of physics in an explanation of the material universe constructed in the manner of the atomists and the Cartesians; in one of his writings[3] he did not hesitate to adopt the following thought of Huygens: "The causes of all natural phenomena are conceived through mechanical reasons, unless we wish to give up all hope of understanding anything in physics."

Being a disciple of Moutier, it was as a convinced partisan of mechanism that we approached the courses in physics pursued at the École Normale. There we were to come under influences very different from those we had experienced until then; the jesting skepticism of Bertin struck in vain against the constantly reborn and constantly abortive attempts of the mechanists. Without going as far as to have the agnosticism and empiricism of Bertin, most of our teachers shared his mistrust regarding hypotheses about the intimate nature of matter. Past masters in experimental manipulation, they saw in experiment the only source of truth; when they accepted physical theory it was on condition that it rest entirely on laws drawn from observation.

Whereas the physicists and chemists rivalled one another in praising the method that Newton had formulated at the end of his book of *Principia*, those who taught us mathematics, especially Jules Tannery, worked to develop and sharpen in us a critical sense and to make our reason infinitely difficult to satisfy when it had to judge the rigor of a demonstration.

The tendencies which the instruction of the experimenters had produced in our mind and the lessons that the mathematicians had fixed in us concurred in making us conceive physical theory to be of quite a different type from what we had imagined it to be until then. This ideal theory, the supreme goal of our efforts, we wished to see resting solidly on laws verified by experiment and completely exempt from those hypotheses about the structure of matter which Newton had condemned in his immortal *General Scholium*; but at the same time we wished theory to be constructed with that logical rigor which the algebraists had taught us to ad-

[3] J. Moutier, "Sur les attractions et les répulsions des corps électrisés au point de vue de la théorie mécanique de l'électricité," *Annales de Chimie et de Physique*, 4th Series, Vol. XVI.

mire. It was to the model of such a theory that we tried hard to make our lessons conform when we were given the first opportunity to teach.

We soon had to recognize how vain our efforts were. We had the good fortune to teach before an elite audience in the Faculty of Sciences at Lille. Among our students, many of whom are today colleagues of ours, the critical sense was hardly asleep; requests for clarification and embarrassing objections indefatigably indicated to us the paradoxes and vicious circles which kept reappearing in our lessons despite our care. This harsh but salutary test did not take long to convince us that physics could not be constructed on the plan we had undertaken to follow, that the inductive method as defined by Newton could not be practised, that the proper nature and true object of physical theory had not yet been exhibited with complete clarity, and that no physical doctrine could be expounded in a fully satisfactory manner so long as this nature and object had not been determined in an exact and detailed manner.

This necessity to take up again the analysis of the method by which physical theory can be developed, down to its very foundations, appeared to us in circumstances of which we retain a very vivid recollection. Little satisfied with the exposition of the principles of thermodynamics that they had encountered "in books and among men," several of our students asked us to edit for them a small treatise on the foundations of that science. While we tried hard to satisfy their desire, the radical impotence of the methods then known for constructing a logical theory came home to us more persistently each day. We then had an intuition of the truths which since that time we have continually affirmed: we understood that physical theory is neither a metaphysical explanation nor a set of general laws whose truth is established by experiment and induction; that it is an artificial construction manufactured with the aid of mathematical magnitudes; that the relation of these magnitudes to the abstract notions emergent from experiment is simply that relation which signs have to the things signified; that this theory constitutes a kind of synoptic painting or schematic sketch suited to summarize and classify the laws of observation; that it may be developed with the same rigor as an algebraic doctrine, for in imitation of the latter it is constructed wholly with the aid of combinations of magnitudes that we have ourselves arranged in our own manner. But we also understood that the requirements of mathematical rigor are no longer relevant when it comes to comparing a theoretical construction with the experimental laws which it claims to represent, and to judging the degree of resemblance between the picture and the object, for this comparison and judgment do not arise from the faculty by which we can unwind a series of clear and rigorous syllogisms. We realized that in order to judge this resemblance between theory and empirical data, it is *not* possible to dissociate the theoretical construction and to submit each of its parts in isolation to the test of facts, for the slightest experimental verification puts into play the most diverse chapters of theory, and we realized that any comparison between theoretical physics and experimental physics consists in an alliance of theory taken in its entirety with the total teaching of experiment.

It was thus through the necessities of teaching, under their urgent and constant pressure, that we were led to produce a conception of physical theory markedly different from what had been current till then. These same necessities led us through the years to develop our first thoughts, to make them more precise, to explain and to correct them. It was through these necessities that our system concerning the nature of physical theory was affirmed in our conviction, thanks to the ease with which it enabled us to connect into a coherent exposition the most diverse chapters of science. And may we be pardoned for insisting here on indicating the quite special authority conferred on our principles by this test to which we have submitted them in the course of many long years? There are many persons today who write about the principles of mechanics and physics, but if someone proposed to them that they give a complete course in physics which would still agree in all particulars with their doctrine, how many of them would accept the challenge?

Our ideas about the nature of physical theory are, therefore, rooted in the practice of scientific research and in the exigencies of teaching. Deeply as we have gone into our examination of our intellectual conscience, it is impossible for us to recognize an influence exerted on the genesis of these ideas by any religious preoccupation whatever. And how could it be otherwise? How could we have imagined that our Catholic faith was interested in the evolution undergone by our opinions as a physicist? Have we not known Christians, as sincere as they were enlightened, who firmly believed in the mechanical explanations of the material universe? Have we not known some of them to be ardent partisans of the inductive method of Newton? Was it not a glaring fact to us, as to any man of good sense, that the object and nature of physical theory are things foreign to religious doctrines and without any contact with them? And, furthermore, as though better to mark to what little extent our manner of viewing these questions was inspired by our religious beliefs, have not the most numerous and liveliest attacks against this manner of viewing things come from those who profess the same faith as we do?

Our interpretation of physical theory is, therefore, essentially positivist in its origins. Nothing in the circumstances which suggested this interpretation can justify the distrust of anyone who does not share our metaphysical convictions or religious beliefs.

3. Our Physical System Is Positivist in Its Conclusions

Our reflections on the meaning and scope of physical theories were induced by preoccupations in which metaphysics and religion had no part; they terminated in conclusions which have nothing to do with metaphysical doctrines and nothing to do with religious dogmas.

Certainly we have relentlessly fought physical theories which claim to reduce the study of the material world to mechanics; we have insisted that the physicist should admit primary qualities into his systems. Now, the doctrines which proclaimed that everything in the material world reduced to matter and motion are metaphysical; some proclaimed that every quality is essentially complex, and that it can and should always be resolved into quantitative elements. It seems that our conclusions are really in opposition to these doctrines; our manner of viewing things cannot be admitted without rejecting by that very fact these metaphysical systems, and, therefore, it seems that our physics underneath its positivistic appearances is, after all, a metaphysics. And that is what M. Rey imagines when he says: "It really seems that M. Duhem has succumbed to a common temptation: he has been metaphysical. He had an idea in the back of his head, a preconceived idea about the validity and scope of science, and about the nature of knowledge."[4]

If this were so—let us repeat it loudly—we should have completely failed in the attempt in which we made every effort: we should not have succeeded in defining a theoretical physics for whose progress positivists and metaphysicians, materialists and spiritualists, nonbelievers and Christians may work with common accord.

But it is not so.

With the help of essentially positivistic methods we have tried hard to distinguish sharply the known from the unknown; we never intended to draw a line of demarcation between the knowable and

[4] A. Rey, *op.cit.*, p. 733.

the unknowable. We have analyzed the procedures through which physical theories were constructed and sought to conclude from this analysis the exact meaning and proper scope or range of the propositions formulated by these theories; our inquiry concerning physics has not led us either to affirm or deny the existence and legitimacy of methods of investigations foreign to this science and appropriate for attaining truths beyond its means.

So we fought against mechanism; but on what terms? Have we postulated at the base of our reasoning some proposition not provided by the method of the physicist? Starting from such postulates have we unwound a series of deductions whose conclusion might be of the following form: mechanism is an impossibility; it is certain that we can never construct an acceptable representation of physical phenomena by means of masses and motions subject only to the laws of dynamics? By no means. What we did do was to submit to a minute examination the systems proposed by the various mechanistic schools and to note that none of these systems offered the characteristics of a good and sound physical theory, for none of them represented with a sufficient degree of approximation an extensive group of experimental laws.[5]

Here is how we expressed ourselves regarding the legitimacy or illegitimacy of the very principle of mechanism:

"For the physicist the hypothesis that all natural phenomena may be explained mechanically is neither true nor false, but has no meaning.

"Let us explain this proposition which might appear paradoxical.

"Only one criterion permits one in physics to reject as false a judgment which does not imply logical contradiction, and that is the noting of a flagrant disagreement between this judgment and the facts of experiment. When a physicist affirms the truth of a proposition, he affirms the fact that this proposition has been compared with the data of experiment, that among these data there were some whose agreement with the proposition under examination was not a priori necessary, but that, nevertheless, the deviations between these data and the proposition remained less than the experimental errors.

"By virtue of these principles we do not state a proposition which physics can hold as erroneous when we advance the view that all

[5] We beg the reader to refer, in our book on the evolution of mechanics (*Evolution de la Mécanique* [Paris, 1903]), to the first part: "Les explications mécaniques," and particularly to Chapter xv: "Considérations générales sur les explications mécaniques."

the phenomena of the inorganic world may be explained mechanically, for experiment cannot inform us of any phenomena not surely reducible to the laws of mechanics. However, neither is it legitimate to say that this proposition is physically true; for the impossibility of running down a formal and irresolvable contradiction between it and the results of observation is a logical consequence of the absolute indetermination allowed by invisible masses and hidden motions.

"So it is impossible for one who holds to the procedures of experimental method to declare the following proposition true: All physical phenomena are explained mechanically. It is just as impossible to declare it false. *This proposition transcends physical method.*"

To assert, therefore, that all the phenomena of the inorganic world are reducible to matter and motion is to be metaphysical; to deny that this reduction is possible is again to be metaphysical. But our critique of physical theory refrained from making such an affirmation or denial. What it affirmed and proved is that there did not exist *at the time* any acceptable physical theory which was in conformity with the requirements of mechanism, and that it was possible *at the time*, by refusing to be subject to these requirements, to construct a satisfactory theory; but in formulating these assertions we were doing the work of a physicist, not that of a metaphysician.

In order to construct this physical theory not reduced to mechanism, we had to make certain mathematical magnitudes correspond to certain qualities, and among these qualities there are some which we did not decompose into simpler qualities but treated as primary qualities. Was it by virtue of a metaphysical criterion that we regarded such qualities as primary? Did we have some means of recognizing a priori whether they were or were not reducible to simpler qualities? By no means. All that we asserted about such qualities was what the procedures proper to physics could teach us: we asserted that we did not know *at the time* how to decompose them, but that it was not absurd to seek their further resolution into simpler elements. We said:

"Physics will reduce the theory of the phenomena presented by inanimate nature to the consideration of a certain number of qualities, but will seek to make this number as small as possible. Each time a new effect presents itself, physics will try in every way to reduce it to qualities already defined; only after recognizing the impossibility of making this reduction will it resign itself to put into its theories a new quality and introduce into its equations a new kind of variable. Thus the chemist discovering a new body tries hard to decompose it into one of the elements already known; only when he has exhausted in vain all the means of analysis at the disposal of laboratories will he decide to add a name to the list of simple bodies.

"The name simple is not given to a chemical substance by virtue of a metaphysical argument proving that it is by nature indecomposable; it is given to it by virtue of a fact, because it has resisted all attempts at decomposition. This epithet (simple) is an admission of present inability, and is nothing definitive and irrevocable; a body that is simple today will cease being so tomorrow if some chemist, more fortunate than his predecessors, succeeds in dissociating it; potash and soda, simple bodies for Lavoisier, were compounds beginning with the work of Davy. So it is with the primary qualities admitted in physics. By calling them primary we do not prejudge them to be irreducible by nature; we simply confess that we do not know how to reduce them to simpler qualities, but this reduction which we cannot effect today will, perhaps, be an accomplished fact tomorrow."[6]

Therefore, in rejecting mechanical theories and proposing instead a qualitative theory, we have in no way been guided by "a preconceived idea about the validity and scope of science and about the nature of the knowable"; we have not made any appeal, consciously or unconsciously, to metaphysical method. We have made use exclusively of the procedures belonging to the physicist; we have condemned theories which did not concord with the laws of observation; we have acknowledged a theory which gave a satisfactory representation of these laws; in short, we have scrupulously respected the rules of positive science.

4. Our System Eliminates the Alleged Objections of Physical Science to Spiritualistic Metaphysics and the Catholic Faith

Led by the positivistic method as practised by the physicist, our interpretation of the meaning and scope of theories has not undergone any influence either of metaphysical opinions or of religious beliefs. This interpretation is by no manner or means the scientific philosophy of a believer; the nonbeliever may admit every article of it.

[6] *ibid.*, Part 2, Ch. I: "La Physique de la Qualité." Cf. Part II, Ch. II of the present volume, on primary qualities.

Does it follow from this that the believer has nothing to gain from this critique of physical science, and that the results to which it leads have no interest for him?

It has been fashionable for some time to oppose the great theories of physics to the fundamental doctrines on which spiritualistic philosophy and the Catholic faith rest; these doctrines are really expected to be seen crumbling under the ramming blows of scientific systems. Of course, these struggles of science against faith impassion those who are very poorly acquainted with the teachings of science and who are not at all acquainted with the dogmas of faith; but at times they preoccupy and disturb men whose intelligence and conscience are far above those of village scholars and café physicists.

Now, the system we have expounded gets rid of the alleged objections that physical theory would raise to spiritualistic metaphysics and Catholic dogma; it makes them disappear as easily as the wind sweeps away bits of straw, for according to this system these objections are, and can never be anything but, misunderstandings.

What is a metaphysical proposition, a religious dogma? It is a judgment bearing on an objective reality, affirming or denying that a certain real being does or does not possess a certain attribute. Judgments like "Man is free," "The soul is immortal," "The Pope is infallible in matters of faith" are metaphysical propositions or religious dogmas; they all affirm that certain objective realities possess certain attributes.

What will be required for the possibility that a certain judgment, on one side, is in agreement or disagreement with a proposition of metaphysics or theology, on the other side? Of necessity it will be required that this judgment have certain objective realities as its subject, and that it affirm or deny certain attributes concerning them. In effect, between two judgments not having the same terms but bearing on the same subjects, there can be neither agreement nor disagreement.

The facts of experience—in the current meaning of the words, and not in the complicated meaning these words take on in physics —and empirical laws—meaning the laws of ordinary experience which common sense formulates without recourse to scientific theories—are so many affirmations bearing on objective realities; we may, therefore, without being unreasonable, speak of the agreement or disagreement between a fact or law of experience, on the one hand, and a proposition of metaphysics or theology, on the other.

If, for example, we noticed a case in which a Pope, placed in the conditions provided by the dogma of infallibility, issued an instruction contrary to the faith, we should have before us a fact which would contradict a religious dogma. If experience led to the formulation of the law, "Human acts are always determined," we should be dealing with an empirical law denying a proposition of metaphysics.

That being settled, can a principle of theoretical physics be in agreement or disagreement with a proposition of metaphysics or of theology? Is a principle of theoretical physics a judgment involving objective reality?

Yes, for the Cartesian and the atomist, and for anyone who makes of theoretical physics a dependency or a corollary of metaphysics; a principle of theoretical physics is a judgment which bears on a reality. When the Cartesian affirms that the essence of matter is extension in length, breadth, and thickness or when the atomist declares that an atom moves with uniform rectilinear motion so long as it does not hit another atom, the Cartesian and the atomist really mean to assert that matter is objectively just what they say it is, that it really possesses the properties they attribute to it, and that it is deprived of the properties they refuse to give it. Consequently, it is not meaningless to ask whether a certain principle of Cartesian or atomistic physics is or is not in disagreement with a certain proposition of metaphysics or of dogma; it may reasonably be doubted that the law imposed by atomism on the motion of atoms is compatible with the action of the soul on the body; it may be maintained that the essence of Cartesian matter is irreconcilable with the dogma of the real presence of the body of Jesus Christ in the Eucharist.

Yes, also, for the Newtonian; a principle of theoretical physics is a judgment involving objective reality for one who, like the Newtonian, sees in such a principle an experimental law generalized by induction. Such a person will see, for instance, in the fundamental equations of dynamics a universal rule whose truth experiment has disclosed and to which all the motions of objectively existing bodies are subject. He will be able to speak without illogicality of the conflict between the equations of dynamics and the possibility of free will, and investigate whether this conflict is resolvable or not.

Thus, the defenders of the schools of physics that we have put in combat may legitimately speak of agreement or disagreement between the principles of physical theory and metaphysical or religious doctrines. This will not be the case with those whose reason has accepted the interpretation of physical theory we proposed, for they will never speak of a conflict between the principles of physical theory and metaphysical or religious doctrines; they understand, in fact, that metaphysical and religious doctrines are judgments touching on objective reality, whereas the principles of physical theory are propositions relative to certain mathematical signs stripped of all objective existence. Since they do not have any common term, these two sorts of judgments can neither contradict nor agree with each other.

What indeed is a principle of theoretical physics? It is a mathematical form suited to summarize and classify laws established by experiment. By itself this principle is neither true nor false; it merely gives a more or less satisfactory picture of the laws it intends to represent. It is these laws which make affirmations concerning objective reality, and which may, therefore, be in agreement or disagreement with some proposition of metaphysics or theology. However, the systematic classification that theory gives them does not add or take away anything concerning their truth, their certainty, or their objective scope. The intervention of the theoretical principle summarizing and ordering them can neither destroy the agreement between these laws and metaphysical or religious doctrines when such agreement existed before the intervention of this principle, nor reinstate such agreement if it did not exist previously. *In itself and by its essence, any principle of theoretical physics has no part to play in metaphysical or theological discussions.*

Let us apply these considerations to an example:

Is the principle of the conservation of energy compatible with free will? That is a question often debated and resolved in different ways. Now, does it even have a meaning such that a man conscious of the exact import of the terms it employs can reasonably think about answering it with either yes or no?

Of course, this question has a meaning for those who make of the principle of the conservation of energy an axiom applicable in all strictness to the real universe, either when they draw this anxiom from a philosophy of nature or when they arrive at it by starting from experimental data with the help of a broad and powerful induction. But we do not accept either side. For us the principle of the conservation of energy is by no means a certain and general affirmation involving really existent objects. It is a mathematical formula set up by a free decree of our understanding in order that this formula, combined with other formulas postulated analogously, may permit us to deduce a series of consequences furnishing us a satisfactory representation of the laws noted in our laboratories. Neither this formula of the conservation of energy nor the formulas that we associate with it can be said, properly speaking, to be true or false, since they are not judgments bearing on realities; all that we can say is that the theory composing a group of laws is a good one if its corollaries represent these laws we intend to classify with a sufficient degree of exactness, and that the theory is a bad one in the contrary case. It is already clear that the question, Is the law of the conservation of energy compatible with free will or not? cannot have any meaning for us. If it had any, in effect it would be the following: Is the objective impossibility of free acts a consequence of the principle of the conservation of energy, or not? Now the principle of the conservation of energy has no objective consequence.

And furthermore, let us insist on this.

How would one go about deriving from the principle of the conservation of energy and from other analogous principles the corollary, "Free will is impossible"? We should observe that these various principles are equivalent to a system of differential equations ruling the changes of state of the bodies subject to them; that if the state and motion of these bodies are given at a certain instant, their state and motion would then be determined unambiguously for the whole course of time; and we should conclude from this that no free movement can be produced among these bodies, since a free movement would be essentially a movement not determined by previous states and motions.

Now, what is such an argument worth?

We selected our differential equations or, what comes to the same thing, the principles they translate, because we wished to construct a mathematical representation of a group of phenomena; in seeking to represent these phenomena with the aid of a system of differential equations, we were presupposing from the very start that they were subject to a strict determinism; we were well aware, in fact, that a phenomenon whose peculiarities did not in the least result from the initial data would rebel at any representation by such a system of equations. We were therefore certain in advance that no place was reserved for free actions in the classification we had arranged. When we note afterwards that a free action cannot be included in our classification, we should be very naïve to be astonished by it and very foolish to conclude that free will is impossible.

Imagine a collector who wishes to arrange sea shells. He takes seven drawers that he marks with seven colors of the spectrum, and you see him putting the red shells in the red drawer, the yellow shells in the yellow drawer, etc. But if a white shell appears, he will not know what to do with it, for he has no white drawer. You would, of course, feel very sorry for his reason if you heard him conclude in his embarrassment that no white shells exist in the world.

The physicist who thinks he can deduce from his theoretical principles the impossibility of free will deserves the same feeling. In manufacturing a classification for all phenomena produced in this world, he forgets the drawer for free actions!

5. Our System Denies Physical Theory Any Metaphysical or Apologetic Import

That our physics is the physics of a believer is said to follow from the fact that it so radically denies any validity to the objections obtained from physical theory to spiritualistic metaphysics and the Catholic faith! But it might just as well be called the physics of a nonbeliever, for it does not render better or stricter justice to the arguments in favor of metaphysics or dogma that some have tried to deduce from physical theory. It is just as absurd to claim that a principle of theoretical physics contradicts a proposition formulated by spiritualistic philosophy or by the Catholic doctrine as it is to claim that it confirms such a proposition. There cannot be disagreement or agreement between a proposition touching on an objective reality and another proposition which has no objective import. Every time people cite a principle of theoretical physics in support of a metaphysical doctrine or a religious dogma, they commit a mistake, for they attribute to this principle a meaning not its own, an import not belonging to it.

Let us again explain what we are saying by an illustration.

In the middle of the last century, Clausius, after profoundly transforming Carnot's principle, drew from it the following famous corollary: The entropy of the universe tends toward a maximum. From this theorem many a philosopher maintained the conclusion of the impossibility of a world in which physical and chemical changes would go on being produced forever; it pleased them to think that these changes had had a beginning and would have an end; creation in time, if not of matter, at least of its aptitude for change, and the establishment in a more or less remote future of a state of absolute rest and universal death were for these thinkers inevitable consequences of the principles of thermodynamics.

The deduction here in wishing to pass from the premises to these conclusions is marred in more than one place by fallacies. First of all, it implicitly assumes the assimilation of the universe to a finite collection of bodies isolated in a space absolutely void of matter; and this assimilation exposes one to many doubts. Once this assimilation is admitted, it is true that the entropy of the universe has to increase endlessly, but it does not impose any lower or upper limit on this entropy; nothing then would stop this magnitude from varying from $-\infty$ to $+\infty$ while the time itself varied from $-\infty$ to $+\infty$; then the allegedly demonstrated impossibilities regarding an eternal life for the universe would vanish. But let us confess these criticisms wrong; they prove that the demonstration taken as an example is not conclusive, but do not prove the radical impossibility of constructing a conclusive example which would tend toward an analogous end. The objection we shall make against it is quite different in nature and import: basing our argument on the very essence of physical theory, we shall show that it is absurd to question this theory for information concerning events which might have happened in an extremely remote past, and absurd to demand of it predictions of events a very long way off.

What is a physical theory? A group of mathematical propositions whose consequences are to represent the data of experiment; the validity of a theory is measured by the number of experimental laws it represents and by the degree of precision with which it represents them; if two different theories represent the same facts with the same degree of approximation, physical method considers them as having absolutely the same validity; it does not have the right to dictate our choice between these two equivalent theories and is bound to leave us free. No doubt the physicist will choose between these logically equivalent theories, but the motives which will dictate his choice will be considerations of elegance, simplicity, and convenience, and grounds of suitability which are essentially subjective, contingent, and variable with time, with schools, and with persons; as serious as these motives may be in certain cases, they will never be of a nature that necessitates adhering to one of the two theories and rejecting the other, for only the discovery of a fact that would be represented by one of the theories, and not by the other, would result in a forced option.

Thus the law of attraction in the inverse ratio of the square of the distance, proposed by Newton, represents with admirable precision all the heavenly motions we can observe. However, for the inverse square of the distance we could substitute some other function of the distance in an infinity of ways such that some new celestial mechanics represented all our astronomical observations with the same precision as the old one. The principles of experimental method would compel us to attribute exactly the same logical validity to both these different celestial mechanics. This does not mean that astronomers would not keep the Newtonian law of attraction in preference to the new law, but they would keep it on account of the exceptional mathematical properties offered by the inverse square of the distance in favor of the simplicity and elegance that these properties introduced into their calculations. Of course, these motives would be good to follow; yet they would constitute nothing decisive or definitive, and would be of no weight the day when a phenomenon would be discovered which the Newtonian law of attraction would be inept to represent and of which another celestial mechanics would give a satisfactory representation; on that day astronomers would be bound to prefer the new theory to the old one.[7]

That being understood, let us suppose we have two systems of celestial mechanics, different from the mathematical point of view, but representing with an equal degree of approximation all the astronomical observations made until now. Let us go further: let us use these two celestial mechanics to calculate the motions of heavenly bodies in the future; let us assume that the results of one of the calculations are so close to those of the other that the deviation between the two positions they assign to the same heavenly body is less than the experimental errors even at the end of a thousand or even ten thousand years. Then we have here two systems of celestial mechanics which we are bound to regard as logically equivalent; no reason exists compelling us to prefer one to the other, and what is more, at the end of a thousand or ten thousand years, men will still have to weight them equally and hold their choice in suspense.

It is clear that the predictions from both these theories will merit equal degrees of confidence; it is clear that logic does not give us any right to assert that the predictions of the first theory, but not those of the second theory, will be in conformity with reality.

[7] This is what they did, in fact, the day when, by introducing the idea of molecular attraction, they complicated the formula of Newtonian attraction in order to be able to represent the laws of capillarity.

In truth these predictions agree perfectly for a lapse of a thousand or ten thousand years, but the mathematicians warn us that we should be rash to conclude from this that this agreement will last forever, and by concrete examples they show us to what errors this illegitimate extrapolation could lead us.[8] The predictions of our two systems of celestial mechanics would be peculiarly discordant if we asked these two theories to describe for us the state of the heavens at the end of ten million years; one of them might tell us that the planets at that time would still describe orbits scarcely different from those they describe at present; the other, however, might very well claim that all the bodies of the solar system will then be united into a single mass, or else that they will be dispersed in space at enormous distances from one another.[9] Of these two forecasts, one proclaiming the stability of the solar system and the other its instability, which shall we believe? The one, no doubt, which will best fit our extra-scientific preoccupations and predilections; but certainly the logic of the physical sciences will not provide us with any fully convincing argument to defend our choice against an attacking party and impose it on him.

So it goes with any long-term prediction. We possess a thermodynamics which represents very well a multitude of experimental laws, and it tells us that the entropy of an isolated system increases eternally. We could without difficulty construct a new thermodynamics which would represent as well as the old thermodynamics the experimental laws known until now, and whose predictions would go along in agreement with those of the old thermodynamics for ten thousand years; and yet, this new thermodynamics might tell us that the entropy of the universe after increasing for a period of 100 million years will decrease over a new period of 100 million years in order to increase again in an eternal cycle.

By its very essence experimental science is incapable of predicting the end of the world as well as of asserting its perpetual activity. Only a gross misconception of its scope could have claimed for it the proof of a dogma affirmed by our faith.

[8] See above, Part II, Ch. III, particularly the third section of that chapter.

[9] Thus the trajectories of the planets under the simultaneous action of Newtonian attraction and capillary attraction might very well not differ over a period of ten thousand years to any appreciable extent from the trajectories of the same bodies subject only to Newtonian attraction; and yet, we could suppose without absurdity that the effects of capillary attraction accumulating over a period of 100 million years might appreciably disturb a planet from the path which Newtonian attraction alone would have made it follow.

6. The Metaphysician Should Know Physical Theory in Order Not to Make an Illegitimate Use of It in His Speculations

There you have, then, a theoretical physics which is neither the theory of a believer nor that of a nonbeliever, but merely and simply a theory of a physicist; admirably suited to classify the laws studied by the experimenter, it is incapable of opposing any assertion whatever of metaphysics or of religious dogma, and is equally incapable of lending effective support to any such assertion. When the theorist invades the territory of metaphysics or of religious dogma, whether he intends to attack them or wishes to defend them, the weapon he has used so triumphantly in his own domain remains useless and without force in his hands; the logic of positive science which forged this weapon has marked out with precision the frontiers beyond which the temper given it by that logic would be dulled and its cutting power lost.

But does it follow from the fact that sound logic does not confer on physical theory any power to confirm or invalidate a metaphysical proposition that the metaphysician is entitled to distrust the theories of physics? Does it follow that he can pursue the construction of his cosmological system without any concern for the set of mathematical formulas by means of which the physicist succeeds in representing and classifying the set of experimental laws? We do not believe so; we are going to try to show that there is a connection between physical theory and the philosophy of nature; we are going to try to show precisely in what this connection consists.

But first, in order to avoid any misunderstanding, let us make a remark. This question, Does the metaphysician have to take account of the statements of the physicist? applies absolutely only to the theories of physics. The question is not to be applied to the facts of experiment or to experimental laws, for the answer cannot be doubtful; it is clear that the philosophy of nature has to take account of these facts and of these laws.

Indeed, the propositions which state these facts and formulate these laws have an objective import which is not possessed by merely theoretical propositions. The former may then be in agreement or disagreement with the propositions constituting a cosmological system; the author of this system does not have the right either to be indifferent to this agreement, which brings valuable confirmation to his intuitions, or to this disagreement, which condemns his doctrines beyond appeal.

The judgment of this agreement or disagreement is generally easy when the facts considered are facts of everyday experience and when the laws aimed at are the laws of common sense,[10] for it is not necessary to be a professional physicist to grasp what is objective in such facts or in such laws.

On the other hand, this judgment becomes infinitely delicate and thorny when it comes to a scientific fact or scientific law. In fact, the proposition which formulates this fact or law is generally an intimate mixture of experimental observation endowed with objective import and theoretical interpretation, a mere symbol devoid of any objective sense. It will be necessary for the metaphysician to dissociate this mixture in order to obtain as pure as possible the first of the two elements forming it; in that element, indeed, and in that observational element alone, can his system find confirmation or run into contradiction.

Suppose, for instance, that it is a question of an experiment on the phenomena of optical interference. The report of such an experiment contains statements bearing surely on the objective characteristics of light, for example, a certain assertion that an illumination which seems constant is in reality the manifestation of a property varying very rapidly from one instant to the next in a periodic manner. But these assertions are, through the very language used to express them, intimately bound up with the hypotheses bearing on optical theory. In order to express them the physicist speaks of the vibrations of an elastic ether or of the alternating polarity of a dielectric ether; now, we must not attribute offhand complete and entire objective reality either to vibrations of an elastic ether or to polarization of a dielectric ether, for they are really symbolic constructions imagined by theory in order to summarize and classify the experimental laws of optics.

And there you have the first reason why the metaphysician should not neglect the study of physical theories. He must know physical theory in order to be able to distinguish in an experimental report what proceeds from theory and has only the value of a means of representation or sign from what constitutes the real content or objective matter of the experimental fact.

Let us not go ahead and imagine, furthermore, that a wholly superficial acquaintance with theory would be enough for that purpose. Very often in the report of a physical experiment, the real and objective matter and the merely theoretical and symbolic

[10] See above, Part II, Chs. IV and V.

form interpenetrate each other in so intimate and complicated a manner that the geometric mind with its clear and rigorous procedures, too simple and inflexible however to be penetrating, may not suffice to separate them. There we need the insinuating and looser methods of the subtle mind with finesse; it alone, by slipping in between this matter and this form, can distinguish them; it alone can surmise that the latter is an artificial construction created of whole cloth by theory and without any value for the metaphysician, whereas the former, rich in objective truth, is suited to instruct the cosmologist.

Now, the subtle mind here, as everywhere else, is sharpened by long practice; it is by profound and detailed study of theory that one will obtain that sort of flair thanks to which one will discern in a physical experiment what is theoretic symbol, and thanks to which one will be able to separate this form, of no philosophical value, from the genuine empirical teaching which the philosopher should take into account.

Thus, it is necessary for the metaphysician to have a very exact knowledge of physical theory in order to recognize it unmistakably when it crosses the boundaries of its own domain and intends to penetrate into the territory of cosmology; in the name of this exact knowledge he will be entitled to halt the theory and remind it that it cannot gain from his assistance nor challenge his objections. The metaphysician has to make a profound study of physical theory if he wishes to be certain that it will not exert any illogical influence on his speculations.

7. Physical Theory Has as Its Limiting Form a Natural Classification

There are still other and more serious reasons why the teachings of physical theory impose themselves on the attention of the metaphysician.

No scientific method carries in itself its full and entire justification; it cannot through its principles alone explain all these principles. We should therefore not be astonished that theoretic physics rests on postulates which can be authorized only by reasons foreign to physics.

Among a number of these postulates is the following one: Physical theory has to try to represent the whole group of natural laws by a single system all of whose parts are logically compatible with one another.

If we limit ourselves to invoking merely the grounds of pure logic, of that logic which allows us to determine the object and structure of physical theory, it is impossible to justify this postulate;[11] it is impossible to condemn a physicist who would claim to represent by several logically incompatible theories either diverse sets of experimental laws or even a single group of laws; all that can be required of him is not to mix up two incompatible theories, that is, not to combine a major premise obtained from one of these theories with a minor premise supplied by the other.

This conclusion, viz., the right of the physicist to develop a logically incoherent theory, is indeed one arrived at by those who analyze the method of physics without recourse to any principle foreign to this method. For them the representations of theory are only convenient summaries and only artificial devices aimed at facilitating the work of discovery. Why should we forbid the worker the successive employment of disparate instruments when he finds that each one of them is well adapted to a certain task and not well adapted to another job?

However, this conclusion greatly shocks a good number of those striving for the progress of physics; some of them wish to see in this scorn for theoretic unity the prejudice of a believer desiring to exalt dogma at the expense of science; and to support this opinion it is observed that the brilliant galaxy of Christian philosophers grouped around Edouard Le Roy readily hold physical theories to be merely recipes. In so reasoning it is too often forgotten that Henri Poincaré was the first to proclaim and teach in a formal manner that the physicist could make use, in succession, of as many theories, incompatible among themselves, as he deemed best; and I do not know that Henri Poincaré shares the religious beliefs of Edouard Le Roy.

It is certain that Henri Poincaré as well as Edouard Le Roy were fully authorized by the logical analysis of physical method to maintain their stand; it is no less certain that this doctrine with its skeptical overtones shocks most of those working for the advance of physics. Although the merely logical study of the procedures they employ does not provide them with any convincing argument in support of their way of viewing things, they feel that this way is the right one; they have an intuition that logical unity is imposed on physical theory as an ideal to which it tends constantly; they feel that any lack of logic, any incoherence in this

[11] See above, Part I, Ch. IV, Sec. 10.

theory, is a blemish, and that the progress of science should gradually remove this blemish.

And this conviction is fundamentally shared even by those who defend the right of theory to logical incoherence. Is there a single one among them who hesitates for an instant to prefer a rigorously coordinated theory to a junk heap of irreconcilable theories, and who in order to criticize an adversary does not strive to discover fallacies and contradictions in him? Therefore, it is not with wholehearted will that they proclaim the right to logical incoherence; like all physicists they regard the physical theory which would represent all experimental laws by means of a single, logically coordinated system as the ideal theory; and if they tend to stifle their aspirations toward this ideal, it is solely because they believe it unrealizable and because they despair of attaining it.

Now, is it right to regard this ideal as utopian? It is up to the history of physics to answer this question; it is up to it to tell us whether men, ever since physics took on a scientific form, have exhausted themselves in vain efforts to unite into a coordinated system the innumerable laws discovered by experimenters; or else, on the other hand, whether these efforts through slow and continuous progress have contributed to fusing together pieces of theory, which were isolated at first, in order to produce an increasingly unified and ampler theory. To our mind that is the great lesson we ought to obtain when we retrace the evolution of physical doctrines, and Abel Rey has very clearly seen that that was the principal lesson we sought in the study of past theories.

When thus interrogated, what answer does history give us? The meaning of this answer is not doubtful, and here is how M. Rey interprets it: "Physical theory by no means presents us with a set of divergent or contradictory hypotheses. On the contrary, it offers us, if we follow its transformations attentively, a *continuous development* and *genuine evolution.* The theory which seems sufficient at a given time in science does not collapse as a whole when the field of science is enlarged. Adequate to explain a certain number of facts, it continues to remain valid for those facts. Only it is not so any longer for the new facts; *it is not ruined; it has become insufficient.* And why? Because our mind cannot grasp the complex except after the simple, the more general except after what is less so. So in order not to get lost in very complicated details masking the exact relations of things, the mind has neglected certain modalities, restricted the conditions of inquiry, and reduced the field of observation and experiment. Scientific discovery, when we really know how to understand it, only gradually enlarges this field, gradually lifts certain restrictions, and reintegrates considerations judged negligible at first."

Diversity fusing into a constantly more comprehensive and more perfect unity, that is the great fact summarizing the whole history of physical doctrines. Why should this evolution, whose law is manifested to us in this history, stop suddenly? Why should not the discrepancies we note today among the various chapters of physical theory be fused tomorrow into a harmonious accord? Why resign ourselves to them as to irremediable vices? Why give up the ideal of a completely unified and perfectly logical theory, when the systems actually constructed have drawn closer and closer to this ideal from century to century?

The physicist, then, finds in himself an irresistible aspiration toward a physical theory which would represent all experimental laws by means of a system with perfect logical unity; and when he asks of an exact analysis of experimental method what the role of physical theory is, he does not find anything in it to justify this aspiration. History shows him that this aspiration is as old as science itself, and that successive physical systems have realized this desire more and more fully from day to day. But the study of the procedures by means of which physical science makes progress does not disclose to him the entire rationale of this evolution. The tendencies directing the development of physical theory are not, therefore, completely intelligible to the physicist if he wishes to be nothing but a physicist.

If he wishes to be nothing but a physicist, and if, as an intransigeant positivist, he regards everything not determinable by the method proper to the positive sciences as unknowable, he will notice this tendency powerfully inciting his own research as it has guided those of all times; but he will not look for its origin, because the only method of discovery which he trusts will not be able to reveal it to him.

If, on the other hand, he yields to the nature of the human mind, which is repugnant to the extreme demands of positivism, he will want to know the reason for, or explanation of, what carries him along; he will break through the wall at which the procedures of physics stop, helpless, and he will make an affirmation which these procedures do not justify; he will be metaphysical.

What is this metaphysical affirmation that the physicist will make, despite the nearly forced restraint imposed on the method he customarily uses? He will affirm that underneath the observable data, the only data accessible to his methods of study, are hidden realities whose essence cannot be grasped by these same methods, and that these realities are arranged in a certain order which physical science cannot directly contemplate. But he will note that physical theory through its successive advances tends to arrange experimental laws in an order more and more analogous to the transcendent order according to which the realities are classified, that as a result physical theory advances gradually toward its limiting form, namely, that of a *natural classification*, and finally that logical unity is a characteristic without which physical theory cannot claim this rank of a natural classification.

The physicist is then led to exceed the powers conferred on him by the logical analysis of experimental science and to justify the tendency of theory toward logical unity by the following metaphysical assertion: The ideal form of physical theory is a natural classification of experimental laws. Considerations of another sort also urge him to formulate this assertion.

Very often a statement representing not an observed law but an observable law can be deduced from a physical theory. If we compare this statement with experimental results, what chance is there that the latter will be in agreement with the former?

If physical theory is nothing but what the analysis of the procedures put into operation by the physicist reveals, there is no sort of chance for the theoretically predicted law to agree with the facts. The statement deduced from the principles of the theory will be, for the physicist anxious to hazard nothing which is not tested by his customary method, exactly as though it were formulated by accident; this physicist will just as soon expect to find this forecast contradicted by observation as to see it confirmed by it; strict logic would disavow formally any preconceived idea regarding the experimental test to which this statement is to be submitted and any anticipated confidence in the success of this test. Indeed, for logic, physical theory is only a system created by a free decree of our understanding in order to classify experimental laws already known. When in this theory we run across an empty compartment, can we conclude from this that there objectively exists an experimental law made to order to fill this compartment? We laughed at the collector who, not having prepared a drawer for white sea shells, deduced accordingly that there are no white sea shells in the world; would it be less ridiculous if from the presence in his conchologist's cabinet of a drawer reserved for the color blue but still empty, he took it upon himself to assert that nature possesses blue sea shells destined to fill the empty drawer?

Now, in what physicist do we ever meet such perfect indifference concerning the result of a test and this absence of any prediction about the meaning of this result when it comes to comparing a law predicted by a theory with the facts? The physicist knows quite well that strict logic absolutely allows him only this indifference and that it authorizes no hope of agreement between theoretical prophecy and the facts; nevertheless, he waits for this agreement, counts on it, and regards it as more probable than the refutation. The probability that he attributes to it is so much the greater as the theory subjected to the test is more perfect; and when he lends his confidence to a theory in which numerous experimental laws have found a satisfactory representation, this probability seems to him to verge on certainty.

None of the rules governing the handling of experimental method justify this confidence in the theory's foreknowledge, and yet this confidence does not seem ridiculous to us. Furthermore, if we harbored some intention to condemn its presumption, the history of physics would surely not take long to compel us to modify our judgment; indeed, it would cite innumerable circumstances in which experiment confirmed down to the smallest details the most surprising predictions of theory.

Why then can the physicist, without exposing himself to ridicule, assert that experiment will disclose a certain law because his theory demands the reality of this law, whereas the conchologist would be ridiculous if the mere presence of an empty compartment in his cabinet drawers devoted to the various colors of the spectrum led him to conclude there are blue sea shells in the ocean? Obviously because the classification of this collector is a purely arbitrary system not taking into account the real affinities among the various groups of mollusks, whereas in the physicist's theory there is something like a transparent reflection of an ontological order.

Everything, therefore, urges the physicist to postulate the following assertion: To the extent that physical theory makes progress, it becomes more and more similar to a natural classification which is its ideal end. Physical method is powerless to prove this assertion is warranted, but if it were not, the tendency which directs the development of physics would remain incomprehensible. Thus, in order to find the title to establish its legitimacy, physical theory has to demand it of metaphysics.

8. There Is an Analogy between Cosmology and Physical Theory

A slave to positive method, the physicist is like the prisoner of the cave:* the knowledge at his disposal allows him to see nothing except a series of shadows in profile on the wall facing him; but he surmises that this theory of silhouettes whose outlines are shadowy is only the image of a series of solid figures, and he asserts the existence of these invisible figures beyond the wall he cannot scale.

So the physicist asserts that the order in which he arranges mathematical symbols in order to constitute a physical theory is a clearer and clearer reflection of an ontological order according to which inanimate things are classified. What is the nature of this order whose existence he asserts? Through what sort of affinity do the essences of the objects coming under his observation approach one another? These are questions he is not allowed to answer. By asserting that physical theory tends toward a natural classification in conformity with the order in which the realities of the physical world are arranged, he has already exceeded the limits of the domain in which his methods can legitimately be exercised; all the more reason why this method cannot disclose the nature of this order or tell what it is. To make out the nature of this order exactly is to define a cosmology; to display it to us is to expound a cosmological system; in both cases it is doing the work not essential to the physicist but to the metaphysician.

The methods by which the physicist develops his theories are without force when it comes to proving that a certain proposition of cosmology is true or false; the propositions of cosmology, on the one hand, and the theorems of theoretic physics, on the other hand, are judgments never bearing on the same terms; being radically heterogeneous they can neither agree with nor contradict one another.

Does it follow that the knowledge of physical theory is useless to anyone working for the progress of cosmology? That is the question we should like to examine now.

First, let us make very clear the precise meaning of this question.

We are not asking whether the cosmologist can without harm be ignorant of physics; the answer to that question would be too obvious, for it is very plain that a cosmological system cannot be reasonably constituted without any knowledge of physics.

* Translator's note: See Plato, *Republic*, Book VII.

The reflections of the cosmologist and the physicist have a common starting point, namely, the experimental laws disclosed by observation applied to the phenomena of the inanimate world. Only the direction they follow after leaving from that point distinguishes the inquiries of the physicist from those of the cosmologist. The former wishes to acquire a knowledge of the laws he has discovered that is increasingly more precise and detailed, but the latter analyzes these same laws in order to lay bare when possible the essential relations they manifest to our reason.

For example, if the physicist and the cosmologist study at the same time the laws of chemical combination, the physicist will wish to know very exactly what the proportion is among the masses of the bodies entering into combination, under what conditions of temperature and pressure the reaction may take place, and how much heat is involved. The preoccupation of the cosmologist will be quite different: observation shows him that certain bodies, viz., the elements in the combination have at least apparently ceased to be, and that a new body, viz., the chemical compound, has appeared; the philosopher will strive to conceive what this change of mode of existence really consists in. Do the elements really subsist in the compound? Or do they persist in it only potentially? Such are the questions he will wish to answer.

Will the details which the physicist will have determined by his numerous and precise experiments all be useful to the philosopher? Undoubtedly not; discovered in order to satisfy a desire for detailed precision, a good number of these details will remain useless in an inquiry solicited by other needs. But will all these details be idle for the cosmologist? It would be odd if this were so, if certain facts did not serve to suggest an answer to some one of the problems which preoccupy the philosopher. When the latter, for instance, attempts to pierce the mystery concealing from him the real state of the elements within the chemical compound, should he not take any account, in his attempts at solution, of certain precise details acquired by the work of the laboratories? Do not laboratory analyses proving that we can always obtain from a compound the elements which went into forming it, without the slightest loss or gain of matter, provide a basis, valuable in its rigor and solidity, for the doctrine which the cosmologist tries to constitute?

There is no doubt then that the knowledge of physics can be useful and even indispensable for the cosmologist. But physical science is composed of an intimate blend of two sorts of elements: one of these is a set of judgments whose subjects are objective realities;

the other is a system of signs serving to transform these judgments into mathematical propositions. The first element represents the share of observation, the second the contribution of theory. Now, if the first of these two elements is manifestly useful to the cosmologist, it may well seem possible that the second is of no use to him, and that he must know it only in order not to confuse it with the first and never to depend on its help prematurely.

This conclusion would certainly be correct if physical theory were only a system of symbols arbitrarily created in order to arrange our knowledge according to a quite artificial order, and if the classification it establishes among experimental laws had nothing in common with the affinities unifying respectively the realities of the inanimate world.

The case is quite different if physical theory has as its limiting form a natural classification of experimental laws. There would be a very exact correspondence between this natural classification or physical theory, after it had reached its highest degree of perfection, and the order in which a finished cosmology would arrange the realities of the world of matter; consequently, the more physical theory, on the one hand, and cosmology, on the other, approach each other in their perfect form, the more clear and detailed should be the analogy of these two doctrines.

Thus, physical theory can never demonstrate or contradict an assertion of cosmology, for the propositions constituting one of these doctrines can never bear on the same terms which the propositions forming the other do, and between two propositions not bearing on the same terms there can be neither agreement nor contradiction. However, between two propositions bearing on terms of different natures it is nevertheless possible that there would be an *analogy*, and it is such an analogy which ought to connect cosmology with theoretic physics.

It is thanks to this analogy that the systems of theoretic physics can come to the aid of progress in cosmology. This analogy may suggest to the philosopher a whole group of interpretations; its clear and tangible presence can increase the thinker's confidence in a certain cosmological doctrine, and its absence put him on guard against another doctrine.

This appeal to analogy forms in many cases a valuable means of investigation or test, but it is well not to exaggerate its power; if at this point the words "proof by analogy" are uttered, it is well to determine their meaning exactly and not to confuse such a proof with a genuine logical demonstration. An analogy is felt rather than concluded; it does not impose itself on the mind with all the weight of the principle of contradiction. Where one thinker sees an analogy, another, more keenly impressed by the contrasts between the terms compared than by their resemblances, may very well see opposition. In order to bring the latter to change his negation into an affirmation, the former cannot use the irresistible force of the syllogism; all he can do by his arguments is to attract the attention of his adversary to the similarities which he judges important and turn him away from the divergencies that he believes negligible. He can hope to persuade the person with whom he is arguing, but he cannot claim to convince him.

Another order of considerations also comes in to limit the range of the proofs in cosmology obtained from the analogy with physical theory.

We said that there ought to be an analogy between the metaphysical explanation of the inanimate world and the perfect physical theory arrived at the state of a natural classification. But we do not possess this perfect theory, and mankind will never possess it; what we possess and what mankind will always possess is an imperfect and provisional theory which by its innumerable gropings, hesitations, and repentances proceeds slowly toward that ideal form which would be a natural classification. Therefore, it is not physical theory as we have it but an ideal physical theory that we must compare with cosmology in order to support the analogy of the two doctrines. Now, for one who knows only what exists, how difficult it is to know what ought to exist! How doubtful and subject to caution his assertions are when he states that *this* doctrine is finally established in the theoretic system and will remain unshakable in the course of time, whereas *that* one is fragile and mutable and will be carried away by the next crop of new discoveries! Of course, in such a matter, we must not be astonished to hear physicists pronounce the most discordant opinions; and in order to choose among these opinions, we must not demand peremptory reasons, but be content with unanalyzable instinctive judgments which the mind of finesse will suggest, whereas the geometric mind will declare itself incapable of justifying them.

These few remarks suffice, we believe, to recommend to the cosmologist that they use with extreme prudence the analogy between the doctrine he professes and physical theory; he should never forget that the analogy he sees most clearly may appear obscure to others to such an extent that they may cease having even a glimpse of it. He should fear above all that the analogy employed in favor of his proposed explanation connects this explanation merely with some provisional and shaky theoretic scaffolding rather than with a definitive and unshakable part of physics. Finally, he should keep in mind that any argument based on an analogy so difficult to judge is an infinitely frail and delicate argument, really incapable of refuting what a direct demonstration would have proven.

Here then are two points we may take as gained: The cosmologist may in the course of his reasoning employ analogy between physical theory and the philosophy of nature; he should employ this analogy only with extreme precautions.

The first precaution that the philosopher should take, before he makes too much of the analogy that his cosmology may have with physical theory, is to become very accurately and minutely acquainted with this theory. If he has merely a vague and superficial acquaintance with it, he will let himself be duped by similarities of detail, by accidental affinities, even by assonances of words which he will take as indications of a real and profound analogy. Only a science capable of penetrating theoretic physics to its most secret arcana and of laying bare its most intimate foundations will be able to put him on guard against these captious errors.

But it is not enough for the cosmologist to know very accurately the present doctrines of theoretic physics; he must also be acquainted with past doctrines. In fact, it is not with the present theory that cosmology should be analogous, but with the ideal theory toward which present theory tends by continual progress. It is not the philosopher's task, then, to compare present-day physics to his cosmology by congealing science at a precise moment of its evolution, but rather to judge the tendency of theory and to surmise the goal toward which it is directed. Now, nothing can guide him safely in conjecturing the path that physics will take if not the knowledge of the road it has already covered. If we perceive in an instant's glance an isolated position of the ball that a tennis player has hit, we cannot guess the end point he aimed at; but if our glance has followed the ball from the moment his hand moved to strike it, our imagination, prolonging the trajectory, marks in advance the point that will be struck. So the history of physics lets us suspect a few traits of the ideal theory to which scientific progress tends, that is, the natural classification which will be a sort of reflection of cosmology.

Consider someone, for instance, who would take physical theory just as we have it, in the year of grace 1905, presented by the majority of those who teach it. Anyone who would listen closely to the talk in classes and to the gossip of the laboratories without looking back or caring for what used to be taught, would hear physicists constantly employing in their theories molecules, atoms, and electrons, counting these small bodies and determining their size, their mass, their charge. By the almost universal assent favoring these theories, by the enthusiasm they raise, and by the discoveries they incite or attribute to them, they would undoubtedly be regarded as prophetic forerunners of the theory destined to triumph in the future. He would judge that they reveal a first draft of the ideal form which physics will resemble more each day; and as the analogy between these theories and the cosmology of the atomists strikes him as obvious, he would obtain an eminently favorable presumption for this cosmology.

How different his judgment will be if he is not content with knowing physics through the gossip of the moment, if he studies deeply all its branches, not only those in vogue but also those that an unjust oblivion has let be neglected, and especially if the study of history by recalling the errors of past centuries puts him on his guard against the unreasoned exaggerations of the present time!

Well, he will see that the attempts at explanation based on atomism have accompanied physical theory for the longest time; whereas in physical theory he will recognize a work produced by the power of abstraction, these attempts at explanation will show themselves to him as the efforts of the mind that wishes to imagine what ought to be merely conceived; he will see them constantly being reborn, but constantly aborted; each time the fortunate daring of an experimenter will have discovered a new set of experimental laws, he will see the atomists, with feverish haste, take possession of this scarcely explored domain and construct a mechanism approximately representing these new findings. Then, as the experimenter's discoveries become more numerous and detailed, he will see the atomist's combinations get complicated, disturbed, overburdened with arbitrary complications without succeeding, however, in rendering a precise account of the new laws or in connecting them solidly to the old laws; and during this period he will see abstract theory, matured through patient labor, take possession of the new lands the experimenters have explored, organize these conquests, annex them to its old domains, and make a perfectly coordinated empire of their union. It will appear clearly to him that the physics of atomism, condemned to perpetual fresh starts, does not tend by continued progress to the ideal form of physical theory; whereas he will surmise the gradually complete realization of this ideal when he contemplates the development which abstract theory has undergone from Scholasticism to Galileo and Descartes; from Huygens, Leibniz and Newton to D'Alembert, Euler, Laplace, and Lagrange; from Sadi Carnot and Clausius to Gibbs and Helmholtz.

9. On the Analogy between Physical Theory and Aristotelian Cosmology

Before proceeding further, let us summarize what we have gained above:

Between the ideal forms toward which physical theory and cosmology slowly travel, there ought to be an analogy. This assertion is by no means a consequence of positive method; although it is imposed on the physicist, it is essentially an assertion of metaphysics.

The intellectual procedure through which we judge the more or less broad analogy existing between a physical theory and a cosmological doctrine is quite distinct from the method through which convincing demonstrations are developed; they do not impose themselves.

This analogy should connect natural philosophy not to the present state of physical theory but to the ideal form toward which it tends. Now, this ideal state is not given in a plain and indisputable manner; it is hinted to us by an infinitely delicate and volatile intuition, whereas the analogy is guided by a profound knowledge of theory and its history.

The sorts of information which the philosopher can obtain from physical theory, either in favor of or against a cosmological doctrine, are therefore scarcely outlined indications; he would be very foolish who would take them as certain scientific demonstrations and be astonished to see them discussed and disputed!

After having thus definitely affirmed how much any comparison between a physical theory and a cosmological demonstration differs from a demonstration proper, after having indicated that it leaves plenty of room for hesitation and doubt, we shall be permitted to indicate the present form of physical theory which appears to us to tend toward the ideal form, and the cosmological doctrine which seems to us to have the strongest analogy with this theory. We do not maintain that this indication is to be given in the name of the positive method belonging to the physical sciences; after what we have said, it is obviously clear that it goes beyond the scope of this method, and that this method can neither confirm nor contradict it. In so doing, in penetrating thereby the domain belonging to metaphysics, we know that we have left the domain of physics behind us; we know that the physicist, after having gone along with us through the latter domain, may very well refuse to follow us into the terrain of metaphysics without violating logically imposed rules.

Which among the various ways, unequally favored by men of science, of dealing with physical theory at present is the one carrying the germs of the ideal theory? Which one already offers us through the order in which it arranges experimental laws something like a sketch of a natural classification? This theory, we have very often said, is in our opinion the one called general thermodynamics.

This judgment is dictated to us by the contemplation of the present state of physics and by the harmonious whole formed by general thermodynamics out of the laws discovered and made precise by experimenters; it is dictated to us, above all, by the history of the evolution which has led physical theory to its present state.

The movement through which physics has evolved may actually be decomposed into two other movements which are constantly superimposed on one another. One of the movements is a series of perpetual alternations in which one theory arises, dominates science for a moment, then collapses to be replaced by another theory. The other movement is a continual progress through which we see created across the ages a constantly more ample and more precise mathematical representation of the inanimate world disclosed to us by experiment.

Now, these ephemeral triumphs followed by sudden collapses making up the first of these two movements are the successes and reverses which have been experienced by the various mechanistic physical systems in successive roles, including the Newtonian physics as well as the Cartesian and atomistic physics. On the other hand, the continual progress constituting the second movement has resulted in general thermodynamics; in it all the legitimate and fruitful tendencies of previous theories have come to converge. Clearly, this is the starting point, at the time we live in, for the forward march which will lead theory toward its ideal goal.

Is there a cosmology which may be analogous to this ideal we glimpse at the end of the road where general thermodynamics engages physical theory? Surely it is not the ancient cosmology of the atomists any more than it is the natural philosophy created by Descartes, or the doctrine of Boscovich inspired by the ideas of Newton. On the contrary, it is a cosmology to which general thermodynamics is unmistakably analogous. This cosmology is the Aristotelian physics; and this analogy is all the more striking for being less anticipated and for the fact that the creators of thermodynamics were strangers to Aristotle's philosophy.

The analogy between general thermodynamics and the physics of the Aristotelian school is marked by many a characteristic whose prominence attracts one's attention from the start.

Among the attributes of substance, equal importance is conferred by Aristotelian physics on the categories of quantity and quality; now, through its numerical symbols, general thermodynamics represents the various magnitudes of quantities and the various intensities of qualities as well.

Local motion was for Aristotle only one of the forms of general motion, whereas the Cartesian, atomistic, and Newtonian cosmologies agree in that the only motion possible is change of place in space. And notice that general thermodynamics deals in its formulas with a host of modifications such as variations in temperature or changes in electrical or magnetic state without in the least seeking to reduce these variations to local motion.

Aristotelian physics is acquainted with transformations still deeper than those for which it reserves the name of motions. Motion reaches only attributes; those transformations, viz., generation and corruption, penetrate to substance itself, creating a new substance at the same time that they annihilate a preexistent substance. Likewise, in the mechanics of chemistry, one of the most important chapters of general thermodynamics, we represent different bodies by masses which a chemical reaction may create or annihilate; within the mass of a compound body the masses of the components subsist only potentially.

These features, and many others, that it would take too long to enumerate, strongly connect general thermodynamics with the essential doctrines of Aristotelian physics.

We say "with the essential doctrines of Aristotelian physics," and we must now emphasize this point.

Experimental science was in its infancy at the time when Aristotle built the impressive monument whose plan has been conserved for us in his *Physics, On Generation and Corruption, On the Heavens*, and *Meteors*; and at the time when his commentators, like Alexander of Aphrodisias, Themistius, Simplicius, Averroes, and innumerable scholastics, strove to chisel down and polish even the slightest portion of this enormous structure. The instruments which so greatly increase the extent, certainty, and precision of our means of knowing were not available to grasp material reality; man had only his naked senses; observable data came to him just as they appear first of all to our perception; no analysis had yet recognized and disentangled a frightful complication; facts, which a more advanced science was to consider as the results of a multitude of simultaneous, interlocked phenomena, were naïvely and hastily taken as the simple and elementary data of natural philosophy. The mark of everything which was incomplete, premature, and childish in this experimental science is necessarily in the cosmology which issues from it. One who hastily runs through the works of the Aristotelians and barely touches the surface of the doctrines expounded in these works notices everywhere strange observations, unimportant explanations, idle and fastidious discussions, in a word, an antique, worn out, deteriorated system in striking contrast with physics at present, so that it is only very remotely possible to recognize in them the slighest analogy with our modern theories.

Quite another impression is experienced by one who digs further. Under this superficial crust in which are conserved the dead and fossilized doctrines of former ages, he discovers the profound thoughts which are at the very heart of the Aristotelian cosmology. Rid of the covering bark which concealed them and at the same time held them in, those thoughts take on new life and movement; as they gradually become animated we see the mask of deterioration which disguised them disappear; soon their rejuvenated look and our general thermodynamics take on a striking resemblance.

He, then, who wishes to recognize the analogy of Aristotelian cosmology with theoretic physics today must not stop at the superficial form of this cosmology, but must penetrate to its deeper meaning.

An illustration may be brought in to clarify our thought and make it precise.

We shall borrow this illustration from one of the essential theories of Aristotle's cosmology, from the theory of the "natural place of the elements"; and we shall consider this theory on the surface, first of all, and, so to speak, from the outside.

In all bodies we always meet, although in various degrees, four qualities: the hot and the cold, the dry and the wet. Each of these qualities characterizes essentially one element: fire is eminently the hot element; air, the cold element; earth, the dry element; and water, the wet one. All the bodies surrounding us are mixtures; to the extent to which each of the four elements, fire, air, water, and earth, enter into the composition of a mixture, it is hot or cold, dry or wet. Beyond these four elements, capable of being transformed into one another by corruption and generation, there exists a fifth essence, incorruptible and nongenerative; this essence forms the celestial orbs and the stars which are condensed portions of these orbs.

Each of the elements has a "natural place"; it remains at rest when it is in this place, but when it is removed from it by "violence," it returns to it by a "natural motion."

Fire is essentially light; its natural place is the concavity of the moon's orb; by natural motion then it rises until it is stopped by this solid vault. Earth is the distinctively heavy element; its natural motion carries it to the center of the world which is its natural place. Air and water are heavy, but less heavy than earth; now, by natural motion the heavier tends to be placed below the lighter; the various elements will therefore be in their natural places when three spherical surfaces concentric with the universe separate water from earth, air from water, and fire from air. What maintains each element in its natural place when it is placed there? What carries it toward this place when it is removed from it? Its substantial form. Why? Because every being tends toward its perfection and in this natural place its substantial form attains its perfection; there it best resists anything which might corrupt it; there it experiences in the most favorable manner the influence of the celestial motions and astral light, the sources of all generation and of all corruption within sublunary bodies.

How childish all this theory of the heavy and the light seems to us! How plainly we recognize the first babblings of human reason trying to give an explanation of falling bodies! How dare we establish the slightest connection between these babblings of an infant cosmology and the admirable development of a science come to full vigor in the celestial mechanics of minds like those of Copernicus, Kepler, Newton, and Laplace?

Of course, no analogy appears between physics today and the theory of natural place, it we take this theory as it appears at first sight with all the details making up its external form. But let us now remove these details and break this mold of outworn science into which the Aristotelian cosmology had to be poured; let us go to the bottom of this doctrine in order to grasp the metaphysical ideas which are its soul. What do we find truly essential in the theory of the natural place of the elements?

We find there the affirmation that a state can be conceived in which the order of the universe would be perfect, that this state would be a state of equilibrium for the world, and what is more, a state of stable equilibrium; removed from this state, the world would tend to return to it, and all natural motions, all those produced among bodies without any intervention of an animated mover, would be produced by the following cause: they would all aim at leading the universe to this ideal state of equilibrium so that this final cause would be at the same time their efficient cause.

Now, opposite this metaphysics, physical theory stands, and here is what it teaches us:

If we conceive a set of inanimate bodies which we suppose removed from the influence of any external body, each state of this set corresponds to a certain value of its entropy; in a certain state, this entropy of the set would have a value greater than in any other state; this state of maximum entropy would be a state of equilibrium and, moreover, of stable equilibrium; all motions and all phenomena produced within this isolated system make its entropy increase; they therefore all tend to lead this system to its state of equilibrium.

And now, how can we not recognize a striking analogy between Aristotle's cosmology reduced to its essential affirmations and the teachings of thermodynamics?

We might multiply comparisons of this kind, and they would authorize, we believe, the following conclusion: If we rid the physics of Aristotle and of Scholasticism of the outworn and demoded scientific clothing covering it, and if we bring out in its vigorous and harmonious nakedness the living flesh of this cosmology, we would be struck by its resemblance to our modern physical theory; we recognize in these two doctrines two pictures of the same ontological order, distinct because they are each taken from a different point of view, but in no way discordant.

It will be said that a physics whose analogy with the cosmology of Aristotle and Scholasticism is so clearly indicated, is the physics of a believer. Why? Is there anything in the cosmology of Aristotle and in that of Scholasticism which implies a necessary adherence to Catholic dogma? May not a nonbeliever as well as a believer adopt this doctrine? And, in fact, was it not taught by pagans, by Moslems, by Jews, and by heretics as well as by the faithful children of the Church? Where then is there that essentially Catholic character with which it is said to be stamped? Is it in the fact that a great number of Catholic doctors, some of the most eminent ones, have worked for its progress? In the fact that a Pope not long ago proclaimed the services that the philosophy of Saint Thomas Aquinas formerly rendered science as well as those that it may render it in the future? Does it follow from these facts that the nonbeliever cannot, without subscribing to a faith not his own, recognize the agreement of Scholastic cosmology with modern physics? Certainly not. The only conclusion that these facts impose is that the Catholic Church has on many occasions helped powerfully and that it still helps energetically to maintain human reason on the right road, even when this reason strives for the discovery of truths of a natural order. Now, what impartial and enlightened mind would dare to testify falsely against this affirmation?

THE VALUE OF PHYSICAL THEORY

CONCERNING A RECENT BOOK[1]

EVER since the most ancient speculations known to us, philosophy had been inseparably linked with the science of nature and with the science of numbers and shapes. A few hundred years ago, this linkage, some thousands of years old in uniting philosophy first to natural philosophy, looked as if it had been weakened to the breaking point. Leaving to the mathematician and the experimenter the task, daily becoming more detailed and more difficult, of working for the advancement of the particular sciences, the philosopher took as the exclusive objects of his reflections the most general ideas of metaphysics, psychology, and ethics; consequently, his thought seemed lighter and more apt to rise to heights which wise men had not been able to reach till then, burdened as they were with so many branches of knowledge alien to their true and noble study.

Rid of mathematics, astronomy, physics, biology, and all the slowly advancing sciences with their complicated techniques and barbarous terminology unintelligible to the uninitiated, philosophy took the form of an easy doctrine, accessible to the multitude, and skillful in formulating its teachings in an eloquent language understandable by all educated men.

The vogue of this separated philosophy did not last long; far-sighted minds did not take long to discern the vicious principle which the seductive externals of this method scarcely concealed. No doubt, this philosophy seemed light and different from the ancient wisdom held down by the enormous weight of scientific detail, but if philosophy now appeared to fly off with the slightest effort, it was not because its wings had become longer and more powerful; it was simply because it had emptied itself of the content to which it owed its solidity, and because it had reduced itself to a vain form deprived of matter.

Numerous soon were the voices crying out in alarm; the reform attempted at the beginning of the nineteenth century imperiled the very future of philosophy; if one did not wish to see it degenerate

[1] The book is Abel Rey's *La Théorie de la Physique chez les physiciens contemporains* (Paris, 1907). This article of ours appeared in the *Revue générale des Sciences pures et appliquées*, XIX (Jan. 15, 1908), 7-19.

into a verbiage whose sound revealed its hollowness, it was necessary to give it the nourishment which had sustained it for so long and had been taken away from it by the claim that it was unnecessary. Very far from separating it from the particular sciences, it was necessary to nourish it with the teachings of these sciences so that it might absorb and assimilate them to itself; it was necessary to merit anew the title which had so long adorned it: *Scientia scientarum* (*Science of sciences*).

The advice was easier to give than to follow. It is easy to break a tradition, but not so easy to renew it. An abyss had been dug between the particular sciences and philosophy; the cable which formerly connected these two continents and established between them a continual exchange of ideas was broken, and the two ends which were to be joined again lay at the bottom of the abyss. Henceforth, deprived of any means of communication, the inhabitants of both shores, the philosophers on one and men of science on the other, were not in a condition to coordinate their efforts toward the union which all felt necessary.

Nevertheless, bold men on both sides took themselves to the task. Among those who had given themselves to the special sciences, several attempted to offer the philosophers in a form which might be agreeable to them the most general and most essential results of their detailed inquiries. Certain philosophers on their side did not hesitate to learn the language of mathematics, physics, and biology, and to become familiar with the technique of these various disciplines so as to be able to borrow from the treasures they had amassed anything which would enrich philosophy.

In 1896 a graduate student of philosophy, formerly a student of the Division of Letters of the Ecole Normale defended a thesis on *The Mathematical Infinite* before the Faculty of Letters in Paris; it was a truly remarkable event, for M. Couturat thus showed to the least attentive the return of philosophy to the study of the sciences and the resumption of the tradition too long abandoned.

In choosing for his doctoral thesis the subject of physical theory among contemporary physicists, Abel Rey has tightened the connection which M. Couturat had renewed. Had he done only that, he would deserve the appreciation of all those concerned with the future of philosophy.

But his work is valuable not only on that acount; it is also valuable for the importance of the problem examined by the author and for the care with which he has prepared the solution he proposes.

I

First, here is how M. Rey poses the problem (p. iii):

"The fideist and anti-intellectualist movement of the nineteenth century, by making of science a utilitarian technique, claims to be supported by a more exact and profounder analysis of physical science than all those that had been made till then. It would express the general spirit of contemporary physics and summarize its necessary conclusions by an impartial examination of its propositions, its methods, and its theories. . . .

"To verify whether these assertions were warranted was the guiding idea which impelled me to undertake this work."

Here is the solution which the author wishes to give this problem (p. 363):

"Yes, science and in particular the physical sciences have a utilitarian value, indeed, one that is considerable. But that is a small matter alongside their value as disinterested knowledge. And to sacrifice this aspect to the former is to bypass the genuine nature of physical science. We may even say that physical science in and of itself has only the value of knowledge."

We may even go further (p. 367): "We shall know *in the strict sense of the word* only what physical science will be capable of attaining, and nothing else. There will be no other means of knowing in the domain which is the object of physics. Thus, however human the measure of physical science may be, we shall be compelled to be content with this science."

Contemporary pragmatism has affirmed that physical theories do not have a value as knowledge, that their role is entirely utilitarian, and that they are in the last analysis only "convenient recipes" enabling us to act "with success" on the external world. To counter this assertion we need only justify the ancient conception of physics: Physical theory does not have merely a practical utility, but also, and above all, has a value as knowledge of the material world. It secures this value not from another method which, applied at the same time to the same objects, would make up for the insufficiencies of the physical method and would confer on its theories a value transcending their own nature. There is no method except the physical method which can serve to study the objects studied by physics; the physical method in itself exhausts the justification of physical theories; it and it alone indicates what these theories are worth as knowledge.

There is the problem stated and the solution formulated. And, so that there be no uncertainty added to throw the debate into confusion, let us recall carefully that the solution does not bear on the whole of physics; experimental facts are outside the argument; nobody except the skeptic whose remarks escape all discussion disputes the documentary value of facts of experiment or denies that they teach us about the external world. The only point of litigation is the value of physical *theory.*

We now know the question which urged the author to compose his work and we know the aim he wished to reach. What road will he follow between the point of departure and the destination?

There is one way which would seem to be the most direct and surest. It consists in weighing one by one and examining carefully the arguments on behalf of pragmatism, and exposing the vulnerable point vitiating them and rendering them improper for justifying the thesis they are intended to prove.

I may perhaps be allowed to regret that the author has not found it to his liking to follow this method. We should have liked to see him attack the doctrine he opposes head on, face to face, and not by way of a detour. Especially should we have liked him to cite and name the champions of this doctrine; the mathematicians and physicists whose names keep coming back each moment in his writing would not be offended by being in such company; philosophers or men of pure science may not share all the opinions of Edouard Le Roy—to mention him only—but he has passed the tests of both sides and both parties regard him as one of their own.

However that may be, let us not lose time in praising the direct route that M. Rey has not wished to follow, and let us walk along with him on the road he has chosen; first, we ask him to indicate this road (pp. ii-iii):

"The method can be only an inquiry among contemporary physicists. And there the task was singularly facilitated by the fact that certain physicists—and some very important ones—are concerned today with the philosophy of physics in giving this subject the nearly positivistic sense of a general, synthetic, and critical point of view on the great problems that a science contains, on its method, and on its processes.

"There remained then, for me to reach my goal, only to seek the opinions maintained at present by physicists about the nature and structure of their science, and to try to present its systematic development by following those who had especially attached themselves to these questions and who seemed to me to have most thoroughly and most clearly expounded them."

To ask of the writings of a certain number of mathematicians, engineers, and physicists what their authors thought of the value of physical theories; to bring together and formulate clearly opinions that are often scattered and remain tacitly understood; to note that all these opinions, despite very often profound differences separating them, are all oriented by a common tendency to converge toward the same proposition; finally, to say that this proposition is the affirmation of a belief in a physical theory whose value is that of *knowledge* and not merely that of practical utility: such is the investigation carried out dutifully by M. Rey, with so much talent that one forgets how laborious it must have been.

But does such an investigation have the import attributed to it by the author? Is it apt to give a convincing solution to the problem posed? It must be observed, first of all, that it is extremely partial and that it could not be otherwise. Naturally, the number of scientists and scholars called to give opinions in this sort of consultation is small in relation to the multitude of those who are not heard. Even if it were more complete and exhaustive, this sort of *referendum* of physicists would still be far from probative, for a question in logic is not resolved by a majority of votes cast. Indeed, may not even those who practice physics with the most success, those whose names are distinguished by the most brilliant discoveries, be deceived, even grossly, about the aim and value of the science to which they have devoted their lives? Did not Christopher Columbus discover America while thinking he had reached India? And is it not one of the favorite themes of pragmatism that men of science most often create illusions about the exact nature of the truths they discover? Does he not subscribe to that formula of Maurice Blondel, so forceful in its odd form: "Science does not know what it knows just as it knows it."

M. Rey, moreover, has understood quite well that in order to learn the true value of physical theory, it is not enough to organize a plebiscite of physicists on this matter; leaving aside the working multitude who people our laboratories, he has taken only the opinions of those who have lived somewhat apart from the din and who have from the heights of "distant hills" been able to discern the general movement of the assault delivered on truth. Thus the author has attached himself exclusively to the opinion of those men who do not cling to the blind confidence of the experimenter in regard to the value of physical theories, but submit this value to a severe critical scrutiny before giving it any credit. That is why the opinions of those men do not simply count as the voice of just any physicist, and why he has attributed a very special weight to these opinions; and whence did this weight come if not from the logical analysis which had transformed an instinctive tendency into a reasoned conviction? That is to say, it is not enough to note the opinion of a logician of physics and to notice that this opinion is favorable to the author's thesis; it is also necessary to examine scrupulously the series of deductions which have served to justify this opinion, for the latter is worth what this reasoning is worth. M. Rey has not been unaware of the necessity of such a critique. Has the latter always been in his work as severe and as careful as he might have been? Has not the joy of welcoming a conclusion conforming to the author's aspirations sometimes prevented him from glimpsing the lacunas separating this conclusion from the premises? We dare not say so.

II

Before gathering the opinion of the physicists or rather, of the logicians of physics, M. Rey has classified them; the mark which serves to designate for each opinion the category into which it falls is supplied by the attitude each has taken with regard to mechanism.

Three attitudes are possible with regard to mechanical theories of matter: a hostile attitude, a simply hopeful or critical attitude, a favorable attitude.

The hostile attitude is one which characterizes first Macquorn Rankine, next Ernest Mach and W. Ostwald, and finally myself.

The simply hopeful and critical attitude is that of Henri Poincaré.

As to the attitude favorable to mechanism, it is more difficult to find those of its representatives who have analyzed, before taking this attitude, their reasons for preferring it to any other, with whom it is a conscious and reflective rather than an instinctive and spontaneous attitude. "It is hardly possible [p. 233] in expounding the mechanistic theory to follow the method which we have followed for the other conceptions of physics. These conceptions, in fact, have been expounded in an explicit fashion by one or another of their adepts. In analyzing the works of these scientists, it is possible to define completely the general spirit which has animated their schools. But with mechanism, it is quite a different matter. First of all, it is a more practical doctrine; we could never expound all its nuances, if we wished. However, this is not an astonishing fact knowing the number of its adepts. Then, there is no one to my knowledge who has proposed to expound and define thoroughly the mechanistic theory of physics. It appears so natural, assisted by tradition, that no one dreams of analyzing it."

And yet an analysis is necessary here, if it is only to make precise in a perfectly clear manner the lines of demarcation drawn by M. Rey among the diverse schools of physicists.

Exactly what do we mean by mechanism?

Shall we define it as a doctrine which proposes to represent all physical phenomena by means of systems moved according to the principles of dynamics or, if we wish to be more precise, according to the equations of Lagrange? We shall then know very exactly what we mean by mechanistic physics, although we can indicate two subdivisions of it. In one, we admit that bodies separated from one another can exert on one another attractive or repulsive forces; this is the mechanistic physics of Newton, Boscovich, Laplace, and Poisson. In the other, we do not admit any force which is not a binding force between two contiguous bodies; this is the mechanistic physics of Heinrich Hertz.

This very exactly delimited meaning of the word mechanism is not the one we must understand in reading M. Rey's work. We see this author rank among the mechanists physicists like J. J. Thomson and Jean Perrin; now, for these men, systems whose motions are to represent the laws of physics are not governed by the equations of dynamics but really by the equations of electrodynamics; such physicists are not *mechanists*, at least in the narrow sense we have just given this word; rather, they are *electrodynamists*.

Consequently, it appears that the word mechanism takes on a very broad meaning with M. Rey. Let us try, however, to delimit it exactly.

If we look for what is common in the very numerous theories, and they are very disparate besides, brought together by M. Rey under the name mechanism, this is what we find: All these theories seek to represent physical laws by means of groups of solid bodies with dimensions close to those that we can see and touch, that can be sculptured in wood or metal; whether they are formed of molecules or atoms, of ions or electrons, the systems whose motions the theorist describes are, despite their extremely small size, conceived as analogous to majestic astronomical systems. All these speculations are alike, therefore, in the following: They wish to reduce all the properties we observe in nature to combinations of shapes and motions subject to expropriation by the imagination. This is clearly shown by the title given by M. Rey to the fourth book of his work: "Les Continuateurs du Mécanisme: Les hypothèses figuratives."

There then is the sharply characterized classification which M. Rey establishes among the various schools of physicists. Permit us to say at once: This classification does not appear to us to be the one that might have been most suitably adopted in view of the problem with which the author instituted his inquiry. It appears, in fact, capable of creating an inextricable confusion between this problem and a different one which, though it is close to the first, is nonetheless essentially distinct from it. The question originally intended to be answered is as follows: Are physical theories simply means for acting on nature, or ought we attribute to them a value as knowledge outside their practical utility? Please don't confuse this problem with another one: Should physics be mechanistic? Or, to speak more precisely, with the following question: Is it necessary for all the hypotheses of physics to be resolved into propositions relative to the motions of small bodies capable of being pictured and imagined? On the other hand, does physics have the right to reason about properties capable of being conceived but irreducible to the motions of systems that can be drawn and sculptured?

There is no doubt that the history of scientific developments and the psychological study of the minds of physicists enable one to establish numerous affinities between the solutions that the various schools have proposed to give to these two problems, but neither is it doubtful that these two problems are essentially independent of one another and that the solution adopted by one physicist for one of them in no way determines by logical necessity the solution that he should adopt for the other.

Does one want examples indicating, clearly enough for all to see, this independence of the two problems?

Is there a physics which has less claim to *knowledge* and which is more clearly and purely utilitarian than that English physics in which theories merely play the roles of *models* without any connection with reality? Is it not that physics which first of all enticed Henri Poincaré when he was studying Maxwell's work and so inspired the famous pages in which physical theories were considered solely as convenient instruments for experimental research? And are not those resounding prefaces of the distinguished professor at the Sorbonne the ones that have in France given birth to the pragmatist critique of physics against which M. Rey protests today? Yet this English physics is entirely mechanical; it employs imaginative hypotheses exclusively.

On the other hand, of all the physical doctrines the one which has most energetically refused to reduce all the properties of bodies to combinations of geometric shapes and local motions is surely the Aristotelian physics. Yet has any one of them more firmly vindicated the name of the science of the real?

We seem therefore to have two logically independent problems in these two questions: Does physical theory have the value of knowledge or not? Should physical theory be mechanistic or not? We have insisted on this independence, for it might easily be missed by the reader of M. Rey's book, even if it has not been by the author. In fact, it seems that M. Rey regards mechanism as a doctrine whose necessary consequence is an absolute confidence in the objective validity of the theories of physics. Let us listen to him (p. 237):

"The question of proving the objectivity of physics does not even pose itself here. The objectivity of physics is the starting point and a necessary postulate. Given the slightest doubt on this point, the least uncertainty, or the smallest share of contingency, and you have left mechanism behind."

Again, he says (pp. 254-256): "The great problem we have had to resolve everywhere in order to maintain the objectivity of physics, the obstacle we have had to overcome with difficulty, but not without leaving an uneasiness sometimes remaining underneath the solution, has been to rejoin two ends of the chain after having broken it.

"Mechanism is not aware of this preoccupation. The problem does not exist for it, since it has simply kept the tradition of the Renaissance and the thought of Galileo, Descartes, Bacon, and Hobbes.

"Mechanism takes the profound unity of the intelligible and experience, of the thinkable and the representable, of the rational and the perceptible, as the solid ground for construction."

Now, has not this deep identity of the real and the intelligible, this *adæquatio rei et intellectus* (adequation of thing and intellect), been the first postulate and almost the essential formula of Aristotelianism, that is to say, the most realistic, the most objective, but at the same time, the least mechanistic and most qualitative of systems of physics?

The indissoluble tie that M. Rey thinks he has established between mechanism and the belief in the objective value of theories seems to us therefore to be a confusion. This confusion engenders some others.

"Mechanism posits [pp. 235-241] as an unshakable base from which all its other characters may be deduced a direct and immediate continuity between experiment and theory. . . . Theory emerges entirely from experiment, and wishes to be a tracing of the object. The empirical object which is its foundation, the model, gives theory its principles, its direction, its step-by-step development, its results, and its confirmation. There is nothing in theoretic physics which is not supported by experiment, which does not stem directly from it, and which is not confirmed by it. At least, that is the claim. And any hypothesis, no matter how hazardous and general, will be based on experiment and will be essentially a *verifiable hypothesis.* . . .

"Thus mechanism repudiates any generalization which is merely a subjective view. Every generalization has to be conceived under the direct and somehow necessary impact of experiment. We have to generalize when experiment does not allow us to do otherwise, when nature almost generalizes for us. A good generalization which is not a dangerous fiction of the imagination will be the natural extension that experiment itself presents when it is made to vary. . . .

"These views have not varied from Newton to Berthelot." And M. Rey recalls the famous statement of Newton on this subject: "Hypotheses non fingo." ("I do not frame hypotheses.")

The method he describes here is really, in fact, the inductive method which Newton canonized in that *General Scholium* with which he ended his book of *Principia.* But is this method, as our author likes to put it, "the unshakable base of mechanism"? When Newton formulated the method, was it as the preface to some treatise on mechanistic physics? Quite the contrary. He stated the rules of inductive physics in order to set them up as an insurmountable barrier for those who reproached him for admitting universal attraction as an "occult quality" and for not explaining it by combinations of shapes and motions. The hypotheses he refuses to feign are mechanical hypotheses about the cause of weight, similar to those imagined by Descartes or Huygens; read this *General Scholium* closely and there will be no doubt about it; there will be still less doubt if you note with the help of Huygens' correspondence what a scandal the method inaugurated by Newton for dealing with physics caused among the mechanists of the time, among men like Huygens, Leibniz, Fatio de Duilliers; and there will be no doubt left at all if you study the admirable development of the *General Scholium* which Cotes inserted as a preface to the second edition of the *Principia.*

A few years ago, a mathematician, prematurely lost to science, formulated anew the rules of the Newtonian inductive method with as much force as clarity. Did Gustave Robin claim he was composing a mechanistic physics by following this method? Not at all; it was a course in thermodynamics from which any mechanical hypothesis was rigorously excluded.

Let us then take it as a genuine truth that there is no necessary connection between the inductive method canonized by Newton and the mechanistic conception of physics. Actually, the mechanists have been seen opposing this method more often than they have maintained it. The purely inductive method can be criticized (we have done it elsewhere); we can strive to prove it is essentially impracticable; but, in any case, this criticism should be sharply distinguished from the criticism of mechanism. The results of one have scarcely any bearing on the other: the rejection of the Newtonian method does not imply the collapse of mechanistic theories; the adoption of the former does not assure in addition the triumph of the latter.

One confusion easily engenders another; from the one we have just dissipated a second has arisen which we are going to try to dissipate in its turn:

"In the mechanistic theory [p. 251], the continuity between experimental physics and theoretical physics is as complete as is conceivable. There is even no longer any room for distinguishing them: experiment and theory imply each other and in the end are identified."

"We know [p. 257] in what the imaginative elements which mechanism puts at the base of theoretic physics integrally consists. Its very name comes from the fact that its elements are those already studied by mechanics and the sciences mechanics presupposes, viz., the science of number and geometry: homogeneous spaces and times, displacements, forces, velocities, accelerations, masses—these are the figures or representations with the aid of which it is intended to render the physical universe intelligible. We have just seen why physics for three centuries has always ended with these very same elements and only with them. . . . There is no knowledge other than what experiment imposes on us. Consequently, it is because experiment has made us fall back on these elements till now, because any representation or any sense perception lets itself be decomposed into these elements and resynthesized starting from these elements, and because analysis and synthesis can be represented objectively with them and only with them, that we have the right to set them down as the primordial elements of physical theory."

It is certain that the ideas by means of which mechanistic theories are constructed, namely, figure and motion, are furnished very directly by experiment. But it is no less certain that experiment just as directly furnishes us with other ideas, for example, light and dark, red and blue, heat and cold. Finally, it is also certain that experiment left to its own resources establishes absolutely no relation between these ideas and the former ones; experiment presents the last ideas to us as radically distinct and essentially heterogeneous from the first ones.

The starting point of mechanistic theories is the following affirmation: The ideas of the first category alone correspond to simple and irreducible objects; those of the second category correspond to complex realities which may and should be resolved into assemblages of shapes and motions.

Such an affirmation obviously transcends experiment; experiment alone can do nothing for or against this affirmation.

In order to be able to establish a contact between such a proposition and experiment, an intermediary is needed. This intermediary is the group of hypotheses which substitutes for the ideas of light, red, blue, heat, etc. more or less complex combinations of ideas furnished by geometry and by mechanics. There is no immediate contact between the immediate data of observation and the statements of mechanistic theory; the transition from one to the other is assured only by the very arbitrary operation which inserts groupings of atoms and molecules and imagines vibrations, paths, and collisions where our eyes see only objects more or less illuminated and variously colored, where our hands apprehend only bodies more or less warm.

Such a theory is much less authorized to offer itself as a direct and inevitable continuation of experiment than a theory such as energetics in which light remains light and heat, heat; the latter theory insists on distinguishing these qualities from shape and motion because observation gives them to us as other than shape and motion, and, without imposing on them a reduction not manifested experimentally, confines itself to grading by means of a numerical scale different intensities of illumination or different temperatures.

This deep fissure, separating directly observable qualities from geometric and mechanical magnitudes to which they are allegedly reduced, marks mechanistic theories with such an essential and evident character that all the adversaries of mechanism have seen in it the weak point, the defect in the armor at which they were to aim their attacks. Their constant reproach against the doctrine they want to destroy is that it is compelled to combine arbitrarily the most complicated agencies and accumulate *hidden masses* and *hidden motions* in order to fill that wide open gap. It was precisely this task that Newton refused to undertake when he enunciated his famous dictum, "I do not frame hypotheses" (*hypotheses non fingo*).

One last confusion seems to me ought to be cleared up; M. Rey says (p. 379):

"Abstract minds are better constituted to put in order what has already been acquired, i.e., well-established knowledge; they clothe science with its logical rigor and rational exactitude. The second sort, imaginative minds, are, on the contrary, better constituted to make discoveries; it is to them that we owe most of the things we have learnt, and the history of the sciences would confirm this easily. We see at once that energetistic theories are generally the work of the first type of mind and will serve particularly in classifying and utilizing acquired science. Mechanistic theories are generally the work of minds with a concrete turn and will serve especially in research and discovery."

The method of energetics would then be essentially a method of exposition; the mechanistic method would be the appropriate one for discovery.

This antithesis has enticed more than one thinker among those who have reflected on physical theory. M. Rey believes it would be easy to justify this by history; the question of knowing whether this antithesis is valid is indeed a question of historical order. We confess that in our opinion history, carefully and impartially consulted, would say that this antithesis is unfounded.

Not that we should wish to maintain that mechanistic theories have never suggested any discovery; it would be easy to refute this claim by illustrations. And, besides, discovery does not let itself be subject to absolute rules. On what strange and unreasonable assumption can one assert that mechanism has never engendered and will never engender any discovery?

We merely mean that mechanism has not in the past had the marked fruitfulness attributed to it. An illusion has been perpetrated: A very great number of discoveries were produced by physicists who adhered firmly to the principles of mechanistic theories, and it is immediately admitted that these principles suggested to them their great discoveries. An attentive study of the work of these physicists nearly always shows that this conclusion is not valid. In general, mechanistic methods are not the ones that have unveiled the truths with which they have enriched science, but the spirit of comparison and generalization and a host of considerations in which the doctrines of mechanism played no role. Very far from its being true that the combinations of shapes and motions facilitated the work of discovery, it was nearly always with great difficulty that they succeeded in operating systems capable of accommodating as well as possible the truths they had discovered despite their mechanistic philosophy. The work of Descartes or of Huygens, very old as it is, could serve us here as an illustration as well as the more recent work of Maxwell or of Lord Kelvin.

Therefore, if one wishes to indicate the advantages of mechanistic method over energetistic method, one should give up invoking either a more perfect continuity with experimental data or a greater aptitude for inciting discovery. There are two and only two advantages for which one can legitimately make a case:

In the first place, and this advantage cannot be contested by anybody, the ideas, assumed to be primitive and irreducible, by means of which mechanism constructs its theories are extremely few, fewer than they are in any energetistic doctrine. Cartesian mechanism employs only shape and motion; atomism admits shape, motion, and mass; Newtonian dynamism adds only force to these.

In the second place, the combinations of small bodies that mechanism substitutes for the qualities directly furnished by experience differ from the purely numerical symbols that energetics employs in grading the intensity of these same qualities in that the former structures can be drawn and sculptured. That is an advantage which is not equally important for all minds; abstract minds hardly prize it, but the more numerous imaginative minds regard it as of first importance.

With these very few ideas easily accessible to minds that are, to use Pascal's language, more ample than strong, mechanism claims to represent the laws of physics as well as energetics can. Is this claim warranted? That is a question of fact to be debated among physicists; whatever opinion one may have concerning the value which physical theory must be accorded as *knowledge* has nothing to do with that debate.

III

Let us, then, leave aside this examination of mechanism and come to the problem which is the essential one for M. Rey's thesis.

Let us begin by formulating this problem sharply as the surest way not to mistake the exact import of the author's arguments.

No one doubts that experience teaches us truths; left to itself, it would suffice to amass a group of judgments about the universe; this group would constitute empirical knowledge.

Theory takes possession of truths discovered by experiment; it transforms them and organizes them into a new doctrine, rational or theoretical physics.

What exactly is the nature of the difference between theoretical physics and empirical knowledge?

Is theory merely an artificial construction which makes the truths of empirical knowledge easier to handle, enabling us to make prompter and more advantageous use of it in our acting on the external world, but teaching us nothing concerning this world which is not already taught us by experiment alone?

Or, on the contrary, does theory teach us something concerning reality which experiment has not taught us and could not possibly teach us, that is, something transcending merely empirical knowledge?

If we must answer this last question affirmatively, we shall be able to say that physical theory is *true* and that it has *value as knowledge*. If we are compelled to say yes to the first question, on the other hand, we shall also have to say that physical theory is not *true* but simply *convenient*, that it has no value as knowledge but only a *practical value*.

In order to cut through this dilemma, M. Rey, we have seen, has instituted an inquiry among men of science who have closely examined physical theory. Let us pursue this inquiry further with him.

The first opinion of those gathered is Rankine's summarized thus (p. 65): "Experiment furnishes the solid and tangible foundations of science and in order to construct a science which is knowledge, employs mathematics so that we may deduce rigorously all the consequences of experiment in order to predict them in a precise manner and to make sure we use all the knowledge acquired in the discovery of new knowledge." These declarations seem clearly to state definitely that the theoretic work accomplished by mathematics is important only as a greater convenience, adding no knowledge to what experience has taught us.

And yet (p. 66), we find in Rankine "a true enthusiasm for science, whose progress he promotes by his work, and an unshakable confidence in the results it has gained and in those it makes him hope for. No trace of skepticism or even of agnosticism is in the work of the British physicist. The objective validity of physics is above all criticism." Now this attitude contrasts strangely with the results of the critical survey through which Rankine assigns a merely utilitarian aim to theoretic mathematics!

Let us now listen to Ernst Mach. Mach's very clear doctrine is entirely summarized in one principle, the principle of economy of thought. The Austrian scientist formulates this principle in the following words: "All science aims to replace experience with the shortest possible intellectual operations." That is why physics first condenses an infinity of real or possible facts into a single law, and why it forms an extremely concentrated synthesis of a multitude of laws in what it calls a theory. "It is a matter [p. 103] of arranging in systematic order the facts presented that have to be reconstructed by thought to form a *system* out of them so that each fact may be recovered and reestablished with the *least intellectual expense.*" It is impossible to state more clearly that the systematic work of theory does not claim to any extent to increase the amount of truth which experiment has dispensed to us, but aims only to make empirical knowledge more easily assimilable and manipulatable by us.

And yet, if the logical criticism which Ernst Mach has pursued with such subtlety and assurance has led him to reduce theory to being no more than an economical tool, almost a technical mnemonic device, he does not appear to wish to be content with this humble role for theory. M. Rey interprets his thought in these terms (p. 105): "The unitary synthesis of physical knowledge at which science aims in its formal development is not, moreover, important simply as economy and harmonious coordination. That synthesis is not an aesthetic crowning of scientific work." And it really seems that Mach sees in it much more than that when he says: "An adequate conception of the world cannot *be given to us*; we have to acquire it; and it is only by leaving the field free to intelligence and experiment wherever they alone ought to decide matters, that we can hope to approach for the good of mankind the ideal of a *unitary* conception of the world which is alone compatible with the ordering of a soundly constituted mind."

After having gathered the opinion of Rankine as well as that of Mach, M. Rey does us the honor of including our opinion, on which we shall not dwell, for it appears plainly, we think, in these pages. However, we shall thank the author for the very great pains he has taken to put in order thoughts that we had scattered in all four quarters of the world. He might have spared himself these pains if, instead of consulting only the various articles in which we had tried out our doctrine, he had read the book in which our opinion about the aim and structure of physical theory sought its complete expression.

After reviewing the adversaries of mechanism, M. Rey goes ahead and consults those who maintain a merely scrutinous attitude with respect to this doctrine; he lets Henri Poincaré speak for them.

M. Rey has striven with a great deal of skill to put a perfect continuity into the statements which M. Poincaré formulated at different times concerning the importance of physical theory. We fear that this unity is more artificial than real. It seems to us that on understanding them well, one sees the opinions of the distinguished mathematician form two groups separated by an abyss. They appear, in the first place, to contradict each other formally; but far from such an attitude being unreasonable, it is thoroughly justified, we believe, by a higher logic, as we shall have occasion to show presently.

The study of the British physicists, in particular from Maxwell on, has led M. Poincaré to scrutinize the principles on which physical theories rest; this scrutiny has led him to conclusions which he has formulated with his customary clarity: "Experience is the sole source of truth; it alone can teach us something new; it alone can give us certainty." The hypotheses on which physical theory rests "are neither true nor false;" they are simply "convenient conventions." It would then be foolish to believe that they add any knowledge whatsoever to purely empirical knowledge.

The logical scrutiny he had made with pitiless rigor drove Henri Poincaré into a corner with the following quite pragmatic conclusion: Theoretic physics is only a collection of prescriptions. Against this proposition he has made a sort of revolution, and he has loudly proclaimed that physical theory has given us something other than the mere knowledge of facts and that it has led us to discover the real relations of things with one another.

Such, it seems to us, is the story, seen in a very foreshortened summary, of the judgments of Henri Poincaré on the value of physical theory.

Now let us see what judgments the continuators of mechanism are going to bring to bear in this same trial.

How does M. Rey define the spirit of modern mechanism, opposed as it is to the spirit of dogmatic mechanism professed by men like Descartes, Huygens, Boscovich, and Laplace?

"Mechanism [p. 225] no longer seeks to give an invariable representation of its object. On the contrary, it offers itself essentially as a method of research, of discovery and progress. All that mechanism insists on is the right to make use of imaginative representations modifiable, of course, as nature is disclosed to us in more complete fashion. . . . Mechanistic physics does not demand the actual unity of a mechanical scheme today; it demands the right to use mechanical schemes for the interpretation and systematization of physicochemical phenomena."

Thus, the mechanist, truly aware of the processes of his own thought, no longer gives us his combinations of shapes and motions as realities underlying qualities directly perceived; he sees in them, following the English school, only *models* which make easier for him an understanding of empirical information already acquired, and facilitate for him the discovery of new facts; he takes them only as fragile and provisional constructions, as scaffolding without essential connection with the monument he is working to complete.

And yet (p. 268): "The conclusion which emerges from the analysis of mechanism is the objectivism of this system. Mechanism, if you like, is the belief in the reality of physical theory (when it has been tested), giving to the words "belief" and "reality" in this definition the same import as in this other definition: Belief in the reality of the external world.

"Mechanism claims, in the midst of inadequate and erroneous conjectures, to be heading for the reproduction of all physical experience. In the end-result we ought to have the complete description of the material universe from the elementary phenomena constituting its warp to the complex details in which it appears to our senses."

M. Rey's inquest stops here. We can, on our part, push it farther and interrogate M. Rey himself; the work he has just accomplished surely confers on him the right to be heard in this debate. What then are the conclusions to which he has been led by his patient research in the writings of others and by his own reflections?

He declares (pp. iv-v) "that all physicists admit a constantly accrued fund of necessary and universal truths, and that this fund of truth is the set of purely experimental results." He admits "that theories are only instruments of work and systematization; which is not to belittle their role, for they thus turn out to be the source of all discovery and of all progress in physical science."

"Physical theory," he says again (p. 354), "has no objective validity independently of experiment. . . . It is a necessary instrument for the physicist; a physicist does not conduct physics without some sort of theory."

Theories (p. 355) "canot claim—at least today—any but a technical value, utilitarian but not objective. Physical theory, or rather, theoretical physics, the set of physical theories of the same form, is only an *organon* (instrument)."

"If physical theories are essentially methods [pp. 357-358], we easily conceive that they may be many. . . . Multiplicity and divergencies do not exist and cannot exist among physicists except in the domain of hypothesis. . . . Hypothesis, in turn, has no other role than as a method of research. Physical theories are multiple and divergent only in that they have a methodological value before all else, and arise from the arbitrary act of the mind in the choice of a hypothesis under whatever name it is disguised."

There are no other truths in physics than the experimental facts; theories are only means of classification and instruments of research. Physics may therefore simultaneously use distinct and incompatible theories; theoretical physics has only a technical and utilitarian value: such are the affirmations to which M. Rey is led logically by his survey of the procedures used in physics and by his examination of the diverse opinions of physicists. What pragmatist could wish for conclusions which were more favorable to his view? Does not the author seem to affirm decidedly the meaning of those who define physical theories as prescriptions claiming to guide our action on nature successfully?

And yet how mistaken we should be about the true thought of the author if we confined ourselves to gathering such affirmations! He would be ranked among the most zealous partisans of the philosophy of action, whereas his book was written precisely in order to reply to pragmatism; the proposition he claims to justify is formulated as follows (p. 359): "The physicochemical sciences have the objective value of *knowledge*. By value of knowledge or theoretic value, I mean their value in relation to a constantly broader and deeper knowledge of nature, and I exclude their value in relation to the practical utilization of natural forces."

The judgments we have gathered in conformity with the text of M. Rey's writings, therefore, express a part but only a part of his thought; they express conclusions he was compelled to enounce in the wake of his inquest and critical study; they are on the surface of his doctrine, very clear and apparent when first inspected, but not connected, it seems, to the very roots of his intellect; they are an adventitious thought imposed, one may almost say, from the outside. Underneath this thought, there is a different one, protruding simultaneously from the most intimate parts of the understanding; and this underlying thought impatiently supports the weight of the one covering it; it protests against the affirmations which logical criticism claims to impose on it, and the formal and precise tone of these affirmations do not succeed in stifling the denials with which nature opposes them.

From the very first pages (pp. iv-v) of his book, M. Rey proclaims "with all physicists that there exists a constantly accrued fund of necessary and universal truths, and that this fund of truths is formed by the set of experimental results." The logician in him, however, knows very well that any experimental result is particular and contingent; but nature protests against logic and cries out to him that the particular and contingent truths disclosed to the physicist by observations are the concrete forms in which necessary and universal truths are manifested to him, although his methods do not permit him to contemplate such truths face to face.

Logical criticism does not succeed in seeing in physical theories anything more than tools. Now, a workingman employs the tool convenient for him, he holds it as he pleases, and he is free to reject it in order to adopt any other. Convenience is decidedly his only guide; provided that his work is well done, what matters the procedure which seemed to him the most appropriate with which to accomplish it! So it goes with physical theories: the physicist may construct them arbitrarily, and change them whenever he sees fit; he may belong to all the schools in succession, today the atomistic, tomorrow the dynamistic, and the day after the energetistic. So long as he discovers new facts, no one has the right to accuse him of inconsistency or reproach him for his palinodes.

And here is how nature protests anew against these critical teachings (p. 354): "Physical theory is not a merely individual suggestion which each scientist may use or reject as he sees fit. . . . If he is faced with several theoretical forms today, they are not opposed to one another as the dream of one individual is to the dream of another, but they are opposed as the conception of one school is to the conception of another, that is to say, as things which claim to be stable and to rally minds to the same road."

By what right does a merely technical procedure insist on imposing itself on a whole school? By what right, above all, does it claim to make itself universally adopted, so that every workingman in the world is obliged to accomplish the same task in the same manner? And yet physical theory does not hesitate to affirm this claim to universal unity, ridiculous as it is if that theory is only a tool or instrument (p. 375): "The present physiognomy of physics is not one which it will always present. Everything leads one to think, on the contrary, that it is due only to relatively transitory contingencies. . . . The divergences or even oppositions that we notice among physical theories will then continue to be attenuated as physics progresses; and they have continued to be attenuated as physics has progressed. These differences are not inherent in the nature of physics; they inhere in the initial phase of its development.

"Hence, as soon as we read the reflections of a physicist, no matter who he is, on physics, we never see him offer the slightest doubt about the profound unity of the science and the final agreement of theories, at least in their general lines. Everybody takes it for granted that the divergences are only temporary."

Let us admit it; let us suppose that all these divergences have been removed, and that we have finally succeeded in constructing that single theory, accepted by all, toward which physicists aspire. This theory will enjoy universal assent; its essence, however, cannot be changed. Now, logical criticism teaches us that physical theory is essentially, only a means of classification, and that it does not contain a morsel of truth which has not been brought to it by experiment. When all physicists have adopted the same theory in which no experimental law will be omitted, what will theoretical physics be? It will still be, and always will be, only empirical knowledge put in order. The order will be extended to all empirical knowledge; the mode of classification from which this order proceeds will be employed by the unanimity of all men of science. Nevertheless, theoretical physics, more conveniently manipulatable and more practical than any brute, inorganic, empirical knowledge, will have no other *value as knowledge* than the latter.

Thus does criticism speak; but nature immediately raises its voice in order to belie it (p. v):

"Theories constitute the domain of hypothesis, that is to say . . . of *successive approximations to the truth*, and this presupposes a truth which they more and more closely approximate. . . . It is legitimate to speak of a homogeneous ideal mind of the physical sciences: it promises at the same time a positive future logic of the physical sciences and a *human* philosophy of matter and its knowledge."

The logical criticism of the method employed by physics and of the testimony of physicists has therefore led M. Rey to the following affirmation: Physical theory is only an instrument suited for increasing empirical knowledge; nothing is true in it except the results of experiment. But nature protests against this judgment; it declares that there exists a universal and necessary truth, and that physical theory through the steady progress which extends it continually while rendering it still more unified gives us from day to day a more perfect insight into this truth, so that it constitutes a veritable philosophy of the universe.

IV

The reading of M. Rey's work has shown us that this author takes in turn two distinct and rather opposite attitudes, a reflective and critical attitude and an instinctive and spontaneous attitude. Critical reflection compels him to declare that theoretical physics knows only experimentally revealed truths, bound to be contingent and particular, and that theory, a mere instrument of classification and discovery, adds no knowledge to purely empirical facts. On the other hand, an instinctive and spontaneous intuition impels him to declare that there exists an absolute and universal truth, consequently one transcending experiment, and that the progress through which physical theory becomes steadily broader and more unified is directed toward a certain insight into this truth, more precise and more complete every day.

Shall we declare these two courses of M. Rey's reasoning, moving in opposite directions, contradictory, and shall we condemn them in the name of logic? Certainly not. We shall not condemn them any more than we condemned the two opposite tendencies we recognized in the thought of the continuators of mechanism, any more than we have accused the propositions formulated by M. Poincaré of incoherence, first for refusing and then for granting objective validity to physical theory. In Mach, Ostwald, and Rankine, and in all those who have scrutinized the nature of physical theory, we were able to note these same two attitudes, one looking like the counterweight to the other. It would be childish to claim that there is only incoherence and absurdity in this; on the contrary, it is clear that this opposition is a fundamental fact essentially connected with the very nature of physical theory, a fact that we must faithfully register and, if possible, explain.

When the physicist, bringing his attention to bear on the science he is constructing, submits to rigorous examination the various procedures he puts to work in constructing it, he discovers nothing which can introduce into the structure of the edifice the least parcel of truth outside of experimental observation. We can say of propositions which claim to assert empirical facts, and only of these, that they are *true* or *false*. Of these and only of these can we affirm that they cannot accommodate any illogicality and that of two contradictory propositions at least one of them must be rejected. As to propositions introduced by a theory, they are neither *true* nor *false*; they are only *convenient* or *inconvenient*. If the physicist judges it convenient to construct two different chapters of physics by means of hypotheses contradicting each other, he is free to do so. The principle of contradiction may be used to judge beyond any appeal between the true and the false; it has no power to decide on the *useful* or the *useless*. Therefore, to oblige physical theory to preserve a rigorous logical unity in its development would be to impose on the physicist's mind an unjust and intolerable tyranny.

When the physicist after submitting his science to this careful examination returns into himself, and when he becomes aware of the course of his reasoning, he at once recognizes that all his most powerful and deepest aspirations have been disappointed by the despairing results of his analysis. No, he cannot make up his mind to see in physical theory merely a set of practical procedures and a rack filled with tools. No, he cannot believe that it merely classifies information accumulated by empirical science without transforming in any way the nature of these facts or without impressing on them a character which experiment alone would not have engraved on it. If there were in physical theory only what his own criticism made him discover in it, he would stop devoting his time and efforts to a work of such meager importance. *The study of the method of physics is powerless to disclose to the physicist the reason leading him to construct a physical theory.*

No physicist, no matter how positivistic, can refuse to admit this. However, his positivism must be very strict, stricter even than that demanded by M. Rey, if he does not go beyond this admission and affirm that his efforts toward a physical theory increasingly more unified and more complete are reasonable, although the logical scrutiny of physical method was not able to discover the reason for it. It will be very difficult for him not to posit this reason in the correctness of the following propositions:

Physical theory confers on us a certain knowledge of the external world which is irreducible to merely empirical knowledge; this knowledge comes neither from experiment nor from the mathematical procedures employed by theory, so that the merely logical dissection of theory cannot discover the fissure through which this knowledge is introduced into the structure of physics; through an avenue whose reality the physicist cannot deny, any more than he can describe its course, this knowledge derives from a truth other than the truths apt to be possessed by our instruments; the order in which theory arranges the results of observation does not find its adequate and complete justification in its practical or aesthetic characteristics; we surmise, in addition, that it is or tends to be a *natural classification*; through an analogy whose nature escapes the confines of physics but whose existence is imposed as certain on the mind of the physicist, we surmise that it corresponds to a certain supremely eminent order.

In a word, the physicist is compelled to recognize that *it would be unreasonable to work for the progress of physical theory if this theory were not the increasingly better defined and more precise reflection of a metaphysics; the belief in an order transcending physics is the sole justification of physical theory.*

The alternating hostile or favorable attitude that any physicist takes in regard to this affirmation is summarized in Pascal's words: "We have an impotence to prove invincible by any dogmatism, and we have an idea of truth invincible by any skepticism."

TRANSLATOR'S INDEX

Abstract, schema of experiment, 133f., 146f.; symbols and facts (q.v.), 151f., 165f.; theories and mechanical models (q.v.), 55-105, 305; type of mind, 56ff., 87f.

Abstraction, vi, 7f., 22, 25f., 55f., 58f., 62, 69f., 73, 75, 88, 97f., 266

Abstractive method, xv, 52f., 55f.; see Rankine

Academy of Sciences, French, viii, xii, 29

Acceleration, vii, 209

Acoustics, v, 7f., 35

Action at a distance, 12f., 16f., 47, 74, 235f.

Adrastus (c. 300 B.C.), 223

Albertus Magnus (1193-1280), 233, 234, 241

Albert of Saxony (1316?-1390), vii, 228, 232, 234, 264

Albumasar (9th cent.), 233f.

Alchemists, 127f.

Alexander of Aphrodisias (2d cent.), 264, 307

Algebra, 76f., 96f., 108f., 112f., 121f., 143, 208

Almagia, Roberto, 233n.

American Journal of Mathematics, 85n.

Ampère, André Marie (1775-1836), 50f., 78, 81, 125f., 148, 154, 177, 196f., 202, 219, 253, 256

Ample type of mind, 55f.

Analogies, in physics, 95, 259f.; of physical theory to cosmology, 299-311

Analytic geometry, vii, 13; see Mechanics

Annales de Philosophie Chrétienne, xvii, 39n., 273n.

Approximation, degree of, 162f., 197; in physics, xvi, 36, 134ff., 168f., 172f., 332f.; mathematics of, 141-143; see Confirmation, Verification

Aquinas, Saint Thomas (1225?-1274?), vii, 41, 223, 233, 310

Arago, Dominique François (1786-1853), 29, 173, 186f., 253

Archimedes (287-212 B.C.), 40, 63, 112, 248

Aristarchus of Samos, pseudonym of Roberval (q.v.)

Aristotelianism, vii, 11f., 14, 17, 40, 43, 111, 123f., 127f., 229, 241, 252, 305-311

Aristotle (384-322 B.C.), vii, 11, 41f., 66, 89, 108, 111, 121n., 123, 222ff., 228, 231, 233f., 239, 245ff., 249n., 263f., 307, 310

Astrology, 234ff.

Astronomy, vii, 39n., 40n.; and physics among Greeks, 40f.; in Aquinas' works, 41f.; methods of, 40f., 169f., 190f.

Atomism, vii, ix, x, xii, xv, 12f., 18, 34f., 51, 73f., 123f.; see Corpuscular

Attraction and repulsion, 11f., 15f., 36, 47ff., 126, 192f., 220-252

Autonomy of physics, xvi, 10f., 19f., 282ff.

Averroes (Ibn Roshd) (1126-1198), 233f., 307

Avicenna (980-1037?), 233f.

Axiomatic method, vi, xi, 43, 63, 206, 208, 259f., 264f.

Bacon, Francis (Lord Verulam) (1561-1626), xi, 65f., 182, 230, 238, 320

Bacon, Roger (1214?-1294), 233f., 244

Balzac, Honoré de (1799-1850), 62

Bartholinus (Barthelsen), Erasmus (died 1680), 34f.

Becquerel, Antoine Henri (1852-1908), 253

Beeckman, I. (1588-1637), 33, 264

Bellantius, Lucius (c. 1500), 234f.

Bellarmino, Cardinal Roberts (1542-1621), 43

Benedetti, G. B. (1530-1590), 224, 264

Bentham, Jeremy (1748-1832), 67

Bernard, Claude (1813-1878), 180, 180n., 218

Bernouilli family, 72, 264n.

Berthelot, Pierre (1827-1907), 321

Bertin, Pierre-Augustin (1818-1884), Duhem's teacher, 276

Bertrand, Joseph (1822-1890), 53n.

Biology, classificatory theories in, 25f.; method of, 57, 180f.

Biot, Jean Baptiste (1174-1862), 29, 160, 187, 189, 253

Blondel, Maurice (1861-1949), 316

Bojanus, organ of, 101

Boltzmann, Ludwig (1844-1906), vi

INDEX

Bordeaux, Duhem at (1893-1916), v, 155
Borelli, Giovanni Alfonso (1609-1679), 248-251
Boscovich, Ruggiero Giuseppe, S.J. (1711-1787), 11f., 14, 36, 49, 306, 318, 328
Bourienne, Louis de (1769-1834), 58
Boussinesq, Joseph (1842-1929), 88f.
Bouvier, system of malacology of, 101
Boyle, Robert (1626-1691), 166, 173f.; see Mariotte
Brahé, Tycho (1546-1601), 42n., 193, 195, 252
Bravais, crystallography of, 35
Broglie, Prince Louis de (1892-), double aspect theory of light, xii; foreword on "Duhem's Life and Work," v-xiii
Bullialdus (Boulliau), Ismaël (1605-1696), 246
Buridan, Jean (1297?-1358?), vii

Cabeus, Nicolaus (Cabeo, Niccolo), S.J. (1585-1650), 11
Caesar, Julius (100-44 B.C.), 62n.
Calcagnini, Caelio (1479-1541), 239
Cardano, Geronimo (1501-1576), 235, 236n., 242, 247, 264
Carnot, Sadi (1837-1894), 255, 287, 305
Cartesianism, 13ff., 34, 44, 51, 66f., 74, 77, 113f., 130, 306f., 325
Cartesians, 48, 123f., 240, 252, 276, 284
Catholicism, Duhem's, xii, xvi, 273-311
Cauchy, Augustin Louis (1789-1857), 13, 77f., 256
Causality, 14f., 45, 47ff.
Cavendish, Henry (1731-1810), 254
Celestial mechanics, vi, 40f., 49f., 141f., 190f., 220-252, 258
Certainty, 42f., 45, 104, 144; inverse to precision (q.v.), 163f., 211, 267
Charles I (1600-1649), 68
Checkers and chess, skill in, 76f.
Chemistry, vi, 28, 74, 127f., 208, 214
Chevrillon, André, 67n., 69n.
Cicero (106-43 B.C.), 233
Classification, laws as natural, x; physical theory as, of laws, 7f., 19f., 23f., 27f., 31f., 38
Clausius, Rudolf (1822-1888), v, vi, 287, 305
Cohn, 91
Colding, August (1815-1888), 255
Collège de France, xii
Columbus, Christopher (1451-1506), 316
Common sense, and scientific knowledge, xvi, 104, 164f., 168f., 204, 209, 259f., 267, 283
Confirmation, of theories, xvi; holistic (q.v.), 183f.; see Verification
Conservation of energy, 285
Contarini, Gasparo (1483-1542), 240
Continental, physicists, 73f.; type of mathematics, 80f., 87f., 89
Continuity, of history of physical theory, 32f., 36, 39f., 177; of traditions, 68f.; see Evolution
Conventionalism, ix; see Poincaré
Copernicus, Nicholas (1473-1543), vii, 41f., 226-231, 239f., 252f., 308
Corneille, Pierre (1606-1684), 64f.
Corpuscular hypothesis, xii, 36f., 72, 122f., 186f., 220f., 247f.; see Waves
Cosmography, 42n.
Cosmology, viii, 11f., 14f., 34, 36, 40f., 73f., 299-311
Cotes, Roger (1682-1716), 49, 321
Coulomb, Charles A. de (1736-1806), 12, 119, 125f., 254
Couturat, Louis (1868-1914), 313
Cromwell, Oliver (1599-1658), 68
Crucial experiment, xi, xii, 188f.
Crystallography, 214f.

D'Alembert, Jean (1717-1783), 305
Darwin, Charles (1809-1882), 67
Davy, Sir Humphry (1778-1829), 128, 253, 282
Deduction, in physics, vi, xi, xv, 19f., 38, 53, 55f., 91, 261; mathematical, 58f., 63f., 79f., 90, 132-143, 208, 267
Definitions, physical theories as, 209f.
Delfino (Delphini), Federico (c. 1550), 242
Descartes, René (1596-1650), vii, 13, 15-18, 22, 33f., 43-49, 51, 65f., 72f., 81, 87f., 113f., 123, 131, 237f., 243n., 247, 251, 264, 305f., 320f., 325, 328
Deville, H. Sainte-Claire (1818-1881), 125
Dickens, Charles (1812-1870), 64
Diderot, Denis (1713-1784), 112
Dielectric, 78, 85f., 129
Dirac, Paul A. M. (1902-), xi
Displacement current, Maxwell's, 78f.
Duhem, Pierre Maurice Marie (1861-1916), bibliographic references, v,

vi, vii, viii, xvii, 24n., 39n., 40n., 55n., 85n., 86n., 98, 103n., 214n., 216n., 273n., 280n., 312n.; historian of science, vii, viii, xv; life and work, v-xiii; philosopher of science, viii, 42n., 216n., 273ff.; positivism (q.v.) and pragmatism (q.v.), in, ix; preface by, xvii; psychologist, 55ff., 60n.
Dulong, Pierre Louis (1785-1838), 173
Duret, Claude (d. 1611), 235, 240ff.
Dynamics, vii; Cartesian (q.v.), 17, 72; Kelvin's molecular, 71ff.

Ecole Normale Supérieure, v, 313
Economy of thought, laws and theories as, ix, 21f., 39f., 48, 55, 327; see Mach
Einstein, Albert (1879-), xii, xv
Elasticity, mechanics of, v, vi; models of, 75f.
Electricity, as fluid, 12, 119; charges of static, 69f., 118f., 203; modern theories of, 70f., 97; Newton on, 47f.; quantities of, 120; see Ampère, Coulomb, Faraday, Maxwell
Electrodynamics, 78f., 90f., 97f., 126, 195f., 252f.
Electromagnetism, vi, xv, 71; Kelvin's model of, 72f.; Maxwell's theory of light as, 79, 120f.
Electrons, vii, x, 304
Eliot, George (1819-1880), 64
Empedocles (500-430? B.C.), 228, 248
Energetics, vi, ix, x, 287ff.
Energy, kinetic, vii; conservation of, 285
English school of physics, xi, 55ff., 63ff.
Entropy, 287ff.
Epicurean philosophy of atoms, 18, 88
Epigraphy, 160
Eratosthenes (276?-195? B.C.), 232
Esprit de finesse, narrow but strong, 57ff.; see Continental
Ether, 9, 26, 47, 82f.
Euclid (c. 300 B.C.), 63, 80, 188, 245, 265
Euler, Leonhard (1707-1783), 261f., 264, 305
Evolution, in biology, 25; of physical theories, 31f., 38f., 103f., 206, 220-252, 253f., 295; of social life, 68
Experiment, and physical theory, vi, x, xi, xii, xv, 3, 7f., 34, 144-164, 180-218; anticipated by theory, 27f., 38, 183; Bacon on, 66f.; fictitious, 201f.; law and, 168f.; see Crucial
Explanation, ix; hypothetical, 19f., 40f., 52f.; physical theory not metaphysical, 5-18, 26f., 31f., 38, 41, 66, 71f., 75f., 103

Facts, and experiment, 163f.; practical and theoretical, 134ff., 149f.; scientific, 149f.; without preconceived ideas, 181f.
Faraday, Michael (1791-1867), 70, 126, 253
Fatio de Duiller, Nicolas (1664-1753), 83, 321
Fermat, Pierre de (1601-1665), 232, 253
Field physics, ix
Force, 69, 80, 126, 194, 264
Foscarini (c. 1615), 43
Foucault, Léon (1819-1868), xi, xii, 187, 189, 258
Fourier, Jean Baptiste, Baron de (1768-1830), 51, 78, 96
Fracastoro, Geronimo (1483-1553), 228f.
Franklin, Benjamin (1706-1790), 12, 29, 119
Fraunhofer, Joseph von (1787-1826), 24
Free will, and determinism, 283-287
French mind, 64ff.
Fresnel, Augustin Jean (1788-1827), xii, 24, 37f., 51, 160, 189, 253, 256

Galen (130-200?), 234
Galilei, Galileo (1564-1642), vii, 33, 39n., 42f., 45, 121f., 224, 227, 239, 243n., 264, 305, 320
Galucci, Giovanni Paolo (1550?- ?), 242
Gamaches, Etienne Simon de (1672-1756), 49n.
Gassendi (Gassend), Pierre (1592-1655), 13, 87f., 121f., 239f., 264
Gauss, Karl F. (1777-1855), 69, 78
Gay-Lussac, Joseph Louis (1778-1850), 29
Geminus, 40
Generalization, 34, 47, 55f., 58
Geometric mind, ample but weak, 57ff., 217; see English, Pascal
Geometry, 13f., 35, 37, 40, 64, 76f., 80, 177, 265f.
German mathematicians and physicists, 69f.

Gibbs, Josiah Willard (1839-1903), v, vi, 95, 168, 207, 305
Gilbert, William (1544-1603), 235
God, 17, 45, 229f., 241; see Religion
Grassmann, Hermann G. (1809-1877), 77
Gravitation, universal, 12, 15, 49f., 83, 192f., 220-252; see Attraction
Greeks, 40
Green, 256
Greisinger, 52
Grimaldi, 24, 212
Grisar, Hartmann (1845-1932), 43
Grisogon (Chrisogogonus), Frederick, of Zara (Dalmatia, c. 1528), 241f.
Grosseteste, Robert (1175?-1253), 233

Hadamard, Jacques (1865-   ), xv, 139, 141f., 216n.
Halley, Edmund (1656-1742), 250f., 255
Hamilton, Sir William Rowan (1805-1865), 38, 77
Haüy, Abbé René Just (1743-1822), 35
Heaviside, Oliver (1850-1925), 91
Helmholtz, Hermann L. F. von (1821-1894), vii, 13, 99f., 305
Hertz, Heinrich (1857-1894), 79f., 90f., 100, 318
Hipparchus, of Rhodes (160?-125? B.C.), 232
History of science, v, vii, viii, xi, xii, xv, xvi, 10f., 14f., 29f., 32f., 39f., 95, 173, 218, 220-252, 295f.; method of, 268-270, 295f.
Hobbes, Thomas (1588-1679), 320
Holistic view of physical theory, xi, xii, 32, 183ff.
Hooke, Robert (1635-1703), 249f., 255
Hume, David (1711-1776), 67
Huygens, Christian (1629-1695), 15f., 34ff., 46f., 72, 96, 171, 189, 250f., 276, 305, 320f., 325, 328
Hydrodynamics, v, vi, 13
Hypotheses, astronomical, 42f., 169f., 190; choice of, 219-270; economy of, 55f.; logical conditions for, 219f.; never isolated, 183f; not deducible from common sense, 259f., not sudden creations, 220-252; principles of deduction from, 20f., 41f., 53, 78

Iceland spar, 34f.; see Polarization of light
Idealism, rejected by Duhem, ix
Imagination, pictorial scientific, x, xi, 55ff., 67f., 79f., 85f., 93, 164
Impetus, medieval theory of, Duhem's work on, vii
Induction, electromagnetic, 71; logic of, 34, 47, 66f., 94, 201, 321
Industrial methods of thinking, 66, 92f.
Inertia, vii; see Galilei, Newton
Instruments, dependent on theory, 153-158, 161f., 182f.; laboratory, 3, 33; theories as research, 319f., 331
International Congress of Philosophy (Paris, 1900), 216
Intuition in physical theory, vi, 27, 37f., 80, 104, 294

Jacobi, Moritz (1801-1874), 63
John of Philopon (5th cent.), 264
Joubert, 201n.
Joule, James (1818-1889), 255

Kelvin, Lord (William Thomson, q.v.)
Kepler, Johannes (1571-1630), 36, 42, 50, 191-196, 227, 230f., 236f., 245-252, 259, 308
Kircher, Athanasius, S.J. (1601-1680), 246
Kirchhoff, Gustav (1824-1887), 53f., 148

Lagrange, Joseph-Louis (1736-1813), v, 112, 305, 318
Laplace, Pierre Simon, Marquis de (1749-1827), 29, 36f., 49f., 78, 81, 119, 142, 154, 158, 171, 173, 174, 187, 189, 253f., 305, 308, 318, 328
La Rive, 253
Lavoisier, Antoine Laurent (1743-1794), 128, 131, 282
Laws, as natural classifications (q.v.) and representations (q.v.), 26f., 74; economy of thought in, 22f.; in English and French nations, 67f.; symbolic relations, 165f., 175f.
Leibniz, Gottfried Wilhelm, Freiherr von (1646-1716), 14f., 18, 49, 122, 247n., 251, 305, 321
Leonardo da Vinci (1452-1519), vii, 226, 252, 264
Le Roy, Edouard Louis E. J. (1870-   ), 144n., 149, 150n., 208f., 267, 294, 315f.
Lesage, Georges Louis (1724-1803), 83
Liénard, 85n.
Light, double aspect of, xii; Descartes'

theory of, 33f.; Fresnel's experiment on diffraction of, 29, 37f.; Newton's laws of, 22f., 35f., 129, 160; polarization of, 35f., 160, 184f.; quantum of, xii; speed of propagation in water of, xi, xii; wave vs. corpuscular theory of, 9f., 37, 71f., 82f.
Lille, Duhem at, v, 277
Living force (*vis viva*), 18, 126
Lloyd, 38
Locke, John (1632-1704), 67
Lodge, Sir Oliver (1851-1940), 69n., 70
Logic, and physics, xv, xvi, xvii, 3, 28, 38, 78, 80, 99, 101, 104, 161, 189f., 205ff., 218, 293f.; hypotheses and, 219f.; inductive, 66f., 204; of astronomy, 38f., 42f.; see Deduction, Mathematics, Method
Lorentz, vii, 98
Loti, Pierre (1850-1923), 64
Louis XIV (1638-1715), 68
Lucretius (96?-55 B.C.), 13
Lycée Henri IV, 156

MacCullagh, James (1809-1847), 83, 256
Mach, Ernst (1838-1916), ix, xv, 21f., 39f., 53f., 264n., 268, 317, 327, 333
Magnetism, 11f.; and electricity (q.v.), 102f.; history of, 225ff.; see Gravitation, Electromagnetism, Tides
Malebranche, Nicolas de (1638-1715), viii, 13, 73, 96
Mansion, P., 40n.
Marchand, J. B., 260
Mariotte, Edme (1620-1684), 166, 173f.; see Boyle
Marsilius van Inghen (d. 1396), 224
Massieu, v
Mathematics, and metaphysics, 10, 46; and physical theory, 13, 19f., 62f., 69f., 72, 76f., 107ff., 121, 132-143, 164, 205f., 215f., 285; in thermodynamics, vi; see Algebra, Euclid, Geometry
Matter, and form, 11, 14; and geometry, 13, 44, 73f., 113ff.; and light, 33f., 73; and qualities (q.v.), 14f., 130; continuous, 74; ideal, 73
Maxwell, James Clerk (1831-1879), vi, vii, 13, 70f., 78ff., 83-86, 89f., 95f., 96, 98, 100, 102, 129f., 190, 319, 325, 328
Mayer, Johann Tobias (1723-1762), 12, 254
Mayer, Julius Robert von (1814-1878), 52, 255
Measurement in physics, xvi, 20f., 63, 108ff., 134ff., 207
Mersenne, Marin (1588-1648), viii, 15, 46, 227, 232, 237, 243n.
Mechanics, analytical, v-vii, xv, 215; chemical, v, vii; history of, viii, 34, 103; philosophy of, 16, 34f., 49, 53f.; statistical, vi; thermodynamics and classical, v, vi; see Dynamics, Statics
Mechanism, as metaphysical explanation, xvi, 16, 34; models of, 69ff., 72f., 319ff.
Metaphysics, and physics, viii, xv, xvi, 5-18, 19f., 38, 41f., 43f., 50f., 63, 74f., 287-290, 291-293, 335
Method, abstractive vs. hypothetical, 52ff.; astronomical, 41f.; Bacon (q.v.) on, 66f.; Descartes (q.v.) on 43ff., 65f., 68; English, 86ff.; mathematical, 62f., 69, 80f., 205f., 265; Newtonian, 190f., 195f., 203; of history, 58f., 62, 68f.; of Scholasticism (q.v.), 120f.; operational, v, vi, xv, xvi, 3, 19f., 104, 207
Microphysics, xi; see Atomism, Electrons
Middle Ages, vii, viii; see Scholasticism
Milhaud, Gaston (1858-1918), 144, 165, 208
Mill, James (1773-1836), 67
Mirandola, Pico della (1463-1494), 234
Models, algebraic, 102; atomic, vi; mechanical, 55-80, 81f,. 93f., 102f.; pictorial, vii, xv, 55f., 69f.
Molanus, 247n.
Molière, Jean-Baptiste Poquelin (1622-1673), 123
Mollien, 59
Moral conditions for scientific progress, 218
Morin, Jean Baptiste (1583-1656), 240, 242
Morin, Paul, 77
Moutier, Jules, teacher of Duhem, 275f.

Napoleon Bonaparte (1769-1821), xi, 57f., 60, 62
National traits of English and French, 63ff.
Natural, classification (q.v.) and physical theory, 19-30, 293-298, 335;

explanation and, 37f.; light of reason, 225; philosophy, 43f., 47, 66; place of elements, 309ff.; science, 25f.
Navier, 75
Necessity, and truth (q.v.), 267
Neumann, Franz Ernst (1798-1895), 78, 148, 184ff., 256
Newton, Sir Isaac (1642-1727), 9, 11f., 23f., 36f., 47-50, 53, 87, 89, 127, 160, 177, 186f., 189-201, 203, 219, 221f., 246, 250ff., 264, 276f., 288, 305-308, 318, 320f., 325
Newtonians, 11, 12, 14f., 36, 284
Nicholas of Cusa (1401-1464), 264
Nifo, Agostino (1473-1546), 224
Noël, Père (c. 1600), Descartes' teacher at La Flèche, 123

Observation, and unobservable reality, xv, 7f., 14f., 18; experiment and, 144f., 158f., 173; without preconceived ideas, 66, 180ff.
Occult causes, 14-16, 48, 73; qualities, 121f.
Oepinus, Ulrich-Theodore (1724-1802), 12, 119
Oersted, Jean Christian (1777-1851), 177, 252-254
Ohm, George Simon (1787-1854), 96, 148
Ontological order, and explanation, x, 7f., 26f., 299ff., 305-311, 335
Operations, algebraic, 76f.; and physical theory, 3f., 19f., 47, 207, 213
Optics, 8f., 22f.; Newton's, 48; see Light
Order, logical and ontological (q.v.), x, 26; of laws in theories, 24f.
Oresme, Nicolas (c. 1320-1382), vii
Osiander, Andreas (1498-1552), 42
Ostwald, Wilhelm (1853-1932), 317, 333

Papin, Denis (1647-1714), 16
Paris, University of, in Middle Ages, vii; Academy of Sciences, viii
Pascal, Blaise (1623-1662), 23, 27, 46, 57, 60, 61n., 62, 76, 87, 104n., 123, 179n., 232, 260, 270, 325, 335
Perrier, Rémy, 101
Perrin, Jean B. (1870-1942), 318
Petrus Peregrinus of Maricourt (fl. 1269), 225n.
Photon, xii
Physical theory, aim of, 5-104; completely developed, 206f.; defined, 19f.; experiment (q.v.) and, 180-218; laws grouped by, 101, 104, 165ff.; mathematics (q.v.) and, 107ff., 205f.; metaphysical explanation irrelevant to, 7-18, 19; natural classification (q.v.) and, 19-30, 104, 293-298; as an organic whole, 187, 204; see Holistic
Picard, Emile, 94, 252
Pictorial, imagination (q.v.) in physical theory, x, xi
Pierre d'Ailly (1350-1420?), 224
Pistophilius, Bonaventura (c. 1525), 239
Plato (427-347 B.C.), 39n., 299n.
Pliny the Elder (23-79), 223, 233
Plutarch (46?-120), 249
Poincaré, Jules Henri (1854-1912), ix, xv, 27, 85f., 91, 101, 142, 149f., 186, 200, 208, 212f., 216n., 294, 317, 319, 328, 333n.
Poisson, Siméon Denis (1781-1840), 12, 29, 69, 75, 81, 119, 125f., 254, 318
Polarization of light, 35; see Light, Electricity, Magnetism
Posidonius (130?-50? B.C.), 40, 232
Positivism, in Duhem, ix, xv, xvi, 7f., 19f., 275-282, 296f.; of Ampère (q.v.), 50f.; of Kirchhoff (q.v.), 53f.; of Mach (q.v.), 53f.; of Le Roy, 209ff.
Pothier, Robert Joseph (1699-1772), 67n.
Pragmatism, of Duhem, ix, xvi, 24f., 330f.
Precision, 152ff.; Ampère's lack of experimental, 198f.; inverse to certainty, 163f., 209; lacking in common-sense knowledge, 210
Prediction, test of theories, xv, 28f., 38
Probability, 37; see Approximation, Certainty, Verification
Psychology, associationist, 67; of physical reasoning, xi, 55f.; of two types of mind, 57ff., 66f.; of national traits, 76f.
Ptolemy (2d cent.), 233, 235, 240
Pyrrhonian skepticism, 27

Qualities, and quantity, 107ff., 113; occult, 14f., 121; of matter, 46f., 113f.; primary, 121-131; sensible, 7f.
Quantity, and measurement, 108-110; and physics, 112ff.; and quality, 107ff., 110-112

Quantum physics, ix
Quaternions, 77

Rabelais, François (1490-1553), 62
Raimondo, Annibale (c. 1589), 242
Rankine, Macquorn W. J. (1820-1872), 52, 53, 55, 317, 326, 327, 333
Raoult, F. M. (1830-1901), 95
Raymond, Hannibal, 241
Regnault, Henri-Victoire (1810-1878), 146, 147, 156, 157, 158, 164, 173, 174
Relativity, Einstein's theory of, xii, xv; in classical and medieval astronomy, 40f.; of "elements" of physical and chemical analysis, 128ff.
Religion, Duhem's, ix, xvi, 273-311; and astronomy (q.v.), 42f.; physical science not opposed to, 282-287
Renaissance science, vii, viii, 121
Representation, function of laws, 26f.; vs. explanation (q.v.), 32f., 43f., 47, 70, 164
Revue de Métaphysique et de Morale, 216
Revue de Philosophie, xvii
Revue des questions scientifiques, 24, 55n.
Revue générale des Sciences pure et appliquées, xvii, 312n.
Rey, Abel (1873-1940), xvi, xvii, 273ff., 295, 312n., 313-334
Rey, Jean (d. 1645), 237
Rheticus, Joachim (1514-1576), 41, 239
Roberval, P. de (1602-1675), 15, 232, 242, 243-251; see Aristarchus of Samos
Römer, Olaus (1644-1710), 34
Robin, Gustave, 201, 202, 208, 321
Rosen, Edward (1908-), 42n.

Saint-Simon, Louis de (1675-1755), 61
Savart, Felix (1791-1841), 253
Scaliger, Julius Caesar (1484-1558), 223, 234, 244, 264
Schiaparelli, Giovanni (1835-1910), 40
Scholasticism, vii, 14f., 40ff., 73, 121f., 240-252, 305, 310; see Aristotelians
Ségur, Monsieur de (1780-1873), 59
Seleucus, Nicator (365?-281? B.C.), 232
Semantics, and syntax of physics, xv, xvi, 20f., 77, 128f., 149f., 164f., 209f.
Sensation, 48
Sévigné, Mme. de (1626-1696), 205n.
Shakespeare, William (1564-1616), 64, 65
Simple, bodies, 127f.; elements, 65, 104, 127f.; properties or qualities (q.v.), 20, 66
Simplicius (6th cent.), 40, 223, 224, 264, 307
Skepticism, and dogmatism, 325
Smith, Sydney (1771-1845), 67n.
Snell, Willebrord (1581-1626), 34
Sorbonne, vii, 234
Spencer, Herbert (1820-1903), 67
Staël, Mme. de (1766-1817), 58
Statistical mechanics, vi, x
Statistics, 70
Stendhal (1783-1842), 58
Strabo (63 B.C.–A.D. 21), 233
Symbolic relations, and physical laws, 165ff., 175f., 195, 293, 301
Sympathy, mutual, cause of gravity, 239f.
System of the world, Duhem's work on, viii, 39n., 42n.; Copernican, 42f., 190ff., 239f., 253; Kepler's, 190f., 251; Newtonian, 47f., 190f., 220-252

Taine, Hippolyte (1828-1893), 58f., 62, 68
Tait, Peter (1831-1901), 72n., 74, 83
Talleyrand-Perigord, De (1754-1838), 61
Tannery, Jules, 276
Tartaglia, Niccolo (1500?-1557), 264
Teaching of physical science, xvi, 3, 92f., 200-205, 257f., 268-270
Themistius (A.D. 4th cent.), 264, 307
Theology, and metaphysics (q.v.), xvi; and physics, 17, 41, 282-290
Theon of Smyrna, 223
Theory, and experiment, vi; aim and scope of physical, x, xvi, 3ff.; structure of physical, 3ff., 20f.; verification of, xi, xii, xvi, 21f., 209
Thermodynamics, v, vi, x, xv, 51f., 72, 97, 287ff.
T(h)imon, the Jew (c. 1516), 224, 234, 241
Thomson, Sir Joseph John (1856-1940), 84n., 98, 99, 318
Thomson, Sir William (Lord Kelvin) (1824-1907), v, x, xi, 13, 71, 74, 75,

80n., 81-84, 89, 91, 97, 98, 99, 103, 325
Tides, Galileo's theory of, 239f.; magnetic theory of, 232f.; medical effects of, 234f.; Newton's theory of, 251f.
Truth, criterion for physical theory, 21, 75, 104, 144, 149, 329f., 335; of physical laws, 168f., 189, 267, 330

Unity of science, 103, 294f., 331
Universals, 165, 329
Ursus, Nicolas Raimarus (c. 1600), 42n.
Utilitarianism, 67, 319, 327

Van der Waals, Johannes D. (1837-1923), 95
Van't Hoff, J. H. (1852-1911), 95
Varignon, Pierre (1654-1722), 262
Vector analysis, 77
Verification of theory as a whole, xi, xii, 20f., 183ff., 209, 216n.
Vortex theories, x, 13, 83; see Descartes, Thomson (Sir William)

Waves, vi, vii, xii; vs. corpuscles in optics, 37f., 189, 218, 256; see Acoustics, Light
Weber, Wilhelm Eduard (1804-1891), 148, 198, 199
Weight, theory of, vii, 12, 46f., 209f., 220-252; see Attraction, Gravitation, Magnetism
Wiener, Otto Heinrich (1862-1927), 184, 185, 186, 258
Wiener, Philip Paul (1905-), translator's preface, xv, xvi
Wilbois, E., 144n.
William of Auvergne, 233
Wren, Sir Christopher (1632-1723), 250, 251, 255

Young, Thomas (1773-1829), 24, 37, 96, 160, 189

Zeeman, Pieter (1865-), 98
Zenker, Wilhelm (1829-1899), 185

The Princeton Science Library

Edwin Abbott Abbott	**Flatland: A Romance of Many Dimensions** With a new introduction by Thomas Banchoff
Friedrich G. Barth	**Insects and Flowers: The Biology of a Partnership**
Marston Bates	**The Nature of Natural History** With a new introduction by Henry Horn
John Bonner	**The Evolution of Culture in Animals**
Paul Colinvaux	**Why Big Fierce Animals are Rare**
Peter J. Collings	**Liquid Crystals: Nature's Delicate Phase of Matter**
Pierre Duhem	**The Aim and Structure of Physical Theory** With a new introduction by Jules Vuillemin
Albert Einstein	**The Meaning of Relativity** Fifth Edition
Niles Eldredge	**Time Frames: The Evolution of Punctuated Equilibria**
Richard P. Feynman	**QED: The Strange Theory of Light**
J. E. Gordon	**The New Science of Strong Materials, or Why You Don't Fall Through the Floor**
Richard L. Gregory	**Eye and Brain: The Psychology of Seeing** Revised, with a new introduction by the author
J.B.S. Haldane	**The Causes of Evolution** With a new preface and afterword by Egbert G. Leigh
Werner Heisenberg	**Encounters with Einstein, and Other Essays on People, Places, and Particles**
Hans Lauwerier	**Fractals: Endlessly Repeated Geometrical Figures**
J. Robert Oppenheimer	**Atom and Void: Essays on Science and Community** With a preface by Freeman J. Dyson

John Polkinghorne	**The Quantum World**
G. Polya	**How to Solve it: A New Aspect of Mathematical Method**
Hazel Rossotti	**Colour, or Why the World isn't Grey**
Henry Stommel	**A View of the Sea: A Discussion between a Chief Engineer and an Oceanographer about the Machinery of the Ocean Circulation**
Hermann Weyl	**Symmetry**